陳雲潮 編著

iLAB FPGA
數位系統設計、模擬測試與硬體除錯

東華書局

國家圖書館出版品預行編目資料

iLAB FPGA 數位系統設計、模擬測試與實體除錯 / 陳雲潮編著. -- 1 版. -- 臺北市：臺灣東華, 2018.09

344 面；17x23 公分.

ISBN 978-957-483-957-5 (平裝)

1. 積體電路 2. 設計

448.62　　　　　　　　　　　　　　107015281

iLAB FPGA 數位系統設計、模擬測試與實體除錯

編 著 者	陳雲潮
發 行 人	陳錦煌
出 版 者	臺灣東華書局股份有限公司
地　　址	臺北市重慶南路一段一四七號三樓
電　　話	(02) 2311-4027
傳　　眞	(02) 2311-6615
劃撥帳號	00064813
網　　址	www.tunghua.com.tw
讀者服務	service@tunghua.com.tw
門　　市	臺北市重慶南路一段一四七號一樓
電　　話	(02) 2371-9320
出版日期	2018 年 9 月 1 版 1 刷
	2020 年 2 月 1 版 2 刷

ISBN　　978-957-483-957-5

版權所有・翻印必究

推薦序

第一次遇到系友陳雲朝先生是在多年前至金門大學電子工程系教學訪視，當時陳老師擔任講座教授。陳老師曾在本校台北工專時代擔任電子科助教、講師、副教授兼任科主任，自 1972 年至美國工作多年，退休後至金門大學擔任教職，近年獲聘回母校電子工程系擔任兼任教師，將其畢生所學傳授給學弟妹。

陳老師將講授「邏輯系統設計」的課程講義編輯成書，讓學生在學習上更能有參考的依據，這本書名為「iLAB FPGA 數位系統設計、模擬測試與硬體除錯」，使用系統模擬軟體為 ModelSim，合成工具使用 Altera Quartus II，實作平台使用 Altera/Terasic DE2-115，以 Digital Pattern Generator/Logic Analyzer 為測試工具，課程中使用 VHDL 來完成數位系統之設計。本書共有十二章，從基本邏輯閘、組合邏輯、順序邏輯、加法器、乘法器至系統電路之設計，這個課程教材設計可以讓學生由基礎設計開始，依序漸近，再進入應用設計階段，相當適合大專院校電機系或電子系相關科系學生所使用的電子實作講義，值得大力推薦給大家。

陳老師在這個進步快速的科技中，仍然要求自己不斷學習精進，更是把一生傳授給學生，並堅持一步一腳印踏實的把教材做好，分享給大家，謹以此序表達感謝之意。

國立臺北科技大學電子工程系主任

李宗演　敬書

2018 年 10 月 1 日

前言

這本實作講義，是我在北科大講授 "邏輯系統設計" 課時，陸續分發給同學們的參考資料的集合。我所採用的課本是 Pedroni 著的 "Circuit Design and Simulation with VHDL. 2nd" MIT Press 2010。實作方面採用 ModelSim 做模擬測試，Quartus II 做電路的合成。DE2-115 為硬體實作的平台，Analog Discovery 的 Digital Pattern Generator 和 Logic Analyzer 為測試工具。

組成 iLAB-FPGA 共 12 章。每章都包括了簡單的介紹，電路系統的組成和範例。還有 ModelSim 模擬測試，Quartus II 的合成，最後才是 DE2-115 平台上的實作和 Digital Pattern Generator / Logic Analyzer 的測試。同學們將會在這門課裡，領悟到模擬測試與硬體實作，在哪些地方有差別，和差別的程度。每章內的範例、實作重點提示、參考資料、和最新的修正等，都登錄在網路上，以便下載。

實作講義部分須借助於教師的重點引導。除了每週三小時的授課時間之外，同學並允許借用 DE2-115 及 Analog Discovery 回家，以完成課外實作，每週約需 8~12 小時。同學除了須繳驗類似工程師筆記之外，同時還有期中及期末的術科考試，藉以反映對整個課程的瞭解和吸收程度。

最後我要謝謝傑出校友宋恭源和六童子賢先生所贈送給電子系 50 套 DE2-115 板子。也感謝友晶創新公司的長時間技術支援。讓這本講義得以在短時間完成，同學們也得以動手用以實作。

台北科技大學電子系

陳雲潮

目次

前言 iii

目次 v

第一章　並發邏輯代碼的電路系統

    1-1　並發邏輯電路的幾個例子 1

    1-2　並發邏輯電路的模擬測試 5

    1-3　並發邏輯電路的 FPGA Synthesis 合成 7

    1-4　DE2-115 的 Expansion Header 8

    1-5　Synthesis 軟體 Quartus II / VHDL 的介紹 11

    1-6　Software Synthesis ALU_simple 電路的例子 13

    1-7　Hardware Synthesis (Programmer) FPGA / ALU_115 電路的例子 26

    1-8　Analog Discovery 測試 FPGA / ALU_115 電路的例子 ... 33

    1-9　課外練習 36

第二章　順序邏輯代碼的電路系統

    2-1　Sequential logic 電路的幾個例子 38

    2-2　順序邏輯電路的 FPGA 合成 (Synthesis) 45

    2-3　Software Synthesis slow_counter 電路的例子 46

    2-4　硬體合成 FPGA / slow_counter 電路的例子 56

    2-5　Analog Discovery 測試 FPGA / slow_counter 電路的例子 63

    2-6　課外練習 67

v

第三章　常數、通用、信號與變數

- 3-1　Signal and Variable 的比較 69
- 3-2　頻率計的例子 72
- 3-3　頻率計電路的合成例子 75
- 3-4　頻率計電路的測試 87
- 3-5　課外練習 89

第四章　測試平台和電路模擬測試

- 4-1　測試平台 (Test Bench) 的種類 91
- 4-2　測試波形的產生 92
- 4-3　第 I 類 TestBench 的寫法 95
- 4-4　第 II 類 TestBench 的寫法 98
- 4-5　第 III 類 TestBench 的寫法 99
- 4-6　第 IV 類 TestBench 的寫法 101
- 4-7　課外練習 104

第五章　有限狀態機

- 5-1　FSM 的模式和它的狀態轉移圖 105
- 5-2　有限狀態機 FSM 的 VHDL 模型 107
- 5-3　Moore machine 的一個例子 108
- 5-4　Mealy machine 的一個例子 111
- 5-5　液晶顯示器的例子 114
- 5-6　DE2-115 液晶顯示器實作 122
- 5-7　課外練習 128

第六章　Intel/Altera 方塊圖電路的設計

- 6-1　VHDL 檔轉變成電路圖形 129

6-2　選用電路庫中的電路 Symbols 136
　　　6-3　電路方塊圖完成後的設計工作 147
　　　6-4　課外練習 152

第七章　串行平行乘法器
　　　7-1　串行平行乘法器的設計 153
　　　7-2　串行平行乘法器的模擬測試 158
　　　7-3　串行平行乘法器電路合成的軟體部份 159
　　　7-4　串行平行乘法器合成電路的硬體測試部份 172
　　　7-5　課外練習 174

第八章　平行乘法器
　　　8-1　並行乘法器的設計 175
　　　8-2　並行乘法器系統的模擬測試 181
　　　8-3　平行乘法器電路合成的軟體部份 182
　　　8-4　平行乘法器合成電路的硬體測試部份 194
　　　8-5　課外練習 196

第九章　乘法-累加電路
　　　9-1　乘法-累加電路的設計 197
　　　9-2　乘法-累加電路的模擬測試 201
　　　9-3　乘法-累加電路的合成 203
　　　9-4　乘法-累加合成電路的硬體測試部份 217
　　　9-5　課外練習 219

第十章　有限脈衝響應數位濾波器
　　　10-1　有限脈衝響應數位濾波器的設計 221
　　　10-2　FIR 數位濾波器的模擬測試 224

- 10-3　FIR 數位濾波器電路的合成 226
- 10-4　課外練習 240

第十一章　Intel/Altera Qsys 系統與 NIOS-SoC 電路的設計

- 11-1　簡介 241
- 11-2　Altera DE2-115 教育板 242
- 11-3　數字硬件系統的例子 243
- 11-4　Altera 的 Qsys 產生 HDL 的軟體工具 245
- 11-5　如何將 Nios II 系統納入 Quartus II Project 263
- 11-6　課外練習 270

第十二章　Intel/Altera "監控程序" 的介紹

- 12-1　簡介 271
- 12-2　使用 Altera 的監控程序，來下載所設計完成的電路，再啟動應用程序 271
- 12-3　Nios II 的組合語言程式 273
- 12-4　Nios II 的 C 語言程式 284
- 12-5　課外練習 295

附錄

- 附錄 A　Altera 版模擬測試軟件 ModelSim VHDL Simulator 的介紹 297
- 附錄 B　測試儀器 Analog Discovery 與其控制軟件 Waveforms 的介紹 313

第一章　並發邏輯代碼的電路系統

並發邏輯 (Concurrent logic) 電路系統，在定義上為：它的輸出完全由它的輸入而定，該系統中既沒有儲存元件也沒有訊號的回授。如圖 1-1 所示。

圖 1-1　並發邏輯同時進行的邏輯系統

1-1　並發邏輯電路的幾個例子

　　VHDL 代碼，基本上是同時進行，不分先後的代碼。而純並發邏輯聲明僅限於 WHEN、SELECT、和 GENERATE，這類的聲明祇能夠存在於**順序碼** (sequential code) 之外。並發碼祇適用於**組合電路** (combinational circuits) 的設計中。圖 1-2 為將 WHEN 和 SELECT 用在**復用器** (Multiplexer) 電路的設計上。

Example： Multiplexer implemented with WHEN and SELECT

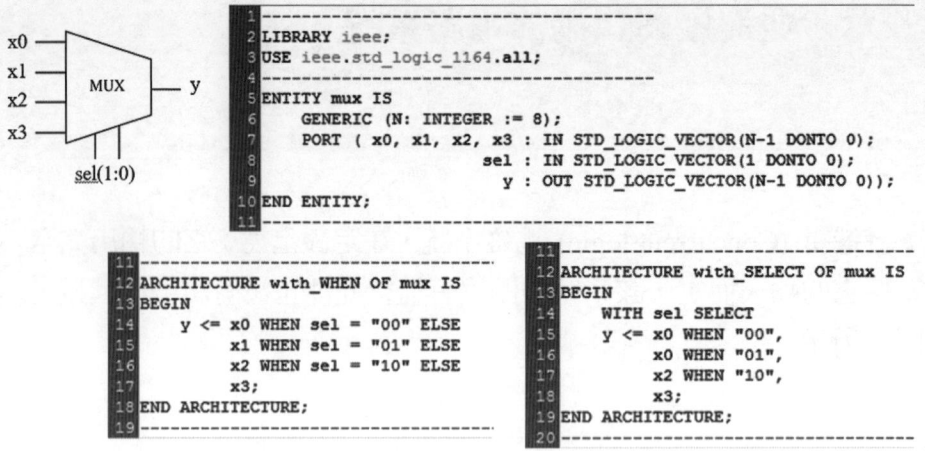

◎ 圖 1-2 ◎　　　將 WHEN 和 SELECT 用在 Multiplexer 電路的設計

圖 1-3 的簡單 ALU 電路方塊圖與其運算關係圖，也可以用 WHEN 和 SELECT 來組成如圖 1-4 所示。

Example 2：ALU implemented with WHEN and SELECT

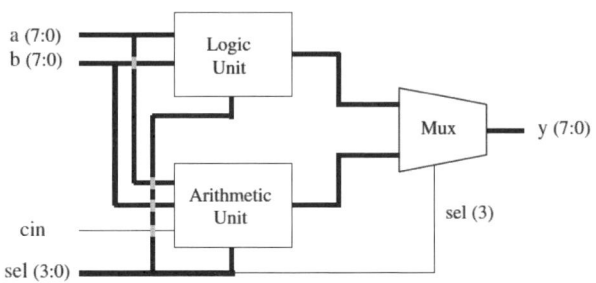

sel	Operation	Function	Unit
0000	y <= a	Transfer a	
0001	y <= a+1	Increment a	
0010	y <= a-1	Decrement a	
0011	y <= b	Transfer b	Arithmetic
0100	y <= b+1	Increment b	
0101	y <= b-1	Decrement b	
0110	y <= a+b	Add a and b	
0111	y <= a+b+cin	Add a and b with carry	
1000	y <= NOT a	Complement a	
1001	y <= NOT b	Complement b	
1010	y <= a AND b	AND	
1011	y <= a OR b	OR	Logic
1100	y <= a NAND b	NAND	
1101	y <= a NOR b	NOR	
1110	y <= a XOR b	XOR	
1111	y <= a XNOR b	XNOR	

圖 1-3　簡單 ALU 電路方塊圖與其運算關係圖

```vhdl
1  -- alu_115.vhd ------------------------------------
2  LIBRARY ieee;
3  USE ieee.std_logic_1164.all;
4  USE ieee.numeric_std.all;
5  ---------------------------------------------------
6  ENTITY alu_115 IS
7    GENERIC (N: INTEGER := 8); --word bits
8      PORT (a, b: IN STD_LOGIC_VECTOR(N-1 DOWNTO 0);
9            cin: IN STD_LOGIC;
10           opcode: IN STD_LOGIC_VECTOR(3 DOWNTO 0);
11              y: OUT STD_LOGIC_VECTOR(N-1 DOWNTO 0));
12 END ENTITY;
13 ---------------------------------------------------
14 ARCHITECTURE alu_115 OF alu_115 IS
15   SIGNAL a_sig, b_sig: SIGNED(N-1 DOWNTO 0);
16   SIGNAL y_sig: SIGNED(N-1 DOWNTO 0);
17   SIGNAL y_unsig: STD_LOGIC_VECTOR(N-1 DOWNTO 0);
18   SIGNAL small_int: INTEGER RANGE 0 TO 1;
19 BEGIN
20      ------Logic unit:--------------
21      WITH opcode(2 DOWNTO 0) SELECT
22          y_unsig <=   NOT  a WHEN "000",
23                       NOT  b WHEN "001",
24                       a AND b WHEN "010",
25                       a OR  b WHEN "011",
26                       a NAND b WHEN "100",
27                       a NOR  b WHEN "101",
28                       a XOR  b WHEN "110",
29                       a XNOR b WHEN OTHERS;
30      ------Arithmetic unit:---------
31      a_sig <= SIGNED(a);
32      b_sig <= SIGNED(b);
33      small_int <= 1 WHEN cin='1' ELSE 0;
34      WITH opcode(2 DOWNTO 0) SELECT
35          y_sig <= a_sig           WHEN "000",
36                   b_sig           WHEN "001",
37                   a_sig + 1       WHEN "010",
38                   b_sig + 1       WHEN "011",
39                   a_sig - 1       WHEN "100",
40                   b_sig - 1       WHEN "101",
41                   a_sig + b_sig   WHEN "110",
42          a_sig + b_sig + small_int WHEN OTHERS;
43      ------Mux:---------------------
44      WITH opcode(3) SELECT
45          y <= y_unsig WHEN '0',
46               STD_LOGIC_VECTOR(y_sig) WHEN OTHERS;
47 END ARCHITECTURE;
48 ---------------------------------------------------
```

圖 1-4　簡單 ALU 電路與其相對之 VHDL

圖 1-5 是並發邏輯代碼中的 generate 用於 address_decoder 的一個例子，它可以在任何輸入 bits 改變下保持 VHDL 代碼的大小不變。

Example 3：address_decoder implemented with GENERATE

```
1  -- Example 5.4 Generic address Decoder with GENERATE----------
2  LIBRARY ieee;
3  USE ieee.std_logic_1164.all;
4  USE ieee.std_logic_unsigned.all;
5  ----------------------------------------------------------------
6  ENTITY address_decoder IS
7      GENERIC (N: NATURAL := 3);  -- number of address bit
8      PORT (address: IN STD_LOGIC_VECTOR(N-1 DOWNTO 0);
9            ena: IN STD_LOGIC;
10           word_line: OUT STD_LOGIC_VECTOR(2**N-1 DOWNTO 0));
11 END ENTITY;
12 ----------------------------------------------------------------
13 ARCHITECTURE addres_decoder oF address_decoder IS
14     SIGNAL addr: NATURAL RANGE 0 TO 2**N-1;
15 BEGIN
16     addr <= conv_integer(address);  -- ieee.std_logic_unsigned
17     gen: FOR i IN word_line'RANGE GENERATE
18         word_line(i) <= '0' WHEN i=address AND ena='1' ELSE '1';
19     END GENERATE;
20 END ARCHITECTURE;
21 ----------------------------------------------------------------
```

圖 1-5 generate 用於 address_decoder 的產生

1-2 並發邏輯電路的模擬測試

Altera 在網路上有免費下載的 ModelSim-Altera Edition software 可供 VHDL 電路的模擬測試之用。這裡選用圖 1-4 簡單 ALU 電路之 VHDL device file 來做測試，它的 VHDL testbench file 如圖 1-6 所示。

```vhdl
--TstBench.vhd ----------------------------
library ieee;
use ieee.std_logic_1164.all;

entity TstBench is
  GENERIC (N: INTEGER := 8); --word bits
end TstBench;

use work.all;

architecture stimulus of TstBench is
--First, declare lower-level entity that to be test
component alu_115
  PORT (a, b: IN STD_LOGIC_VECTOR(N-1 DOWNTO 0);
        cin: IN STD_LOGIC;
     opcode: IN STD_LOGIC_VECTOR(3 DOWNTO 0);
          y: OUT STD_LOGIC_VECTOR(N-1 DOWNTO 0));
end component;

--Next, declare TstBench's SINGNALs
SIGNAL a, b: STD_LOGIC_VECTOR(N-1 DOWNTO 0) := (others =>'0');
SIGNAL cin: STD_LOGIC := '0';
SIGNAL opcode: STD_LOGIC_VECTOR(3 DOWNTO 0) := "0000";
SIGNAL y: STD_LOGIC_VECTOR(N-1 DOWNTO 0) := (others =>'0');

begin
  DUT: alu_115 port map ( a, b, cin, opcode, y);

  -- Concurrent Code for Periodical waveforms
  a <= "00010001";     b <= "00001111";
  cin <= '1' after 800 ns;
  -- Sequential Code for Non-periodical waveform
  process
  begin
    opcode <= "0000"; wait for 100 ns;
    opcode <= "0001"; wait for 100 ns;
    opcode <= "0010"; wait for 100 ns;
    opcode <= "0011"; wait for 100 ns;
    opcode <= "0100"; wait for 100 ns;
    opcode <= "0101"; wait for 100 ns;
    opcode <= "0110"; wait for 100 ns;
    opcode <= "0111"; wait for 100 ns;
    opcode <= "1000"; wait for 100 ns;
    opcode <= "1001"; wait for 100 ns;
    opcode <= "1010"; wait for 100 ns;
    opcode <= "1011"; wait for 100 ns;
    opcode <= "1100"; wait for 100 ns;
    opcode <= "1101"; wait for 100 ns;
    opcode <= "1110"; wait for 100 ns;
    opcode <= "1111"; wait;
  end process;
------------------------------------------
end stimulus;
```

圖 1-6　testbench.vhd 測試圖 1-4 簡單 ALU 電路之 VHDL device file

測試結果之 Timing Diagram 如圖 1-7 所示。

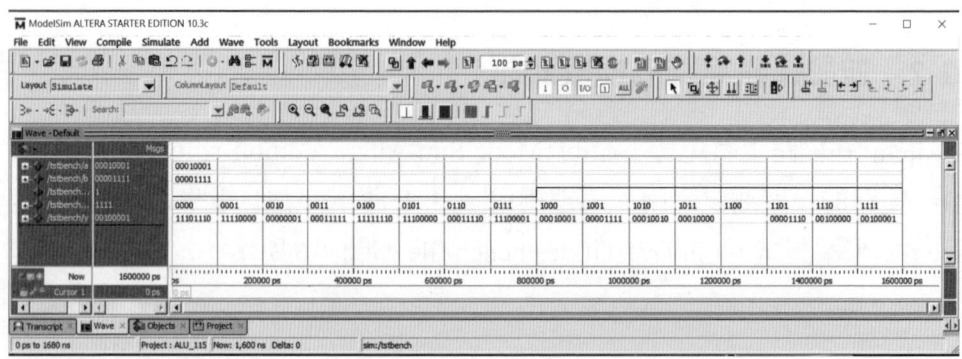

圖 1-7　ALU 115 測試結果之 Timing Diagram

Timing Diagram 之解讀：輸入 a = "00010001"，輸入 b = "00001111"。

@100 ns，Opcode = "0000" y = NOT a = "11101110"；
@200 ns，Opcode = "0001" y = NOT b = "11110000"；
……
@900 ns， Opcode = "1000" y = signed a = "00010001"；
@1000 ns，Opcode = "1001" y = signed b = "00001111"；
……
@1500 ns，Opcode = "1111" y = add a AND b with carry= "00100001"；

1-3　並發邏輯電路的 FPGA Synthesis 合成

　　本章將使用由硬體供應商 Altera 所提供的 Quatus II，實作平台使用的是台灣生產的 Altera/Terasic DE2-115。圖 1-8 是 DE2-115 的照相，上面有許多 Slide - switches、Leds、SSD 和 LCD 等輔助測試用的裝置。為了更能觀察邏輯電路間的 Timing 關係，本章將使用 Analog Discovery Module 中的 Logic Analyzer 和 Logic Pattern Generator 來替代。

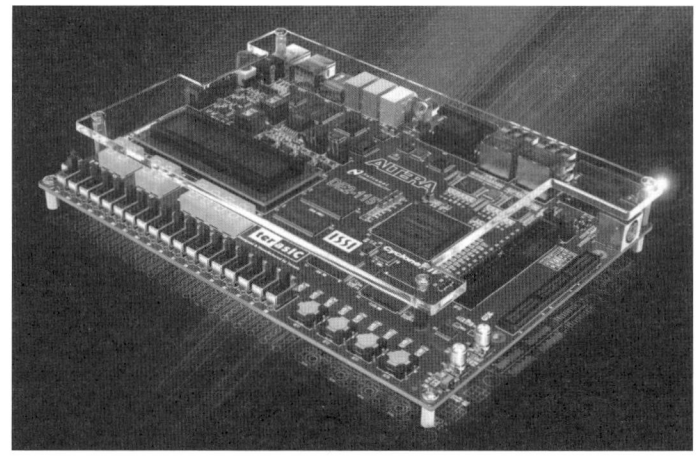

圖 1-8　　實作平台 DE2-115 全貌

1-4　DE2-115 的 Expansion Header

為了配合 Analog Discovery Module 以便觀測邏輯電路不同訊號間的 Timing 關係，與 DE2-115 最容易的連接點，當為其 JP4 Expansion Header 和 JP5 Expansion Header。

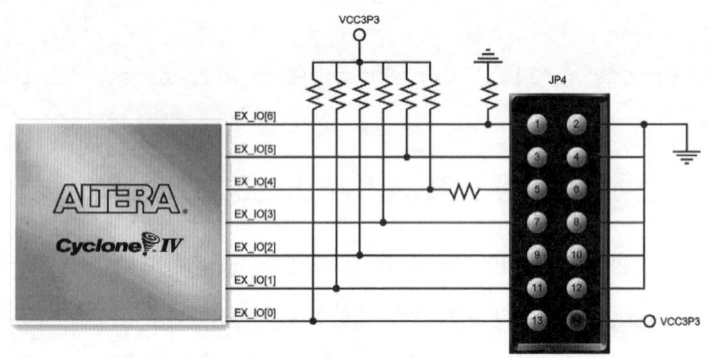

圖 1-9　JP4 Expansion Header 與 FPGA 間的關係

JP4 Expansion Header，如圖 1-9 所示。它的訊號名稱和在 FPGA 上的接腳編號，如圖 1-10 所示。

Signal Name	FPGA Pin No.	Description	I/O Standard
EX_IO[0]	PIN_J10	Extended IO[0]	3.3V
EX_IO[1]	PIN_J14	Extended IO[1]	3.3V
EX_IO[2]	PIN_H13	Extended IO[2]	3.3V
EX_IO[3]	PIN_H14	Extended IO[3]	3.3V
EX_IO[4]	PIN_F14	Extended IO[4]	3.3V
EX_IO[5]	PIN_E10	Extended IO[5]	3.3V
EX_IO[6]	PIN_D9	Extended IO[6]	3.3V

圖 1-10　JP4 Expansion Header 的訊號名稱和在 FPGA 上接腳編號

JP5 Expansion Header，如圖 1-11 所示。它的訊號名稱和在 FPGA 上的接腳編號，如圖 1-12 所示。其中的 5 V 和 3.3 V 電源，可以提供的最大電流，如圖 1-13 所示。

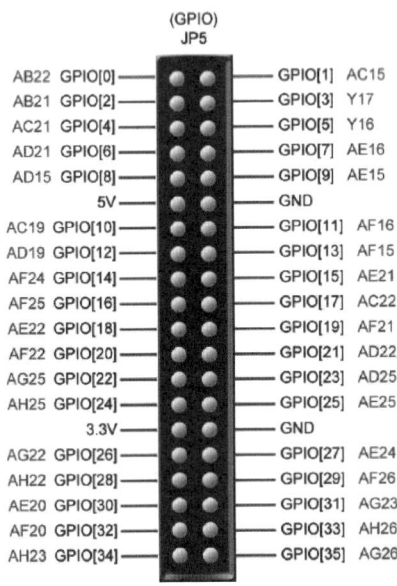

圖 1-11　JP5 Expansion Header 與 FPGA 間的關係

Supplied Voltage	Max. Current Limit
5V	1A
3.3V	1.5A

圖 1-12　5 V 和 3.3 V 電源可以提供的最大電流

Signal Name	FPGA Pin No.	Description	I/O Standard
GPIO[0]	PIN_AB22	GPIO Connection DATA[0]	Depending on JP6
GPIO[1]	PIN_AC15	GPIO Connection DATA[1]	Depending on JP6
GPIO[2]	PIN_AB21	GPIO Connection DATA[2]	Depending on JP6
GPIO[3]	PIN_Y17	GPIO Connection DATA[3]	Depending on JP6
GPIO[4]	PIN_AC21	GPIO Connection DATA[4]	Depending on JP6
GPIO[5]	PIN_Y16	GPIO Connection DATA[5]	Depending on JP6
GPIO[6]	PIN_AD21	GPIO Connection DATA[6]	Depending on JP6
GPIO[7]	PIN_AE16	GPIO Connection DATA[7]	Depending on JP6
GPIO[8]	PIN_AD15	GPIO Connection DATA[8]	Depending on JP6
GPIO[9]	PIN_AE15	GPIO Connection DATA[9]	Depending on JP6
GPIO[10]	PIN_AC19	GPIO Connection DATA[10]	Depending on JP6
GPIO[11]	PIN_AF16	GPIO Connection DATA[11]	Depending on JP6
GPIO[12]	PIN_AD19	GPIO Connection DATA[12]	Depending on JP6
GPIO[13]	PIN_AF15	GPIO Connection DATA[13]	Depending on JP6
GPIO[14]	PIN_AF24	GPIO Connection DATA[14]	Depending on JP6
GPIO[15]	PIN_AE21	GPIO Connection DATA[15]	Depending on JP6
GPIO[16]	PIN_AF25	GPIO Connection DATA[16]	Depending on JP6
GPIO[17]	PIN_AC22	GPIO Connection DATA[17]	Depending on JP6
GPIO[18]	PIN_AE22	GPIO Connection DATA[18]	Depending on JP6
GPIO[19]	PIN_AF21	GPIO Connection DATA[19]	Depending on JP6
GPIO[20]	PIN_AF22	GPIO Connection DATA[20]	Depending on JP6
GPIO[21]	PIN_AD22	GPIO Connection DATA[21]	Depending on JP6
GPIO[22]	PIN_AG25	GPIO Connection DATA[22]	Depending on JP6
GPIO[23]	PIN_AD25	GPIO Connection DATA[23]	Depending on JP6
GPIO[24]	PIN_AH25	GPIO Connection DATA[24]	Depending on JP6
GPIO[25]	PIN_AE25	GPIO Connection DATA[25]	Depending on JP6
GPIO[26]	PIN_AG22	GPIO Connection DATA[26]	Depending on JP6
GPIO[27]	PIN_AE24	GPIO Connection DATA[27]	Depending on JP6
GPIO[28]	PIN_AH22	GPIO Connection DATA[28]	Depending on JP6
GPIO[29]	PIN_AF26	GPIO Connection DATA[29]	Depending on JP6
GPIO[30]	PIN_AE20	GPIO Connection DATA[30]	Depending on JP6
GPIO[31]	PIN_AG23	GPIO Connection DATA[31]	Depending on JP6
GPIO[32]	PIN_AF20	GPIO Connection DATA[32]	Depending on JP6
GPIO[33]	PIN_AH26	GPIO Connection DATA[33]	Depending on JP6
GPIO[34]	PIN_AH23	GPIO Connection DATA[34]	Depending on JP6
GPIO[35]	PIN_AG26	GPIO Connection DATA[35]	Depending on JP6

圖 1-13 JP5 Expansion Header 的訊號名稱和在 FPGA 上接腳編號

JP5 接腳的輸入或輸出電位，可以經由改變 JP6 的 Jumper 的位置而改變如圖 1-14 所示。

JP6 Jumper Settings	Supplied Voltage to VCCIO4	IO Voltage of Expansion Headers (JP5)
Short Pins 1 and 2	1.5V	1.5V
Short Pins 3 and 4	1.8V	1.8V
Short Pins 5 and 6	2.5V	2.5V
Short Pins 7 and 8	3.3V	3.3V (Default)

圖 1-14　JP6 Jumper 的設定與 JP5 接腳的輸入或輸出電位的關係

1-5　Synthesis 軟體 Quartus II / VHDL 的介紹

Altera 的 Quartus II 軟體，有 Schematic Design、VHDL Design、和 Verilog Design 等三種。為配合先前使用的 Modelsim/VHDL Simulator，本章所使用的是 VHDL Design。不同於 Design 電路，Synthesis 電路比較機械化。它不像 VHDL 電路的設計，而必須按照一定的程序進行，少有使用者可以變動的地方。除了顯得繁雜，但不困難。當電路的 VHDL 檔，通過 Test Bench 檔測試成功後。該 VHDL 檔，就可以用 FPGA 來做 Synthesis，合成電路的程序如下：

1. Starting a New Project
   決定 Project 的名稱和位置，及 VHDL 電路的檔名，使用 FPGA 的編號。
2. First Compilation the Design
   第一次對 Project 的 VHDL 電路的檔 FPGA 接腳的編輯。
3. Pin Assignment
   決定電路輸入和輸出端在 FPGA 接腳上的位置。
4. Second Compilation the Design

第二次對 Project 的 VHDL 電路的檔 FPGA 接腳的編輯。
5. Programming and Configuring the FPGA Device
把完成了的電路 Bit Pattern 直接加入到 FPGA 上。
6. Test the Designed Circuit
使用 Analog Discovery 的 Logic Pattern Generator 和 Logic Analyzer 來測試所設計完成的電路是否工作正常。
7. Active Serial Mode Programming
FPGA 在 Power off 之後將失去程序 5 所有加入的 Bit Pattern，如果要讓 FPGA 在 Power on 之後，還能立刻動作，必須另加 Configuration Device，DE2-115 的 Configuration Device 是編號為 EPCS64 的 EEPROM。EEPROM 在 Power off 之後還能保持內有 Data Pattern 一段很長的時間，當 Power on 的時候，可以很快地存回 FPGA。所以叫做 Active Serial Mode Programming。
8. Third Compilation the Design
選用了 Configuration Device 之後，也需要做一次編輯。
9. Programming and Configuring the FPGA Device
把完成了的電路 Bit Pattern 直接加入到 EEPROM 上。
10. Test the Designed Circuit
使用與程序 6 的裝置設定，來測試所設計完成的電路是否工作正常。

　　從以上 10 點，使人感覺是整個 Synthesis 過程的繁雜，但從經驗上得知，祇要一步步小心地做，多做幾個電路，熟悉之後，也就無所謂繁雜了。

1-6　Software Synthesis ALU_simple 電路的例子

這個電路來自圖 1-6 的簡單 ALU 例子。當 Quartus II 14.1 啟動之後點擊 "New Project Wizard" 進入第一階段 S1。如圖 S1-1。

在合成 ALU_simple 之前，首先要做的是在 Altera 檔案夾底下產生一個名為 ALU_115 的檔案夾。再把通過模擬測試的 ALU_simple.vhd 搬移到 ALU_115 的檔案夾內，並且改名為 ALU_115.vhd。

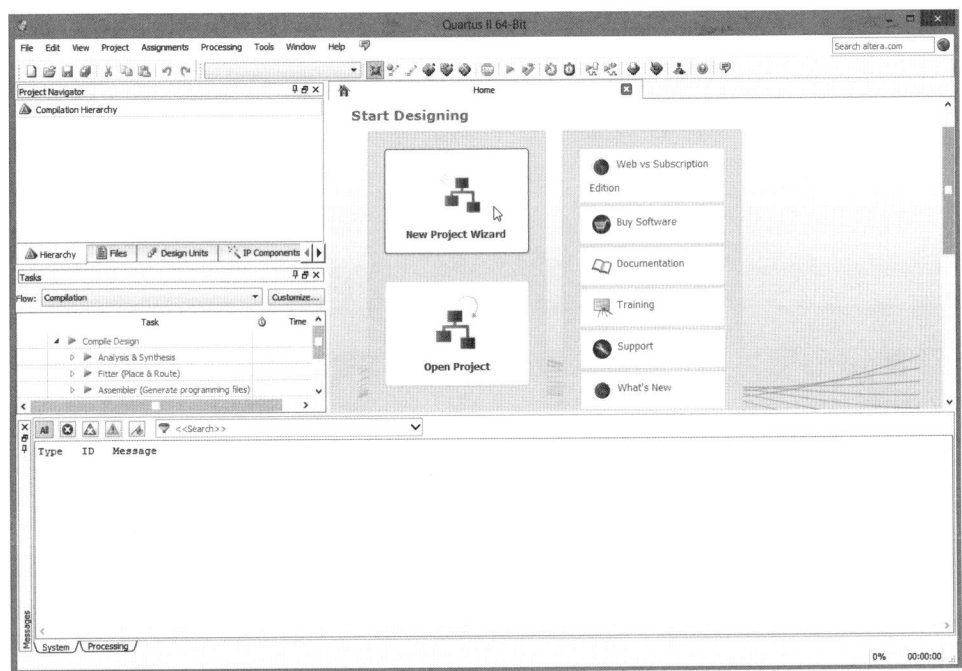

圖 S1-1　進入 New Project Wizard

14　iLAB FPGA 數位系統設計、模擬測試與實體除錯

　　New Project Wizard 視窗下有 Directory, Top-Level Entity，Project Type，和 Add Files 等視窗需要填寫和選用如圖 S1-2 所示。

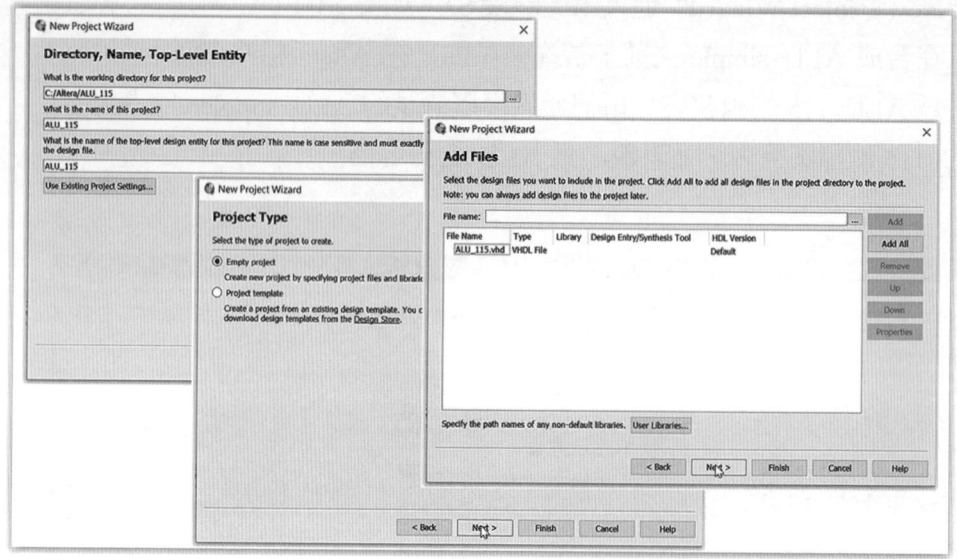

圖 S1-2　　New Project Wizard 視窗下各視窗需要填寫和選用

最為重要的是 Family & Device Setting，DE2-115 用的是 Cyclone IV E Family 的 EP4CE115F29C7，如圖 S1-3 所示。EDA Tool Setting 的 Simulation 的 Tool Name 為 VHDL。最後是 New Project Wizard 的 Summary

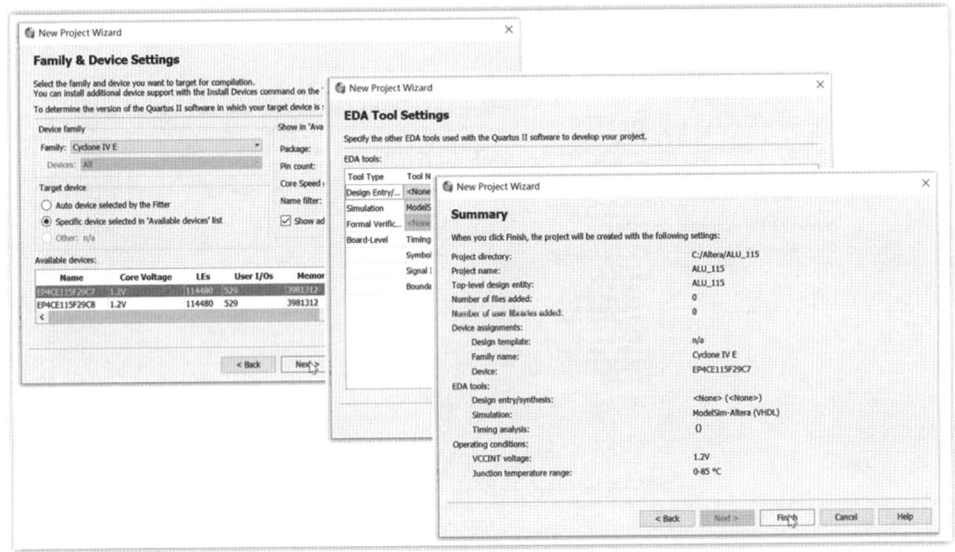

圖 S1-3　DE2-115 Family & Device Setting 和 New Project Wizard 的 Summary

選用 **Project** 視窗中的 **Assigments** > **Settings**...可以得到如圖 S1-4 以檢視 alu_115.vhd 存在於 Project 中。

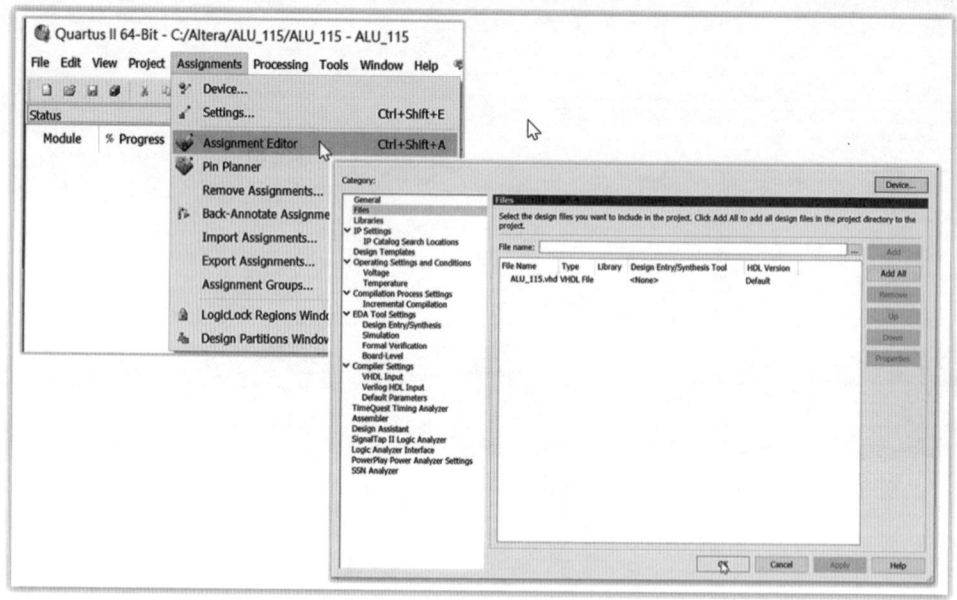

圖 S1-4　Assigments > Settings... 以檢視 alu_115.vhd 存在於 Project 中

最後還需要用 **Processing > Start Compilation** 來對這 S1 階段做檢驗，如圖 S1-5 所示。Compilation 必須做到 0 errors。

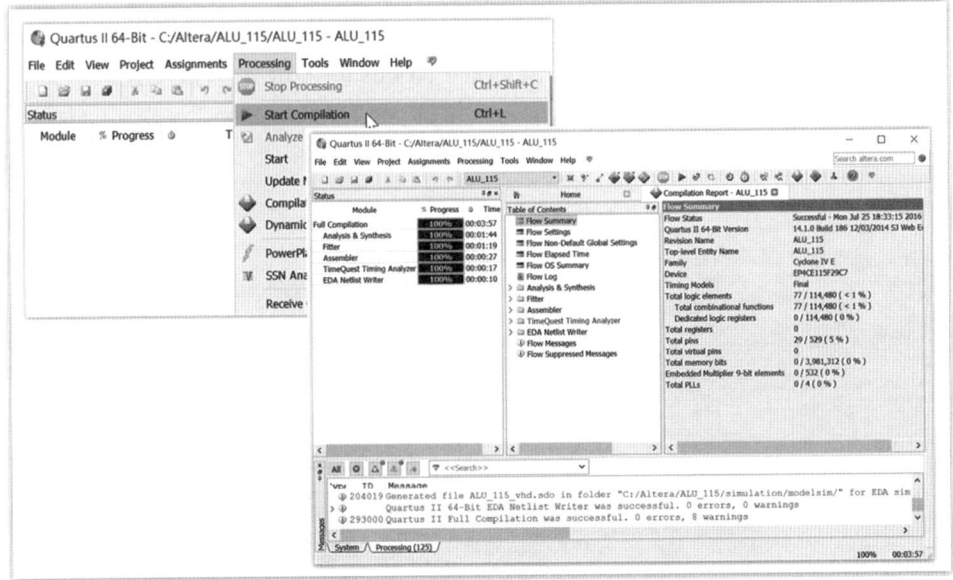

☘ 圖 S1-5 ⌇﹉﹉﹉ 用 Compilation 來對這 S1 階段做檢驗，必須做到 0 errors

S1 階段所產生的 Folder 和 Files 如圖 S1-6 所示。

☘ 圖 S1-6 ⌇﹉﹉﹉ S1 階段所產生的 Folder 和 Files

第 1 階段 Compilation pass 後就可以從 **Assignments** > **Assignment Editor** 進入第 2 階段的 S2：選用 FPGA/EP4CE115F29C7 的接腳，如圖 S2-1 所示。首先是單擊圖中的 **1**，然後是單擊圖中的 **2**，最後是單擊圖中的 **3**。再在圖中的 Nodes Found 內選取 a(0)~a(7), b(0)~b(7), cin, opcode(0)~opcode(3), y(0)~y(7) 經 > 移入到 Selected Nodes 內，最後再按 OK。

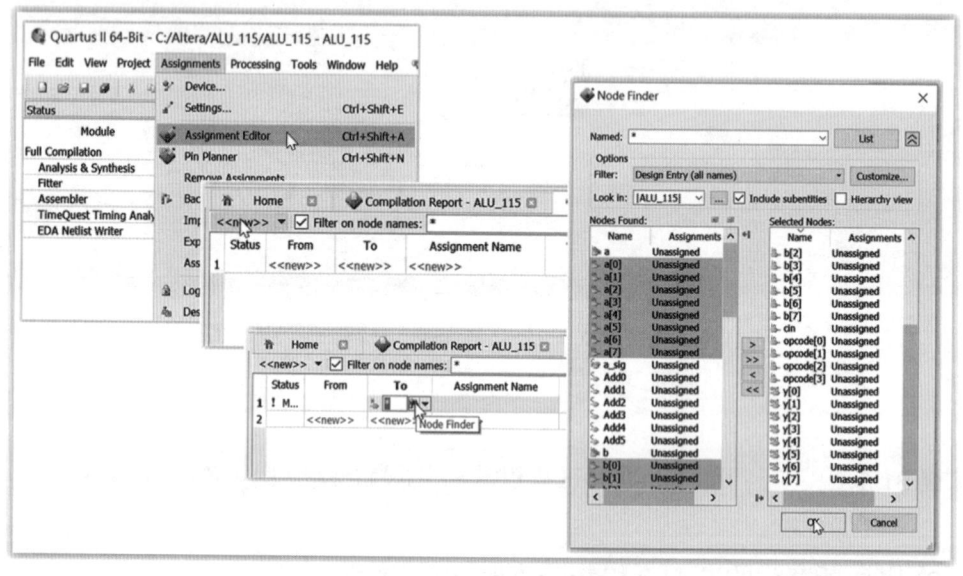

圖 S2-1　FPGA/EP4CE115F29C7 接腳的 Assignment

第一章　並發邏輯代碼的電路系統　19

　　EP4CE115F29C7 的接腳的選用，關係到 ALU_115 電路的 I/O 和 DE2-115 Slide Switches 和 JP5 Expansion Header 間的測試連線。如圖 S2-2 所示。

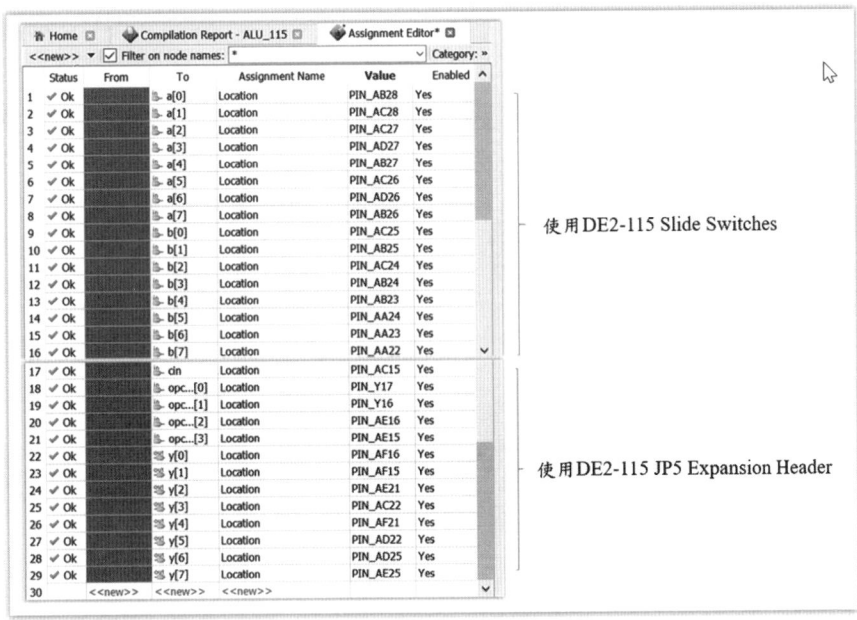

圖 S2-2　　ALU_115 電路的 I/O 和 JP5 間的測試連線關係

ALU_115 電路的 I/O 和 JP5 間的測試連線的 Assignment 過程中是否有錯，可以再一次使用 **Processing > Start Compilation** 如圖 S2-3 加以測試，結果必須為 Full Compilation was successful, 0 errors。

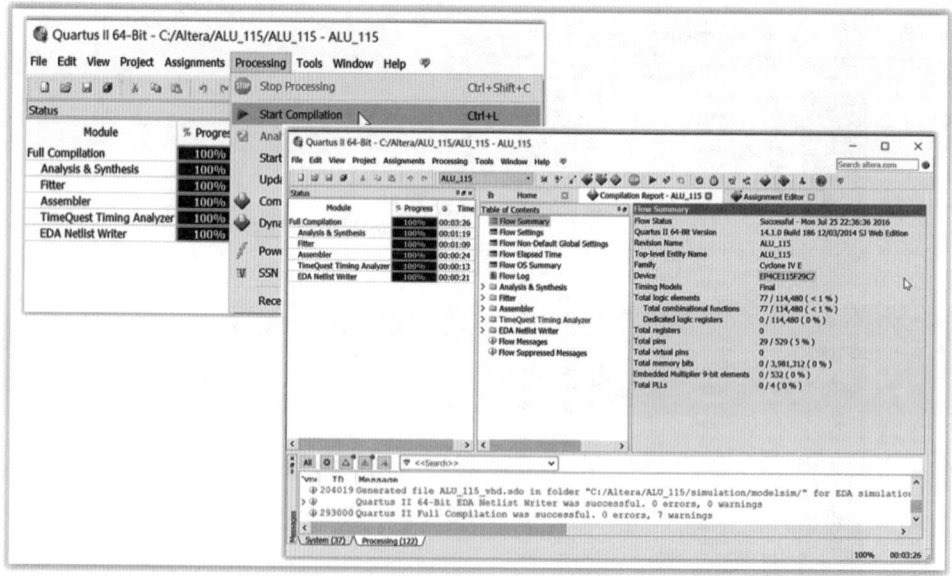

圖 S2-3　ALU_115 電路的 I/O 和 SW 及 JP5 間的連線，通過 Compilation 測試

S2 階段所產生的 Folder 和 Files 如圖 S2-4 所示。其中的 output_files Folder 裏面的 ALU_115.sof 可用來直接 Programming FPGA/EP4CE115F29C7。

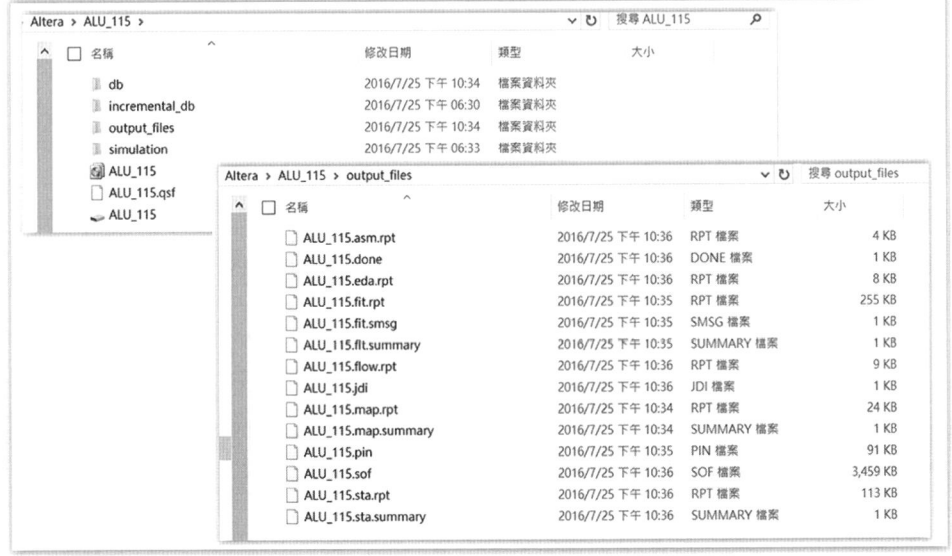

圖 S2-4　S2 階段所產生的 Folder 和 Files

ALU_115.sof 可用來直接 Programming FPGA/EP4CE115F29C7，但是當電源 OFF，Program 立刻消失。如果事先存入 DE2-115 中的 EEPROM/EPCS64，讓它在電源 ON 時立刻再存入 FPGA。圖 S3-1 為 EEPROM 的選取，它從點擊 **Device and Pin Options...** 進入。是為 S3 階段的開始。

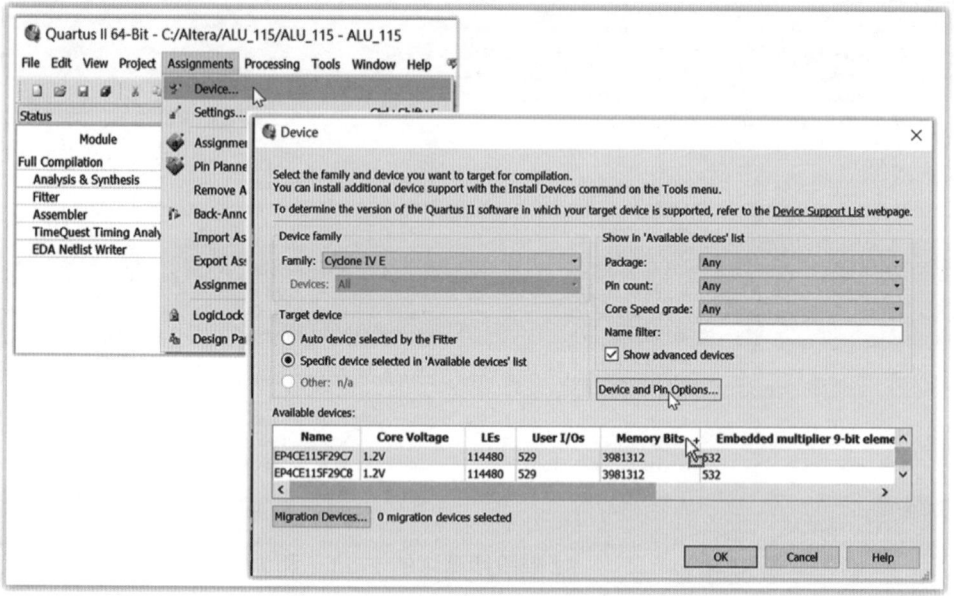

圖 S3-1　DE2-115 中 EEPROM 的選取，從點擊 Device and Pin Options... 進入

當圖 S3-2 的 Device and Pin Options 視窗出現後，選用 **Configuration device** 為 EPCS64。再在 Device 視窗點擊 **OK**。

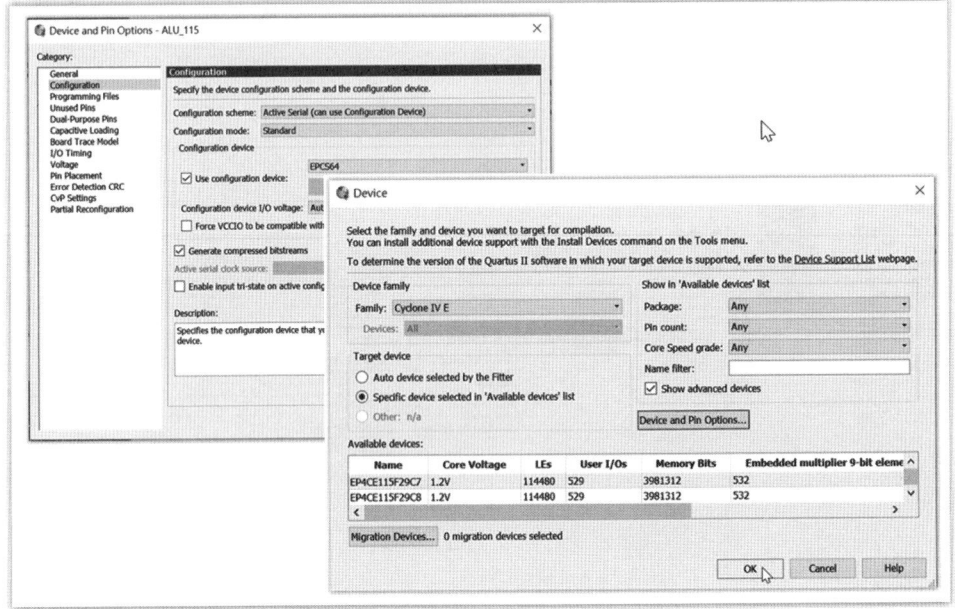

◌ 圖 S3-2 ◌ 　　選用 Configuration device 為 EPCS64。再在 Device 視窗
　　　　　　　　點擊 OK

Configuration device 為 EPCS64 的選用過程中是否有錯,可以再一次使用 **Processing** > **Start Compilation** 如圖 S3-3 加以測試,結果也必須為 Full Compilation was successful, 0 errors。

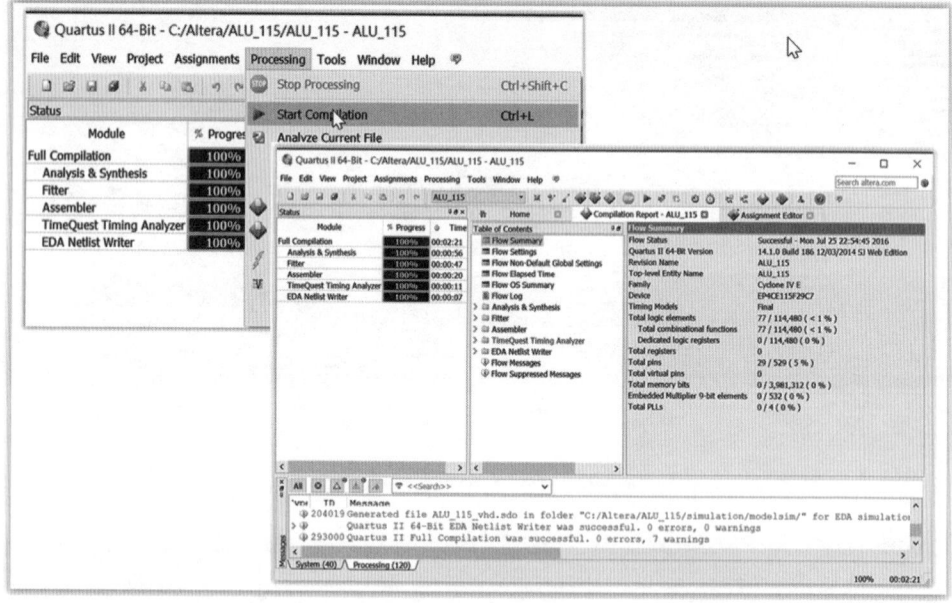

圖 S3-3　使用 Compilation 來測試 EPCS64 的選用過程中是否有錯

S3 階段所產生的 Folder 和 Files 如圖 S3-4 所示。其中的 output_files Folder 中的 ALU_115.pof 可用來間接 Programming DE2-115 中的 EEPROM/ EPCS64。

圖 S3-4　　S3 階段所產生的 Folder 和 Files

1-7　Hardware Synthesis (Programmer) FPGA / ALU_115 電路的例子

硬體合成是將上一節 1-6 的 Software Synthesis ALU_115 電路所獲得的 ALU_115.sof 經 Programmer 置入到 DE2-115 的 FPGA/ EP4CE115F29C7 中或將 ALU_115.pof 經 Programmer 置入到 DE2-115 的 EEPROM/ EPCS64 內。要將任何 *.sof 或任何 *.pof 置入到 DE2-115 上，首先要將 DE2-115 接上 Power 和事先已將 USB-Blaster (USB-0) 設定好，並且連接到 PC 的 USB 上。圖 T1-1 為點擊 File > Open Project 後開啓 ALU_115.qpf。

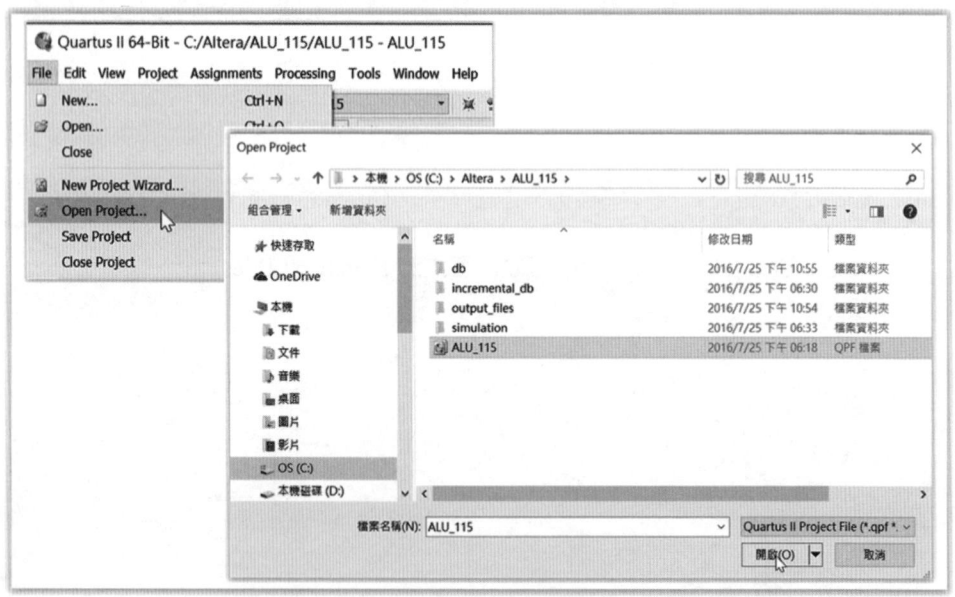

圖 T1-1　Quartus II 開啓 ALU_115.qpf

第一章　並發邏輯代碼的電路系統　27

結果得到圖 T1-2 為 Quartus II C:/Altera/ALU_115 – ALU_115

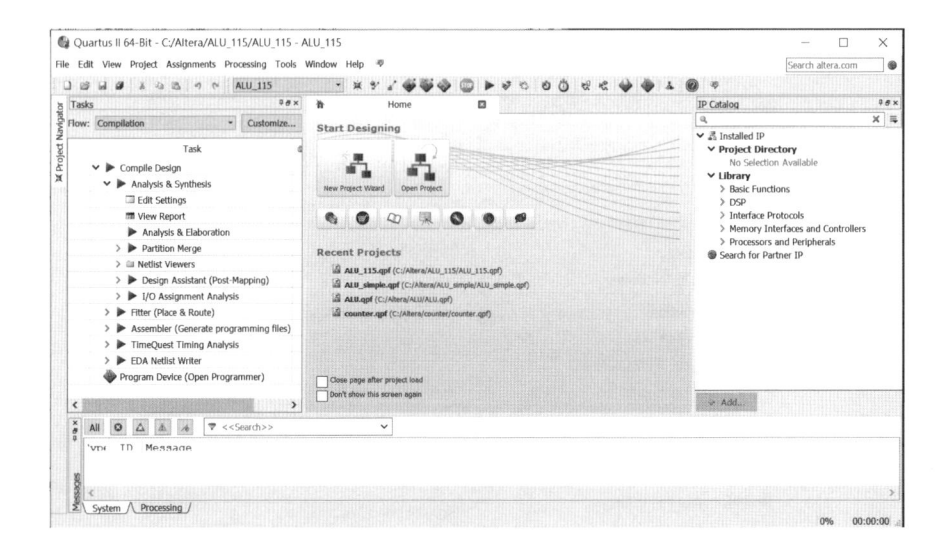

◎ 圖 T1-2 ◎　　Quartus II 開啟 C:/Altera/ALU_simple – ALU_simple

28　iLAB FPGA 數位系統設計、模擬測試與實體除錯

然後點選 **Project** 中的 **Tool** > **Programmer** 如圖 T1-3 所示。

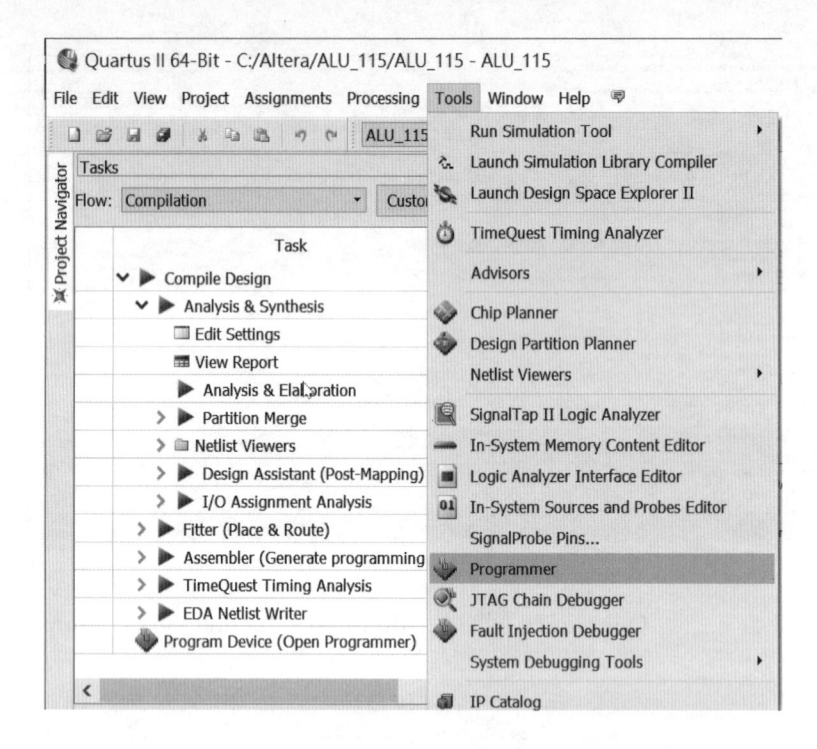

▓ 圖 T1-3　點選 Project 中的 Tool > Programmer

Programmer 視窗，它顯示的是 USB-Blaster (USB-0) Driver 已經設定。電源也已開啟，同時 Cable 已連接 DE2-115 到 PC 上，Program FPGA/EP4CE115F29 需要的 Mode 為 JTAG，點選 **Add File** 並從 output_files 的 folder 中選用 ALU_115.sof 如圖 T1-4 所示。最後才點擊 **Start**，如果沒做錯當顯示 100% (Successful)。

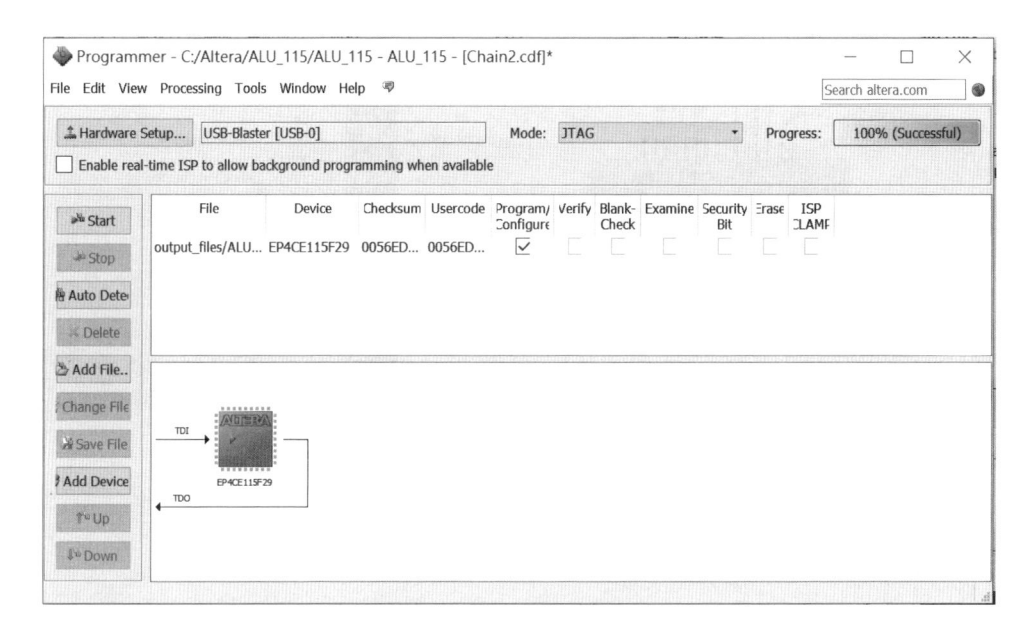

◎ 圖 T1-4 ◎　　Program ALU_115 到 FPGA/EP4CE115F29 的步驟

為了最後能夠用 Analog Discovery 的 Digital Pattern Generator 和 Logic Analyzer 來測試 ALU_115 電路的 Logic Timing Diagram，如圖 T1-5 所示，為 Analog Discovery 和 DE2-115 JP-5 之間的接線關係。

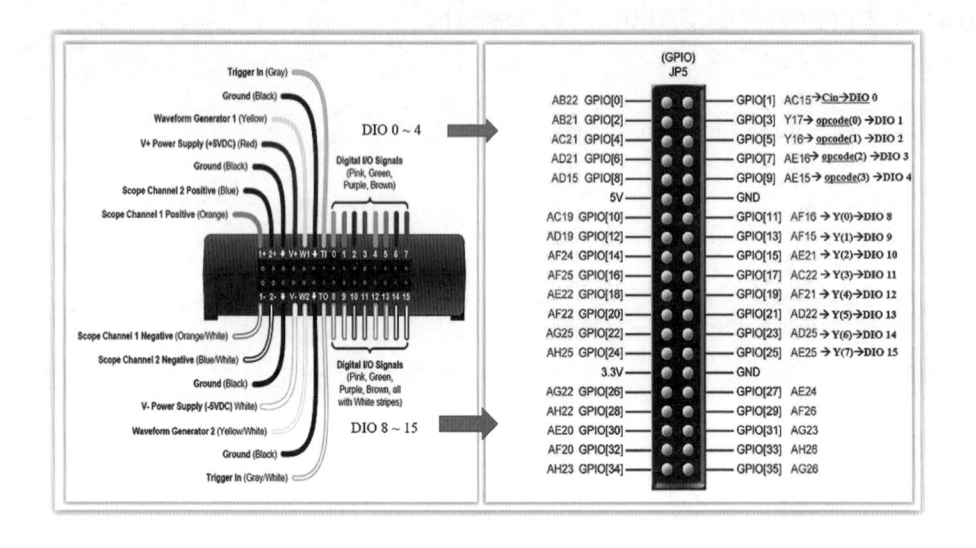

◎ 圖 T1-5　Analog Discovery 和 DE2-115 JP-5 之間的接線關係

其中 a0~a7 和 b0~b7 如果設定為常數可以直接連接 DE2-115 的 Slide Switches 上 以補救 Analog Discovery Digital I/O 僅 16 條連線的不足。如圖 T1-6 所示。

	FPGA Pin No.	Description
a(0) →	PIN_AB28	Slide Switch[0]
a(1) →	PIN_AC28	Slide Switch[1]
a(2) →	PIN_AC27	Slide Switch[2]
a(3) →	PIN_AD27	Slide Switch[3]
a(4) →	PIN_AB27	Slide Switch[4]
a(5) →	PIN_AC26	Slide Switch[5]
a(6) →	PIN_AD26	Slide Switch[6]
a(7) →	PIN_AB26	Slide Switch[7]
b(0) →	PIN_AC25	Slide Switch[8]
b(1) →	PIN_AB25	Slide Switch[9]
b(2) →	PIN_AC24	Slide Switch[10]
b(3) →	PIN_AB24	Slide Switch[11]
b(4) →	PIN_AB23	Slide Switch[12]
b(5) →	PIN_AA24	Slide Switch[13]
b(6) →	PIN_AA23	Slide Switch[14]
b(7) →	PIN_AA22	Slide Switch[15]

圖 T1-6　DE2-115 的 Slide Switches 和 a( )、b( ) 之間的接線關係

Program EEPROM/EPCS64 跟 FPGA 相似，所不同的 Mode 要改成 Active Serial Programming，File 改或 ALU_115.pof，Device 改成 EPCS64。同時要選用 Program configure，Verify，和 Blank check。因此 Program 所需的時間也比較長。如圖 T1-7 所示。

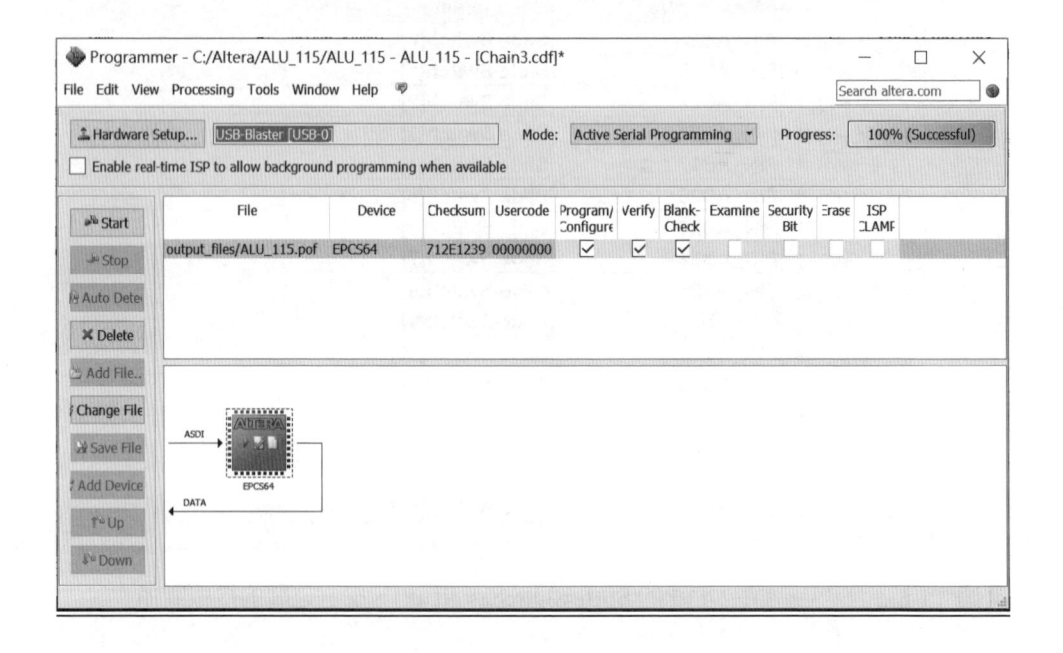

圖 T1-7　Program ALU_115 到 EEPROM/EPCS64 的步驟

1-8　Analog Discovery 測試 FPGA / ALU_115 電路的例子

Analog Discovery 的 Digital Pattern Generator 和 Logic Analyzer 用以測試 ALU_115 電路的動態 Timing Diagram。它跟 DE2-115 間的連線已經在上一節的圖 T1-5 和圖 T1-6 中有詳細的說明。圖 T2-1 是它們的實體照相。

◎ 圖 T2-1　　Analog Discovery 與 DE2-115 對測試 ALU_115 電路間的連線照相

Analog Discovery 測試軟件 Digilent Waveforms 1 中的 Digital out – Patterns 用來產生測試 ALU_115 電路的輸入訊號 cin，和 opcode ( )。Digital in – Analyzer 用來觀測 ALU_115 電路的輸出訊號 Y ( )。如圖 T2-2 所示。

34　iLAB FPGA 數位系統設計、模擬測試與實體除錯

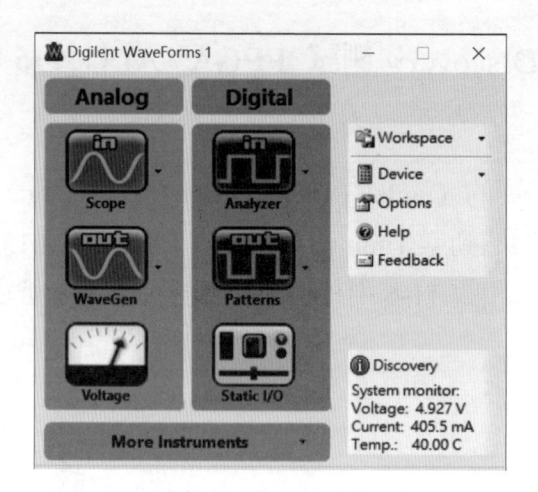

◦ 圖 T2-2 ◦　　　開啟 Analog Discovery 測試軟件 Digilent Waveforms 1

　　Digital out – Patterns 的設定參照圖 1-6：testbench.vhd 測試的波形。其中 a ( ) 和 b ( ) 因為是常數，所以採用 DE2-115 上面的 slide-switches 來完成。Opcode ( ) 為 4 bits，1 ms (1 KHz) 為單位，run 16 ms (20 ms)。Cin 為 '0' 及 '1' 各 8 ms。如圖 T2-3 所示。注意：Digital Pattern Generator 的 Trigger 必須選用 Analyzer。

圖 T2-3：Digital out - Patterns 對測試 ALU-115 電路的輸入設定

第一章　並發邏輯代碼的電路系統　35

　　Digital in – Analyzer 是用來測試電路的波形變化。為了觀察電路的輸入和輸出的相對**時序** (Timing) 關係，因此把輸入和輸出的訊號全部拉到一起，如圖 T2-4 所示。這個測試的結果在 Bus Y 的 binary codes 中顯示出來，跟圖 1-7：simple_ALU 測試結果之 Timing Diagram 完全相同。

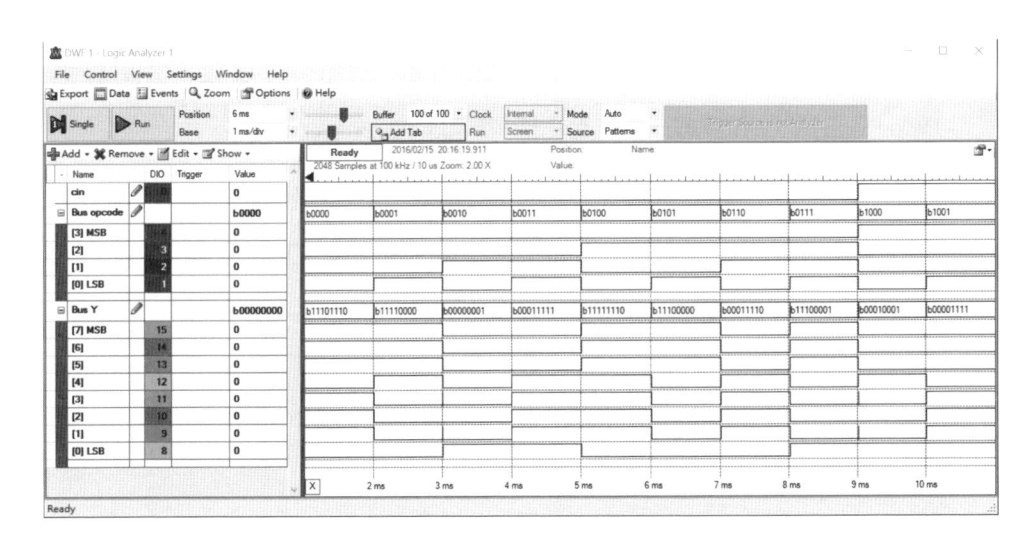

◎ 圖 T2-4　　Digital in - Analyzer 對測試 ALU_115 電路的輸入和輸出訊號 Timing

1-9 課外練習

(1) DE2-115 出廠原裝時，它的 EEPROM/EPCS64 裝有 DE2_115_Default.pof，該 program 每當開啓電源時，會自動測試板子上的所有的顯示器。而在本章的圖 T1-7 為 ALU_115.pof 所取代。試述如何將 DE2_115_Default.pof 裝回去的步驟。

(2) 為了配合 Analog Discovery 的 timing analysis test，每個實驗都會使用 DE2-115 的 J5，試完成一個能夠完整測試 J5 的 Project。

(3) 試述 VHDL 和 C/C++ 在處理 Code 時的最大不同點？

(4) 試述 VHDL 在 Coding 電路的 Simulation 和 Synthesis 的最大不同點？

第二章　順序邏輯代碼的電路系統

順序邏輯 (Sequential logic) 代碼的電路系統，在定義上為：它的輸出不完全由它的輸入而定，該系統中除了有訊號的回授也可能有儲存元件。如圖 2-1 所示。

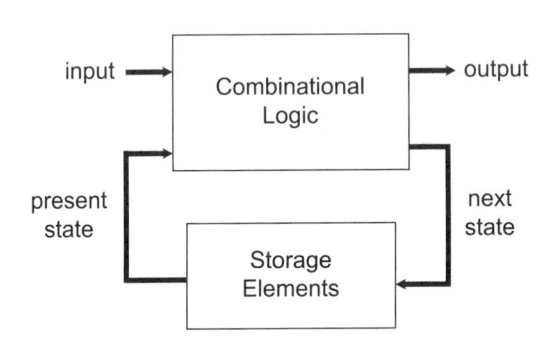

◎ 圖 2-1　　順序邏輯代碼的電路系統

2-1　Sequential logic 電路的幾個例子

　　純順序邏輯聲明為 IF, WAIT, CASE 和 LOOP，這類聲明祇能夠存在於順序代碼的電路系統，也就是包括在 PROCESS, FUNCTION, 和 PROCEDURE 的內部。順序代碼同時也適用於組合電路的設計中。圖 2-2 為將 IF, WAIT, CASE 和 LOOP 用在電路的設計上。

IF

```
IF conditions THEN
   assignments;
ELSIF conditions THEN
   assignments; ...
ELSE
   assignments;
END IF;
```

```
IF x=a AND y=b THEN
   output <= '0';
ELSIF x=a AND y=c THEN
   output <= '1';
ELSE
   output <= 'Z';
END IF;
```

WAIT

```
WAIT UNTIL signal_condition;
```

```
WAIT UNTIL clk'EVENT AND clk='1';
```

CASE

```
CASE identifier IS
   WHEN value => assignments;
   WHEN value => assignments;
   ...
END CASE;
```

```
CASE sel IS
   WHEN 0 => y <= a;
   WHEN 1 => y <= b;
   WHEN OTHERS => y <= c;
END CASE;
```

LOOP

```
[lab:] FOR identifier IN range LOOP
   (sequential statements)
END LOOP [label];
```

```
FOR i IN x'RANGE LOOP
   x(i) <= a(M-i) AND b(i);
END LOOP;
```

圖 2-2　純順序邏輯聲明 IF, WAIT, CASE 和 LOOP 用在電路的設計上

第二章　順序邏輯代碼的電路系統

Example：BCD_counter

圖 2-3 的 BCD_counter 例子，PROCESS 之內包含順序邏輯聲明 IF。

```vhdl
1  -- BCD_counter.vhd --------------------------
2  LIBRARY ieee;
3  USE ieee.std_logic_1164.all;
4  --------------------------------------------
5  ENTITY counter IS
6      PORT (clk, reset : IN STD_LOGIC;
7            digit : OUT INTEGER RANGE 0 TO 9);
8  END counter;
9  --------------------------------------------
10 ARCHITECTURE counter OF counter IS
11 BEGIN
12     PROCESS(clk, reset)
13         VARIABLE temp : INTEGER RANGE 0 TO 10;
14     BEGIN
15         IF (reset = '1') THEN
16             temp := 0;
17         ELSIF (clk'EVENT AND clk='1') THEN
18             temp := temp + 1;
19             IF (temp = 10) THEN temp := 0;
20             END IF;
21         END IF;
22         digit <= temp;
23     END PROCESS;
24 END counter;
```

圖 2-3　PROCESS 之內包含順序邏輯聲明 IF 的一個例子

這個例子的 clk 和 digit 的波形 Timing Diagram 如圖 2-4 所示。

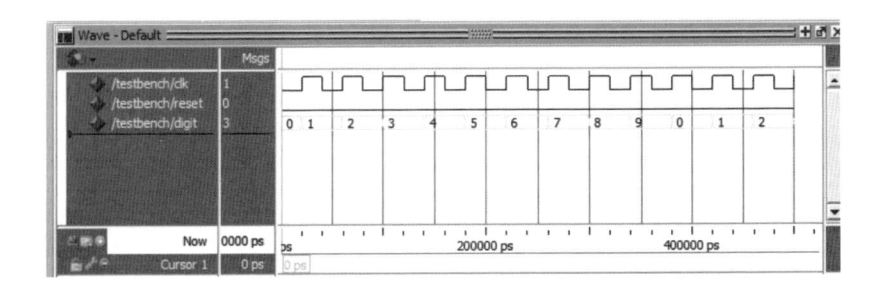

圖 2-4　BCD_Counter 的 clk 和 digit 波形 Timing Diagram

BCD_counter 須經由 BCD to SSG 轉換電路，才能夠推動 SSG 顯示器。轉換電路可由 PROCESS 和 CASE 來獲得，如圖 2-5 所示。

```vhdl
-- BCD2SSD.vhd ---------------------------------
LIBRARY ieee;
USE ieee.std_logic_1164.all;
-----------------------------------------------
ENTITY BCD2SSD IS
    PORT (digit: IN INTEGER RANGE 0 TO 9;
          ssd: OUT std_logic_vector(0 to 6));
END ENTITY;
-----------------------------------------------
ARCHITECTURE BCD2SSD OF BCD2SSD IS
BEGIN
    PROCESS (digit)
    begin
        CASE digit IS
            WHEN 0 => ssd<="0000001";        --"0" on SSD
            WHEN 1 => ssd<="1001111";        --"1" on SSD
            WHEN 2 => ssd<="0010010";        --"2" on SSD
            WHEN 3 => ssd<="0000110";        --"3" on SSD
            WHEN 4 => ssd<="1001100";        --"4" on SSD
            WHEN 5 => ssd<="0100100";        --"5" on SSD
            WHEN 6 => ssd<="0100000";        --"6" on SSD
            WHEN 7 => ssd<="0001111";        --"7" on SSD
            WHEN 8 => ssd<="0000000";        --"8" on SSD
            WHEN 9 => ssd<="0000100";        --"9" on SSD
            WHEN OTHERS => ssd<="0110000";   --"E"rror
        END CASE;
    END PROCESS;
END ARCHITECTURE;
```

圖 2-5　BCD to SSG 轉換電路由 PROCESS 和 CASE 來獲得

這個例子的 digit 和 ssd 的波形 Timing Diagram 如圖 2-6 所示。

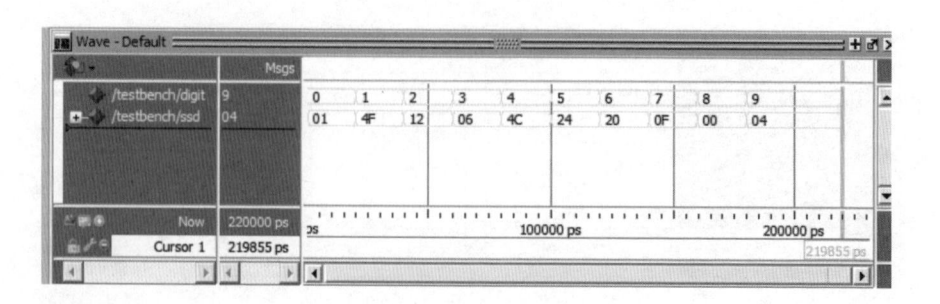

圖 2-6　BCD to SSG 轉換電路的 digit 和 ssd 波形 Timing Diagram

把 BCD_counter 和 BCD to SSG 電路合併在一起，組成 slow_counter 電路的 VHDL 檔如圖 2-7 所示。

```vhdl
1 -- slow_count.vhd Slow 0-9 Counter with SSD ------------
2 LIBRARY ieee;
3 USE ieee.std_logic_1164.all;
4 ------------------------------------------------------
5 ENTITY slow_counter IS
6     PORT ( clk: IN std_logic;
7            reset: IN std_logic;
8            digit: INOUT INTEGER RANGE 0 TO 9;
9            ssd: OUT std_logic_vector(6 DOWNTO 0));
10 END ENTITY;
11 ------------------------------------------------------
12 ARCHITECTURE structure OF slow_counter IS
13
14     component BCD_counter
15         PORT (clk, reset : IN STD_LOGIC;
16               digit : OUT INTEGER RANGE 0 TO 9);
17     END component;
18
19     component BCD2SSD
20         PORT (digit: IN INTEGER RANGE 0 TO 9;
21               ssd: OUT std_logic_vector(6 DOWNTO 0));
22     END component;
23
24 BEGIN
25     U1: BCD_counter port map(clk, reset, digit);
26     U2: BCD2SSD port map(digit, ssd);
27 END structure;
```

◎ 圖 2-7　　BCD_counter 和 BCD to SSG 組成的 slow_counter 電路

　　圖中 U1：BCD_counter 和 U2：BCD2SSD 成為 slow_counter 的 components。port map 是它們連線，U1 的 digit 和 U2 的 digit 名字相同，代表連在一起。

模擬測試 slow_counter 電路的 TestBench.vhd 如圖 2-8 所示。Sequential code 的 PROCESS 和 WAIT 被用來做 reset 時間的控制。WAIT 後面沒列時間，代表無限制地等下去。

```vhdl
1  -- TestBench.vhd -------------------------------
2  LIBRARY ieee;
3  USE ieee.std_logic_1164.all;
4  -----------------------------------------------
5  entity TestBench is
6  end TestBench;
7  -----------------------------------------------
8  use work.all;
9  -----------------------------------------------
10 architecture stimulus of TestBench is
11 -- First, declare component that to be test
12 component slow_counter
13     PORT ( clk: IN std_logic;
14            reset: IN std_logic;
15            digit: INOUT INTEGER RANGE 0 TO 9;
16            ssd: OUT std_logic_vector(6 DOWNTO 0));
17 end component;
18 -- Next, declare TstBench's SINGNALS and DUT
19 SIGNAL   clk: std_logic := '0';
20 SIGNAL reset: std_logic := '0';
21 SIGNAL digit: INTEGER RANGE 0 TO 9;
22 SIGNAL   ssd: std_logic_vector(6 DOWNTO 0);
23 begin
24     DUT: slow_counter port map (clk, reset, digit, ssd);
25 -- Then Testbench signals generation
26     clk <= NOT clk AFTER 20 ns;
27     Process
28     begin
29         reset <='1'; wait for 50 ns;
30         reset <='0'; wait;
31     end process;
32 end stimulus;
```

圖 2-8　Sequential code 的 PROCESS 和 WAIT 被用來做 reset 時間的控制

這個例子的 clk，reset，digit 和 ssd 的波形 Timing Diagram 如圖 2-9 所示。其中 SSD 訊號，使用 Hexadecimal code 來表示。

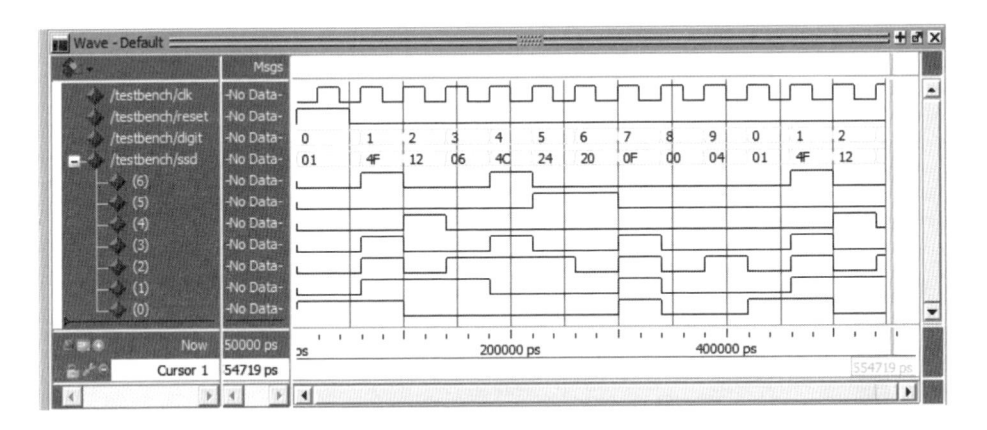

◎ 圖 2-9 ◎　　slow_counter 電路的 clk，reset，digit 和 ssd 的波形 Timing Diagram

Example：Carry–Ripple Adder

圖 2-10 Carry–Ripple Adder 的例子，PROCESS 之內包含 Sequential statements LOOP。使用 LOOP 來做 N 級相同電路的處理上，多能達到省時省力的效果。

```vhdl
-- Carry-Ripple Adder.vhd ----------------------
LIBRARY ieee;
USE ieee.std_logic_1164.all;
--------------------------------------------------
ENTITY carry_ripple_adder IS
    GENERIC (N : INTEGER := 8); --number of bits
    PORT (a, b: IN STD_LOGIC_VECTOR(N-1 DOWNTO 0);
          cin: IN STD_LOGIC;
            s: OUT STD_LOGIC_VECTOR(N-1 DOWNTO 0);
         cout: OUT STD_LOGIC);
END ENTITY;
--------------------------------------------------
ARCHITECTURE structure OF carry_ripple_adder IS
BEGIN
    PROCESS(a, b, cin)
      VARIABLE c: STD_LOGIC_VECTOR(N DOWNTO 0);
    BEGIN
      c(0) := cin;
      FOR i IN 0 TO N-1 LOOP
         s(i) <= a(i) XOR b(i) XOR c(i);
         c(i+1) := (a(i) AND b(i)) OR (a(i) AND c(i)) OR
                   (b(i) AND C(i));
      END LOOP;
      cout <= c(N);
    END PROCESS;
END ARCHITECTURE;
```

圖 2-10　使用 PROCESS 和 LOOP 來完成 Carry-Ripple Adder 的例子

這個例子的 a，b，cin，s，和 cout 的波形 Timing Diagram 如圖 2-11 所示。

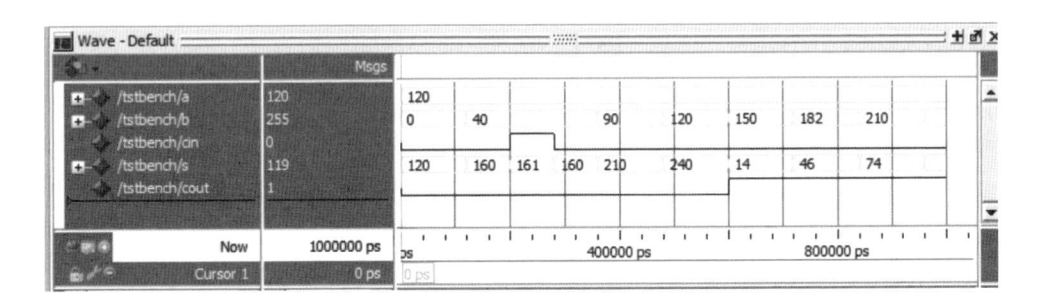

◦ 圖 2-11 ◦ Carry-Ripple Adder 的 Timing Diagram

2-2 順序邏輯電路的 FPGA 合成 (Synthesis)

順序邏輯電路的 FPGA Synthesis 合成和上一章 1-3：並發邏輯電路的 FPGA 合成完全一樣。而且也用到了圖 1-4：DE2-115 的 Expansion Header JP5 和 Slide-Switches。合成軟體 Quartus II / VHDL 的介紹請參照上一章的 1-5 節，為了節省篇幅，不再重複。

2-3　Software Synthesis slow_counter 電路的例子

這個電路來自由 BCD_counter 和 BCD2SSG 電路合併組成的 slow_counter 電路如圖 2-10 所示。首先要產生一個 slow_counter 的 folder，再把 BCD2SSG.vhd，BCD_counter.vhd，和 slow_counter.vhd 置入其中。當 Quartus II 14.1 啟動之後點擊 "**New Project Wizard**" 進入第一階段 S1。後填寫 Directory、Top-Level Entity、和 ProjectType，如圖 S1-1。

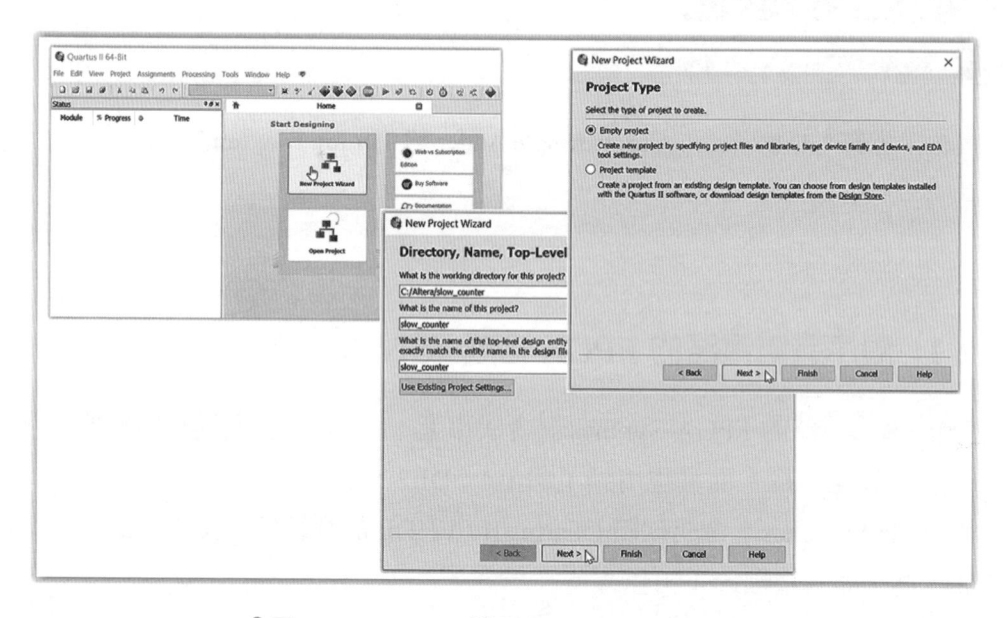

圖 S1-1　　進入 New Project Wizard

第二章　順序邏輯代碼的電路系統　47

接下來單擊 **Add All** 將準備好了的 3 個 VHDL 檔加入到 **Project** 中，並且選用 Cyclone IV E 的 FPGA/EP4CE115F29C7。如圖 S1-2 所示。

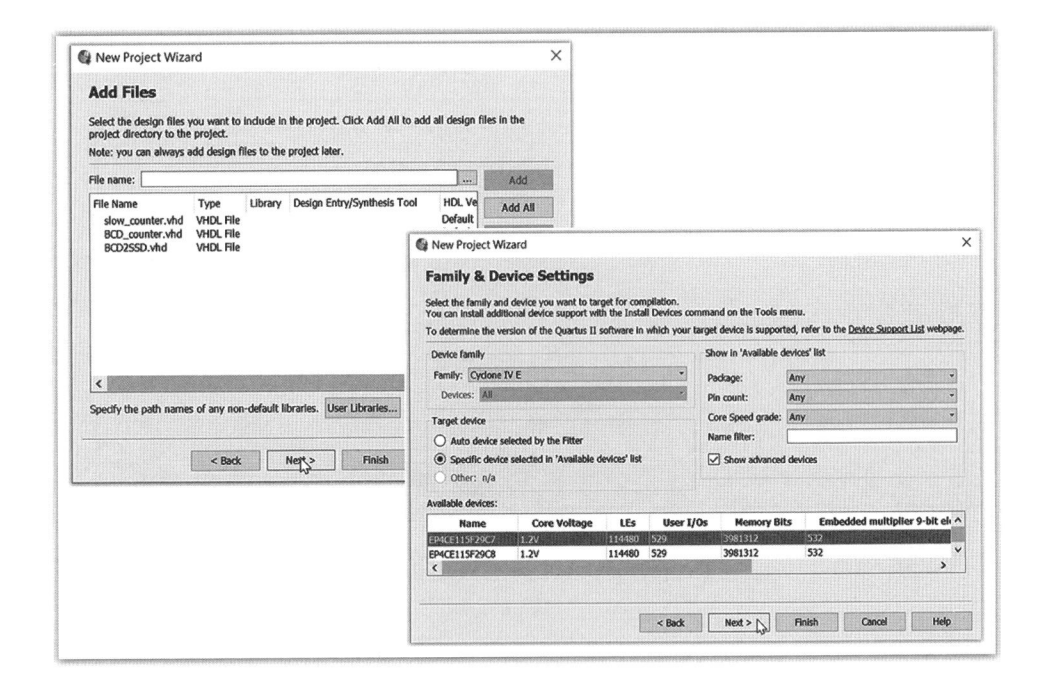

圖 S1-2　加入 Project 所需的 files 和 FPGA/EP4CE115F29C7 的選用

最後顯示 Project EDA 工具的設定和 Summary，如圖 S1-3 所示。

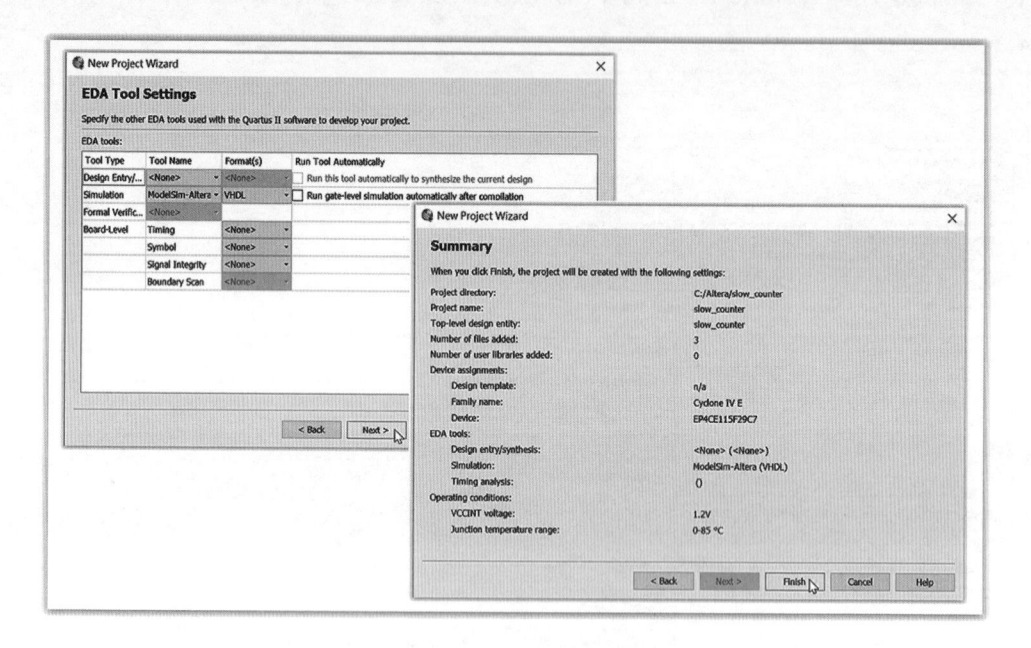

圖 S1-3 Project EDA 工具的設定和 Summary 的顯示

第二章　順序邏輯代碼的電路系統　49

　　選用 **Project** 視窗中的 Assigments > Settings…可以得到如圖 S1-4 以檢視 slow_counter.vhd 存在於 Project 中。

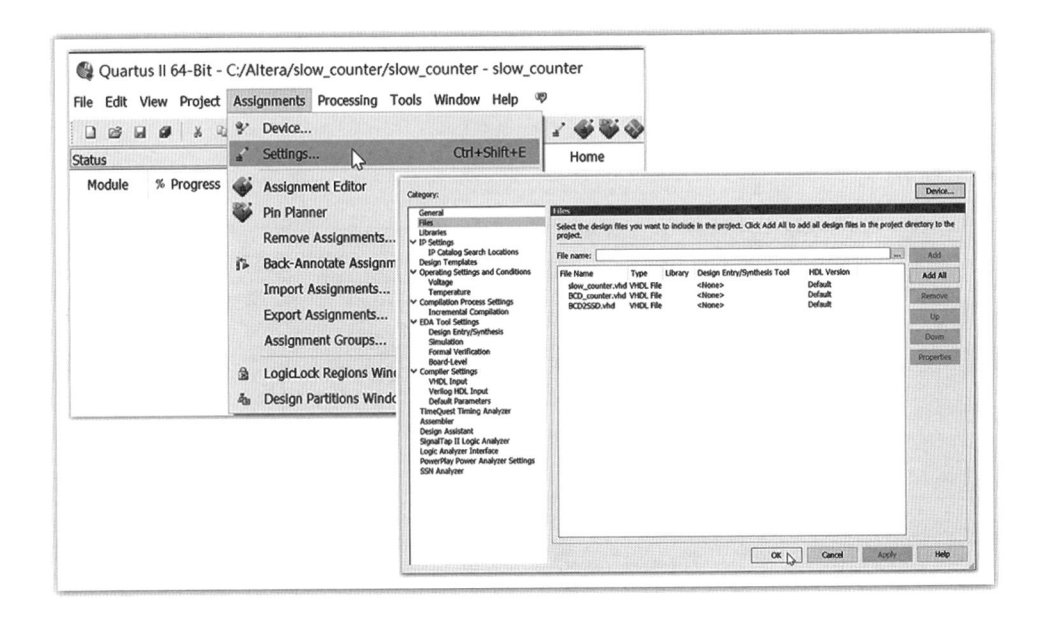

◎ 圖 S1-4 ◎　　Assigments > Settings… 以檢視 slow_counter.vhd 確存於 Project 中

最後還需要用 **Processing > Start Compilation** 來對這 S1 階段做檢驗，如圖 S1-5 所示。Compilation 必須做到 0 errors。

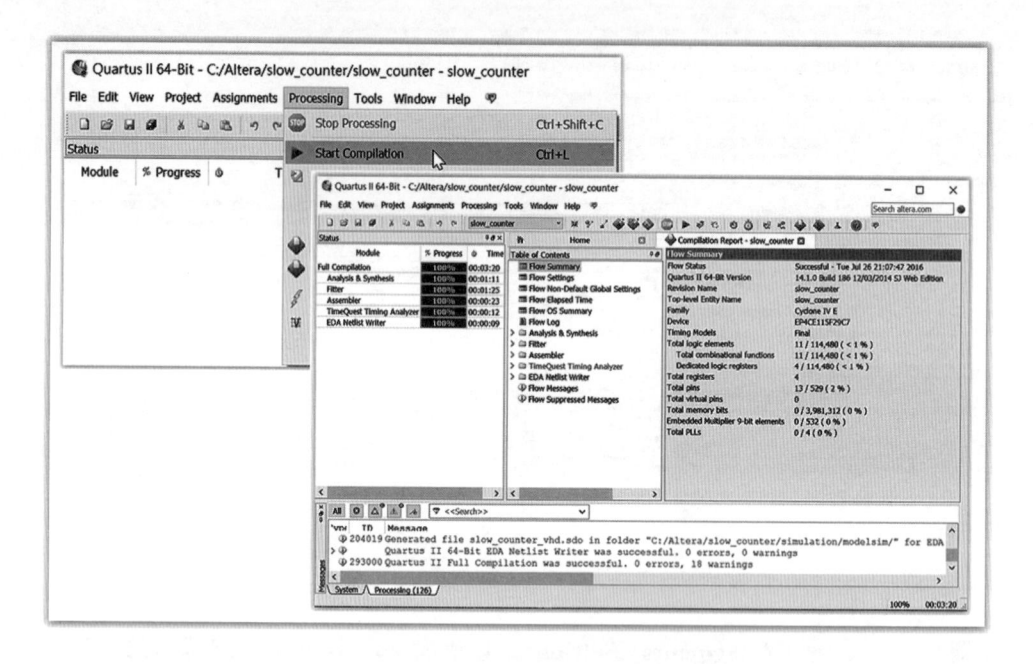

◎ 圖 S1-5 　　用 Compilation 來對這 S1 階段做檢驗，必須做到 0 errors

第 1 階段 Compilation pass 後就可以從 **Assignments > Assignment Editor** 進入第 2 階段的 S2：選用 FPGA/EP4CE115F29C7 的接腳，如圖 S2-1 所示。首先是單擊圖中的 **1**，然後是單擊圖中的 **2**，最後是單擊圖中的 **3**，再單擊圖中的 **4**。在圖中的 Nodes Found 內選取 clk, cin, digit(0)~digit(3), ssd0)~ssd(6) 經 > 移入到 Selected Nodes 內，最後單擊圖中的 **5** 全部完成 **OK**。

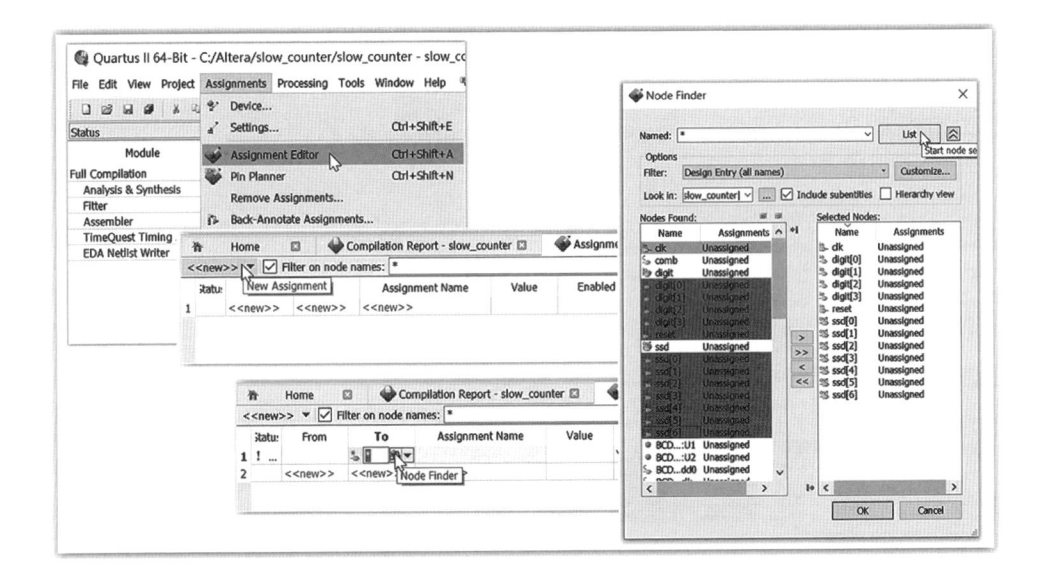

圖 S2-1　FPGA/EP4CE115F29C7 接腳的 Assignment

接下來圖 S2-2 顯示 DE2-115 的 JP5 與 slow_counter 的 I/O 接腳關係。

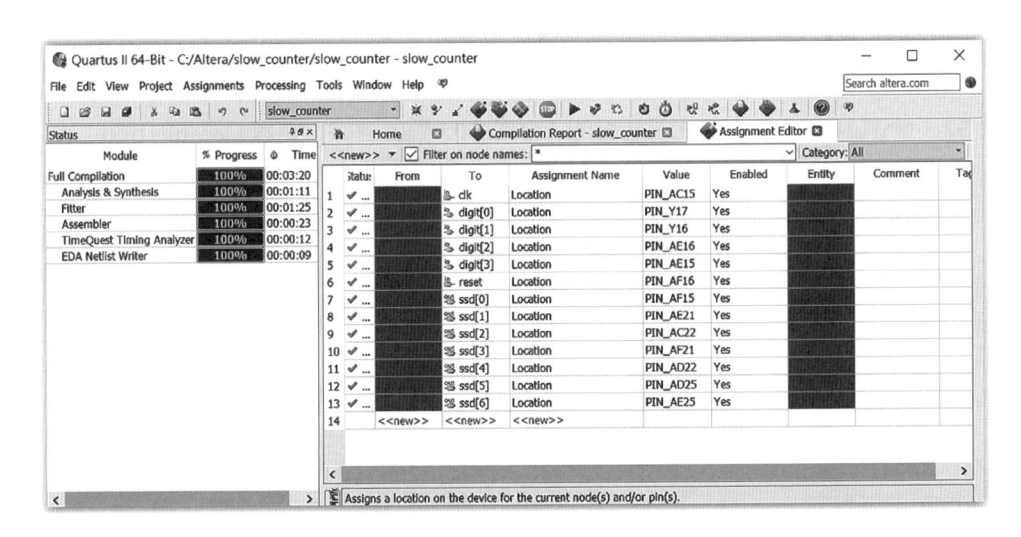

圖 S2-2　DE2-115 的 JP5 與 slow_counter 的 I/O 接腳關係

Slow_counter 電路的 I/O 和 JP5 間的測試連線的 Assignment 過程中是否有錯，可使用 Processing > Start Compilation 如圖 S2-3 加以測試，結果必須為 Full Compilation was successful, 0 errors。

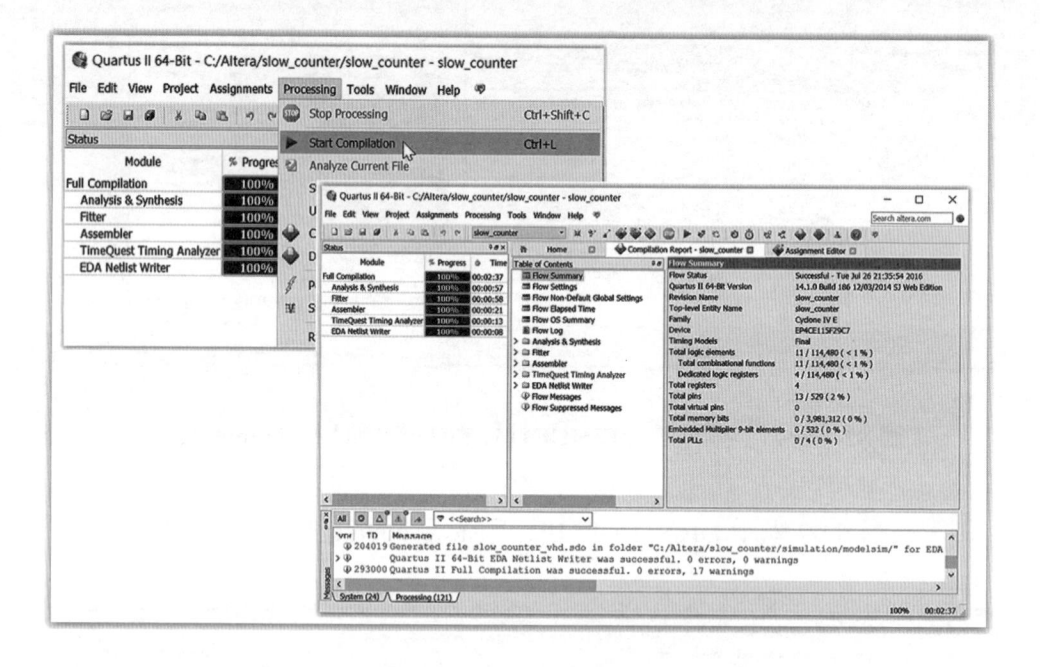

◎ 圖 S2-3 ◎ slow_counter 電路的 I/O 和 JP5 間的連線，通過 Compilation 測試

第二章　順序邏輯代碼的電路系統　53

　　Slow_counter.sof 可用來直接 Programming FPGA/EP4CE115F29C7，但是當電源 OFF，Program 立刻消失。如果事先存入 DE2-115 中的 EEPROM/EPCS64，讓它在電源 ON 時立刻再存入 FPGA。圖 S3-1 為 EEPROM 的選取，它從點擊 **Device and Pin Options...** 進入。是為 S3 階段的開始。

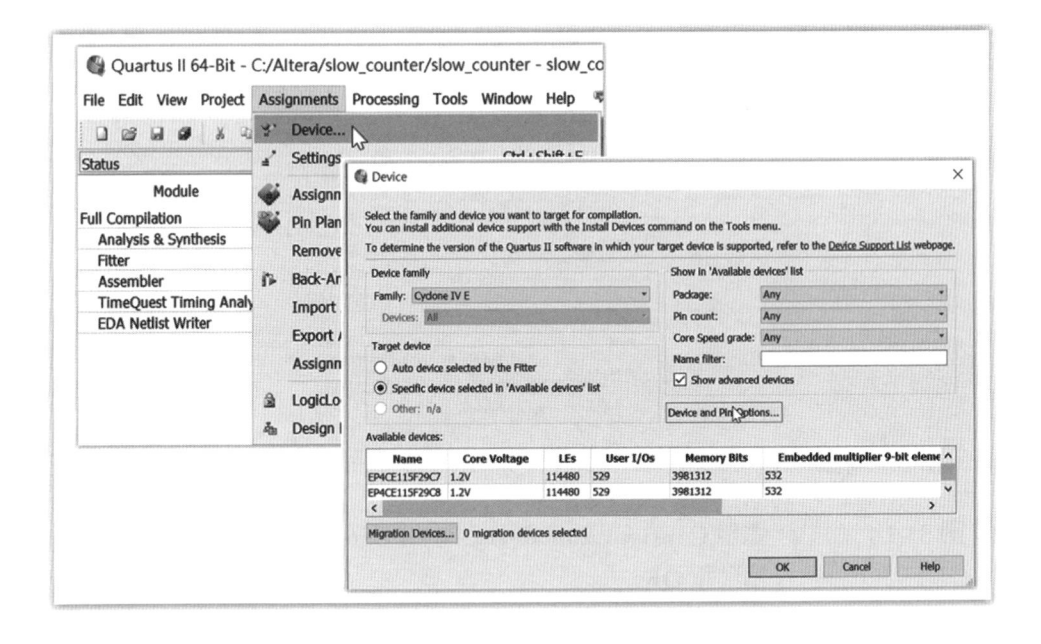

❃ 圖 S3-1 ❃　　DE2-115 中 EEPROM 的選取，從點擊 Device and Pin Options... 進入

當圖 S3-2 的 Device and Pin Options 視窗出現後，選用 **Configuration device** 為 EPCS64。再在 Device 視窗點擊 **OK**。

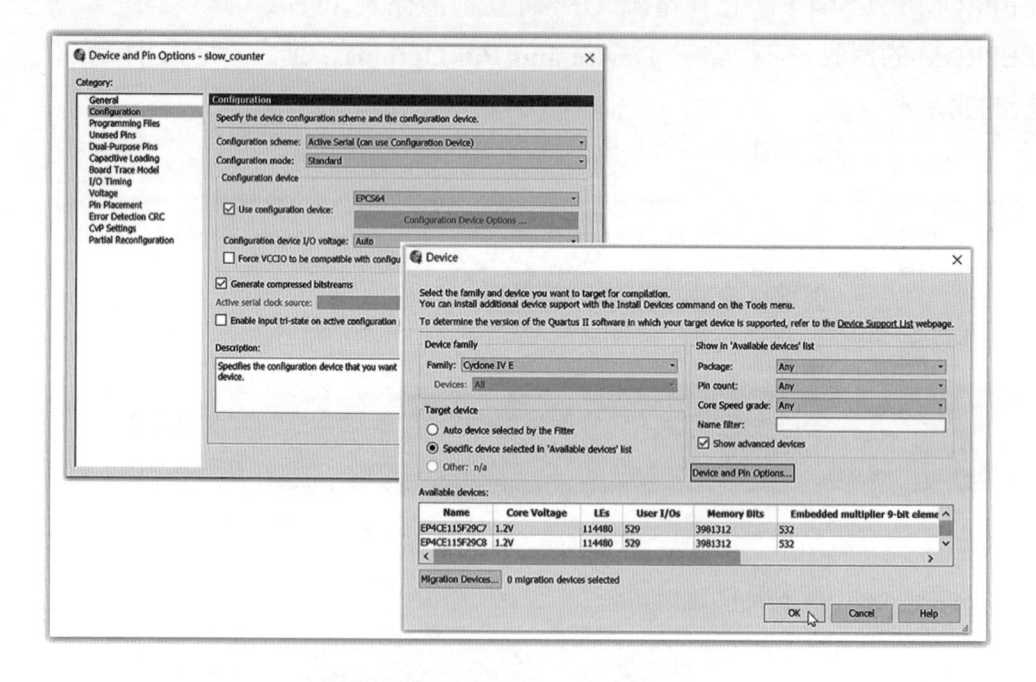

☙ 圖 S3-2 ☙　　選用 Configuration device 為 EPCS64。再在 Device 視窗點擊 OK

Configuration device 為 EPCS64 的選用過程中是否有錯，可以再一次使用 Processing > Start Compilation 如圖 S3-3 加以測試，結果也必須為 Full Compilation was successful, 0 errors。

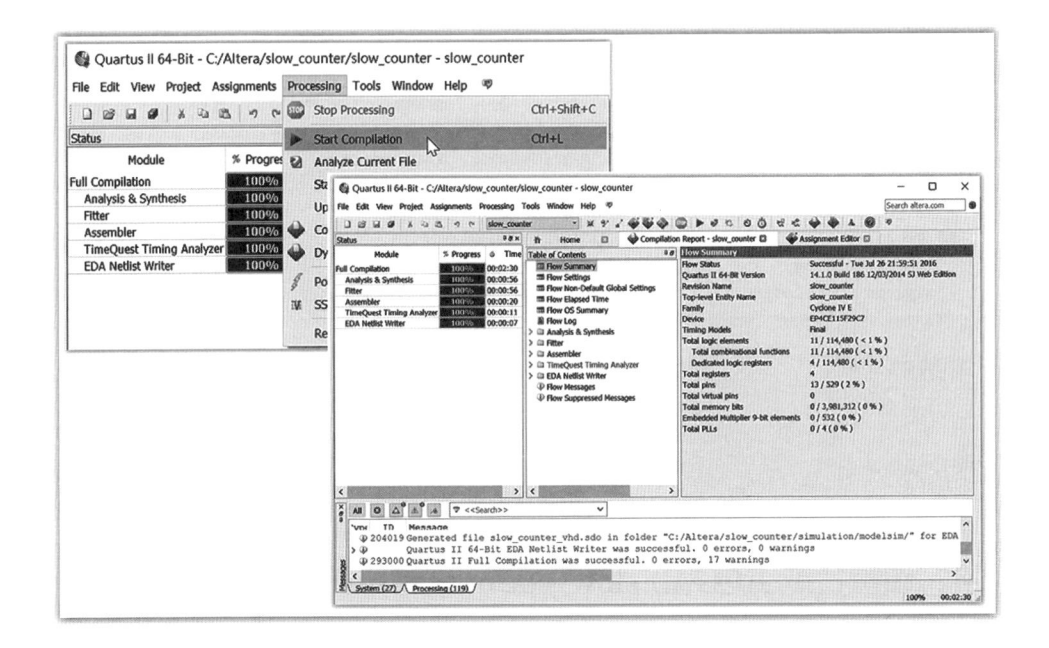

◦ 圖 S3-3 ◦　　　使用 Compilation 來測試 EPCS64 的選用過程中是否有錯

S3 階段 slow_counter 所產生的 output_files Folder 如圖 S3-4 所示，其中的 slow_counter.sof 可以用來直接 Programming DE2-115 中的 EP4CE115F29C7 FPGA。slow_counter.pof 可用來間接 EPCS64 EEPROM。

名稱	修改日期	類型	大小
slow_counter.asm.rpt	2016/7/26 下午 09:59	RPT 檔案	5 KB
slow_counter.done	2016/7/26 下午 09:59	DONE 檔案	1 KB
slow_counter.eda.rpt	2016/7/26 下午 09:59	RPT 檔案	8 KB
slow_counter.fit.rpt	2016/7/26 下午 09:59	RPT 檔案	233 KB
slow_counter.fit.smsg	2016/7/26 下午 09:59	SMSG 檔案	1 KB
slow_counter.fit.summary	2016/7/26 下午 09:59	SUMMARY 檔案	1 KB
slow_counter.flow.rpt	2016/7/26 下午 09:59	RPT 檔案	9 KB
slow_counter.jdi	2016/7/26 下午 09:59	JDI 檔案	1 KB
slow_counter.map.rpt	2016/7/26 下午 09:58	RPT 檔案	25 KB
slow_counter.map.summary	2016/7/26 下午 09:58	SUMMARY 檔案	1 KB
slow_counter.pin	2016/7/26 下午 09:59	PIN 檔案	91 KB
slow_counter.pof	2016/7/26 下午 09:59	POF 檔案	8,193 KB
slow_counter.sof	2016/7/26 下午 09:59	SOF 檔案	3,459 KB
slow_counter.sta.rpt	2016/7/26 下午 09:59	RPT 檔案	103 KB
slow_counter.sta.summary	2016/7/26 下午 09:59	SUMMARY 檔案	1 KB

圖 S3-4　S3 階段 slow_counter 所產生的 output_files Folder

2-4　硬體合成 FPGA / slow_counter 電路的例子

硬體合成 (Programmer) 是將上一節 1-6 的 Software Synthesis slow_counter 電路所獲得的 slow_counter.sof 經 Programmer 置入到 DE2-115 的 FPGA/ EP4CE115F29C7 中或將 slow_counter.pof 經 Programmer 置入到 DE2-115 的 EEPROM/EPCS64 內。要將任何 *.sof 或任何 *.pof 置入到 DE2-115 上，首先要將 DE2-115 接上 Power 和事先已將 USB-Blaster (USB-0) 設定好，並且連接到 PC 的 USB 上。

第二章　順序邏輯代碼的電路系統　57

圖 T1-1 為點擊 File > Open Project 後開啓 slow_counter.qpf。

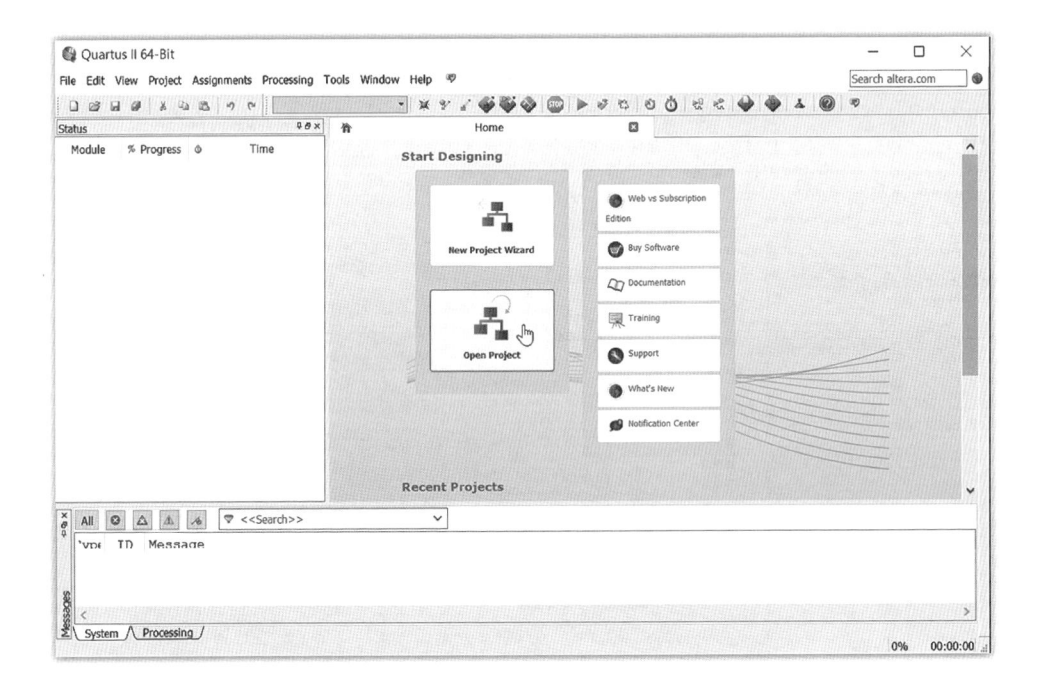

◦ 圖 T1-1 ◦　　Quartus II 開啓 Open Project

58 iLAB FPGA 數位系統設計、模擬測試與實體除錯

結果得到圖 T1-2 為 Quartus II C:/Altera/slow_counter/ slow_counter.qpf

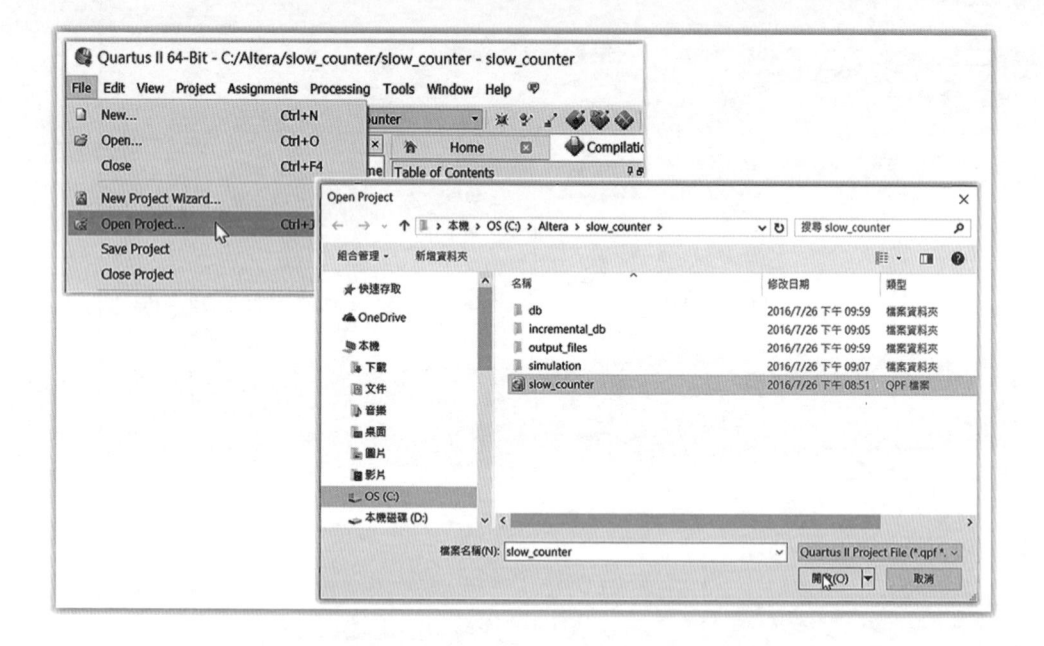

圖 T1-2　Quartus II 開啟 C:/Altera/ slow_counter/ slow_counter.qpf

然後點選 Project 中的 Tool > Programmer 如圖 T1-3 所示。

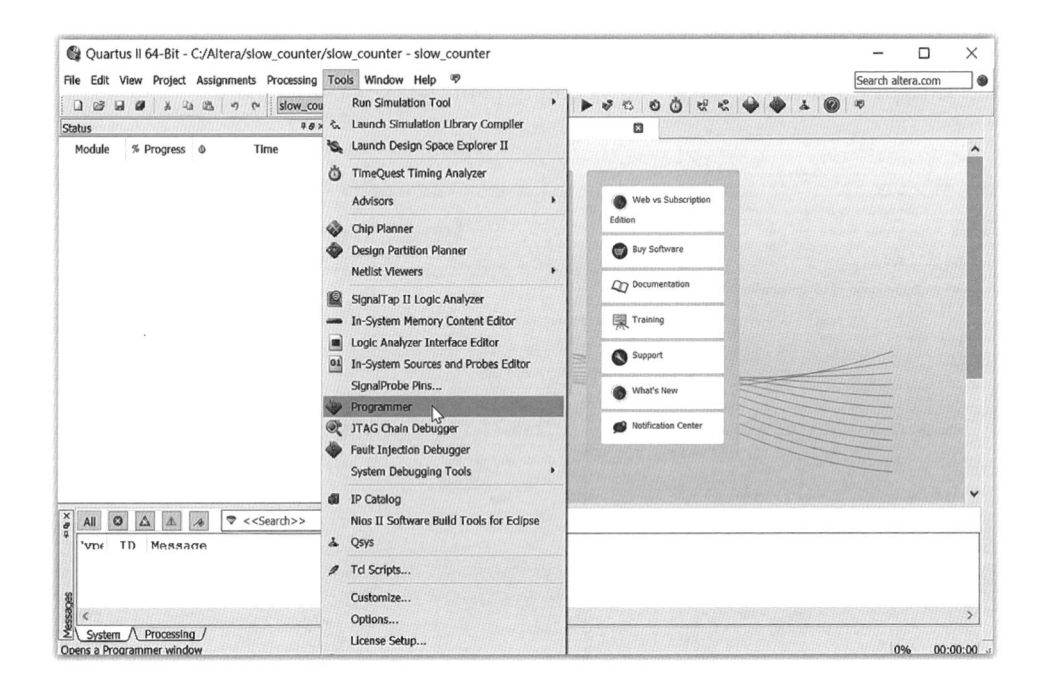

圖 T1-3　點選 Project 中的 Tool > Programmer

Programmer 視窗，它顯示的是 USB-Blaster(USB-0) Driver 已經設定。電源也已開啟，同時 Cable 已連接 DE2-115 到 PC 上，Program FPGA/EP4CE115F29 需要的 Mode 為 JTAG，點選 **Add File** 並從 output_files 的 folder 中選用 slow_counter.sof 如圖 T1-4 所示。

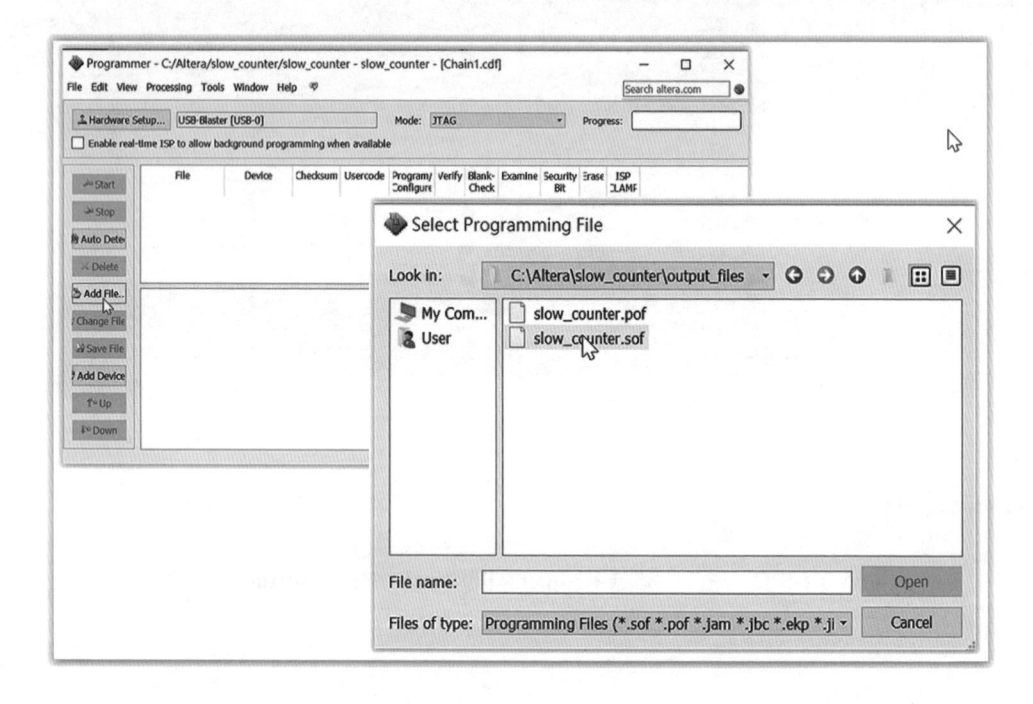

◦ 圖 T1-4 ◦　　　**Program FPGA/EP4CE115F29 的設定及選項**

最後才點擊 **Start**，如圖 T1-5 所示。如果沒做錯當顯示 100% (Successful)。

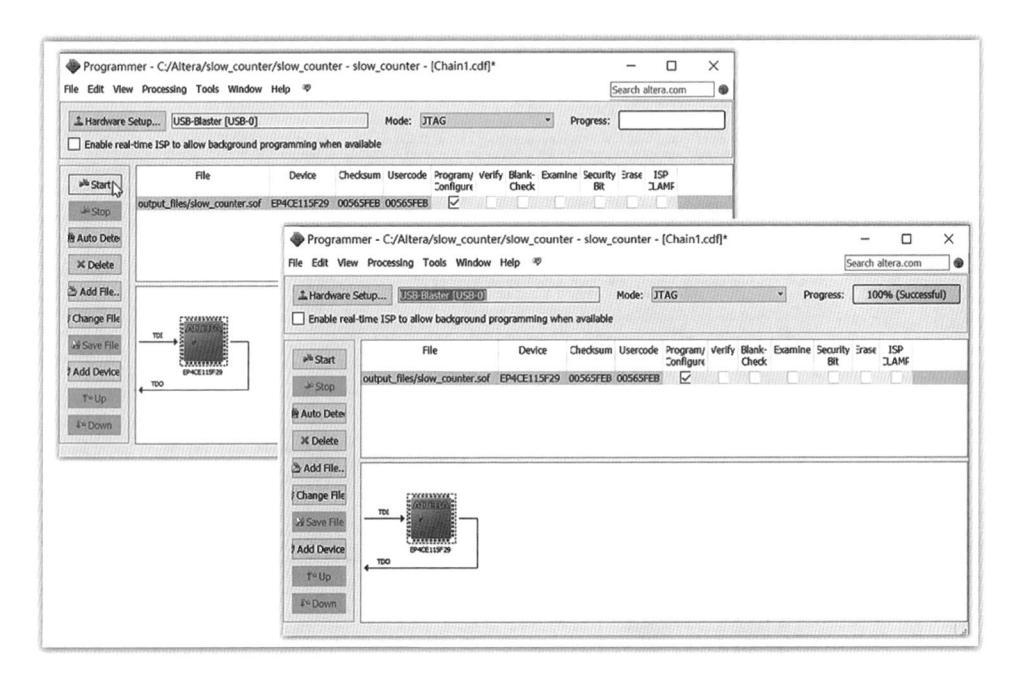

◦ 圖 T1-5 ◦———— 開始 Program FPGA/ EP4CE115F29 並且成功

Program EEPROM/EPCS64 跟 FPGA 相似，所不同的 Mode 要改成 Active Serial Programming，File 改或 slow_counter.pof，Device 改成 EPCS64。同時要選用 **Program configure**，**Verify**，和 **Blank check**。因此 Program 所需的時間也比較長。如圖 T1-6 所示。

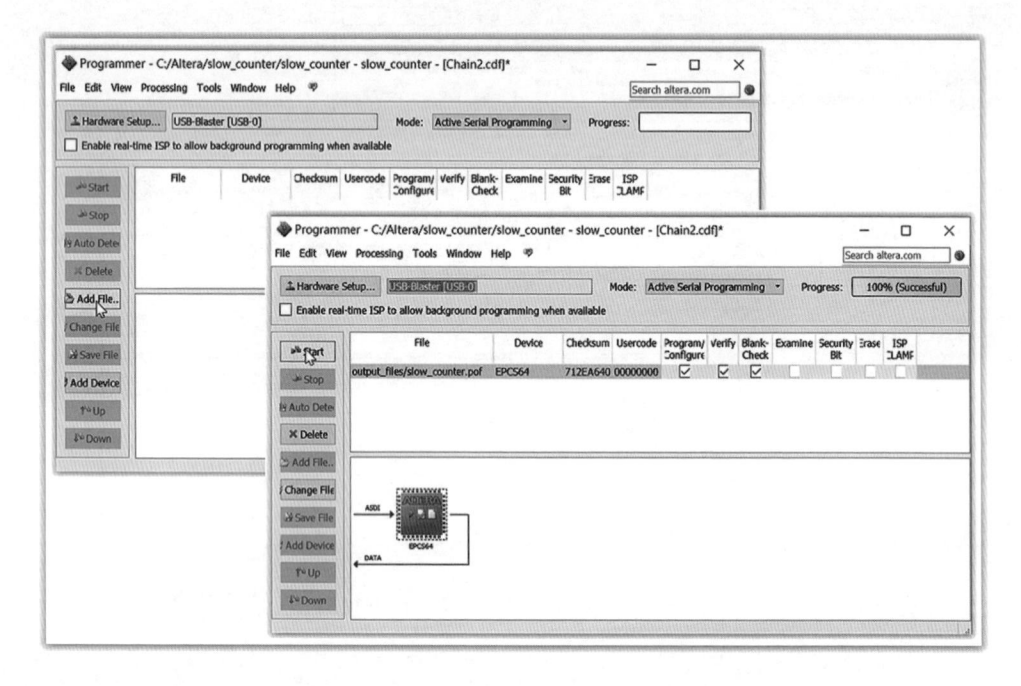

圖 T1-6　　Program slow_counter.pof 到 EPCS64 的步驟

2-5　Analog Discovery 測試 FPGA / slow_counter 電路的例子

　　Analog Discovery 的 Digital Pattern Generator 和 Logic Analyzer 用以測試 slow_counter 電路的動態 Timing Diagram。它跟 DE2-115 間的連線已經在上一節的圖 T1-5 和圖 T1-6 中有詳細的說明。圖 T2-1 是它們的實體照相。

◎ 圖 T2-1 　　Analog Discovery 與 DE2-115 對測試 slow_counter 電路間的連線照相

為了最後能夠用 Analog Discovery 的 Digital Pattern Generator 和 Logic Analyzer 來測試 slow_counter 電路的 Logic Timing Diagram，如圖 T2-2 所示，爲 Analog Discovery 和 DE2-115 JP-5 之間的接線關係。

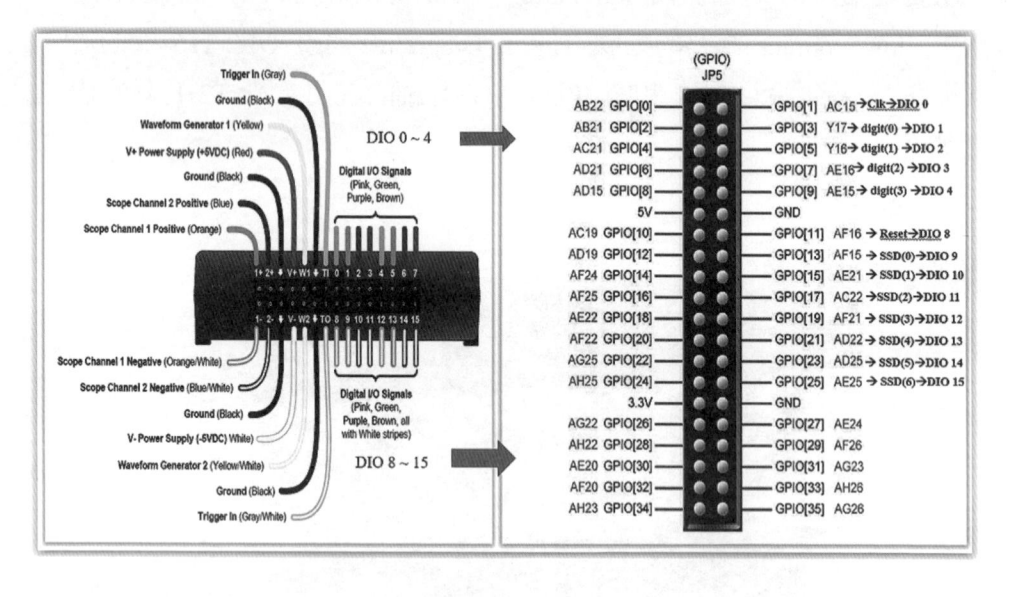

☙ 圖 T2-2 ❧　　Analog Discovery 和 DE2-115 JP-5 之間的接線關係

Analog Discovery 測試軟件 Digilent Waveforms 1 中的 Digital out – Patterns 用來產生測試 slow_counter 電路的輸入訊號 clk，digit ( ) 和 reset。Digital in – Analyzer 用來觀測 slow_counter 電路的輸出訊號 SSD( )。如圖 T2-3 所示。

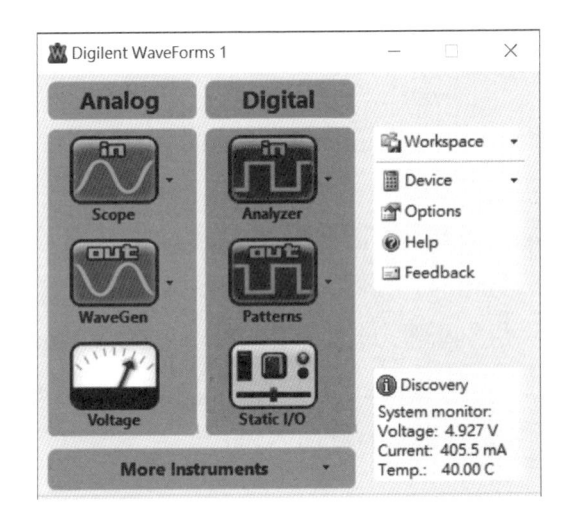

◎ 圖 T2-3　　　開啟 Analog Discovery 測試軟件 Digilent Waveforms 1

Digital out – Patterns 的設定參照圖 2-9：slow_counter 電路的 clk，reset，digit 和 ssd 的 Timing Diagram。其中 clk 為 1 KHz，reset 為 1.5 ms 單一脈波和 digit 採用 BCD code，如圖 T2-4 所示。注意：Digital Pattern Generator 的 Trigger 必須選用 **Analyzer**。輸出，如圖 T2-5 所示。

◎ 圖 T2-4　測試 slow_counter 的 Digital Pattern Generator 設定

Digital in – Analyzer 是用來測試電路的波形變化。為了觀察電路的輸入和輸出的相對**時序** (Timing) 關係，因此把輸入和輸出的訊號全部拉到一起，如圖 T2-5 所示。這個測試的結果由 7 bits SSD( ) binary codes 中顯示出來，跟圖 2-9 的 slow_counter 測試結果之 Timing Diagram 完全相似。(注意：模擬測試和實體測試的時間不同)

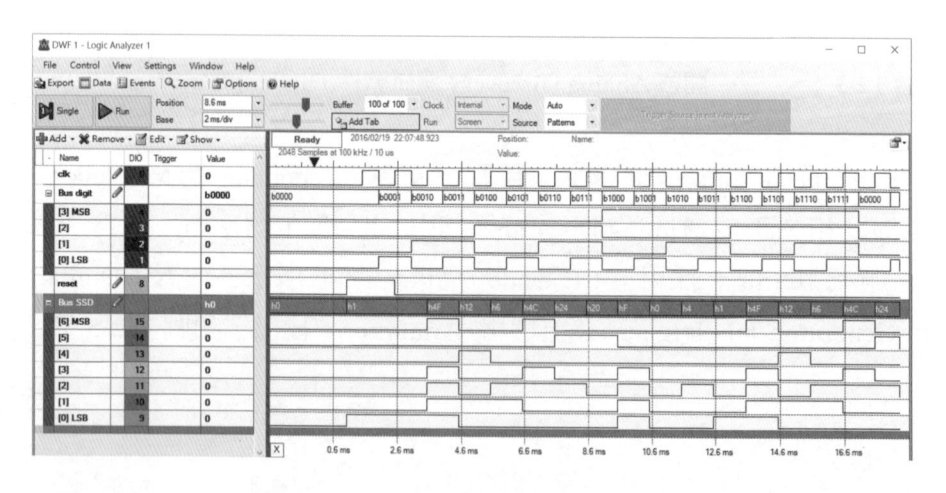

◎ 圖 T2-5　測試 slow_counter 的結果 SSD( ) 由 Logic Analyzer 顯示出來

2-6 課外練習

(1) Loop 代碼內可以用 Exit 的，功能是在條件適合時，立刻退出現在的 Loop。以下為計數每組 8 bits data 從左到右有幾個 '0' 的 VHDL。試將該 LeadingZeros 的 Project 用 DE2-115 完成 LeadingZeros.sof 並用 Analog Discovery 來測試它。

```
1 -- LeadingZeros.vhd ------------------
2 LIBRARY ieee;
3 USE ieee.std_logic_1164.all;
4 ------------------------------------
5 ENTITY LeadingZeros IS
6     PORT ( data: IN STD_LOGIC_VECTOR (7 DOWNTO 0);
7            zeros: OUT INTEGER RANGE 0 TO 8);
8 END LeadingZeros;
9 ------------------------------------
10 ARCHITECTURE behavior OF LeadingZeros IS
11 BEGIN
12     PROCESS (data)
13         VARIABLE count: INTEGER RANGE 0 TO 8;
14     BEGIN
15         count := 0;
16         FOR i IN data'RANGE LOOP
17             CASE data(i) IS
18                 WHEN '0' => count := count + 1;
19                 WHEN OTHERS => EXIT;
20             END CASE;
21         END LOOP;
22         zeros <= count;
23     END PROCESS;
24 END behavior;
25 ------------------------------------
```

(2) 試用 Sequential code (Process) 來完成以下 Combinational 電路的設計。

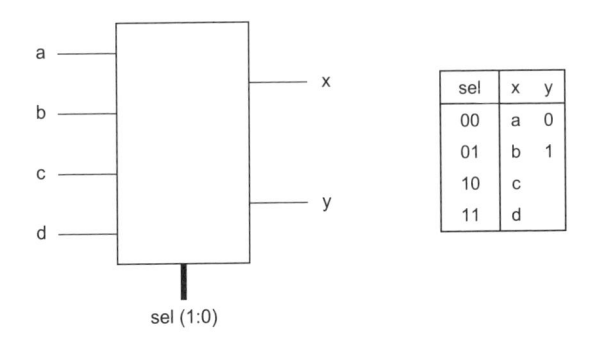

sel	x	y
00	a	0
01	b	1
10	c	
11	d	

(3) 圖 2-5 BCD to SSG 轉換電路為何 SSG 須用 (0 to 6)？

(4) 試述 Concurrent 和 Sequential code 在處理上的不同點？

(5) 試述在混合 Concurrent 在先，Sequential code 在後的處理方式。

第三章　常數、通用、信號與變數

VHDL 的靜態代碼，有 Constant 和 Generic。動態代碼則有 Signal 和 Variable 可供選用。**常數** (Constant)、**通用** (Generic)、**信號** (Signal)、和**變數** (Variable) 可以在並發電路或順序電路的代碼中設定。常數、通用、和信號可供整個 Program 之參用，所以被稱為 Global。變數則不同，它祇能做順序程序的內部設定和參用，因此被稱為 Local。

3-1　Signal and Variable 的比較

信號和變數的選用，並不是一件直截了當的事，它們間的主要異同，如圖 3-1 所示。

Rule	SIGNAL	VARIABLE
1. Local of declaration	ENTITY, ARCHITECTURE, BLOCK, GENERATE, and PACKAGE (declaration in sequential code is forbidden)	Only in sequential units (PROCESS and subprograms), except shared variables (declared in ENTITY, ARCHITECTURE, BLOCK, GENERATE, PACKAGE)
2. Scope (local of use and of modification)	Can be global (used and modified anywhere in the code)	Local (used and modified only inside its own sequential unit), except shared variables (can be global, but modified by only one sequential unit)
3. Update	New value available only at the end of the current cycle	Updated immediately (new value ready to be used in the next line of code)
4. Assignment operator	Values are assigned using "<=" Example: sig<=5;	Values are assigned using ":=" Example: var:=5;
5. Multiple assignments	Only one assignment is allowed	Multiple assignments are fine (because update is immediate)
6. Inference of registers	Flip-flops are inferred when an assignment to a signal occurs at the transition of another signal	Flip-flops are inferred when an assignment to a variable occurs at the transition of a signal and this variable's value eventually affects a signal's value

圖 3-1　信號和變數間的主要異同點

Example：BCD Counter 使用 Signal 和 Variable 的不同結果

圖 3-2 為 BCD counter 電路的 VHDL file，temp1 和 temp2 用同樣的 RANGE。

```
1  -- counter with SIGNAL and VARIABLE ------------------
2  LIBRARY ieee;
3  USE ieee.std_logic_1164.all;
4  ------------------------------------------------------
5  ENTITY counter_s_v IS
6      PORT (clk: IN std_logic;
7            count1, count2: OUT INTEGER RANGE 0 TO 10);
8  END ENTITY;
9
10 ARCHITECTURE dual_count OF counter_s_v IS
11     SIGNAL temp1: INTEGER RANGE 0 TO 10;
12 BEGIN
13     ----- counter_s_v 1: with signal: -----
14     with_sig: PROCESS(clk)
15     BEGIN
16         IF (clk'EVENT AND clk='1') THEN
17             temp1 <= temp1 + 1;
18             IF (temp1 = 9) THEN
19                 temp1 <= 0;
20             END IF;
21         END IF;
22         count1 <= temp1;
23     END PROCESS with_sig;
24     ----- counter_s_v 2: with variable: -----
25     with_var: PROCESS(clk)
26         VARIABLE temp2: INTEGER RANGE 0 TO 10;
27     BEGIN
28         IF (clk'EVENT AND clk='1') THEN
29             temp2 := temp2 + 1;
30             IF (temp2 = 10) THEN
31                 temp2 := 0;
32             END IF;
33         END IF;
34         count2 <= temp2;
35     END PROCESS with_var;
36 END ARCHITECTURE;
```

圖 3-2　選用 Signal 和 Variable 的 BCD counter

同樣為了達到除以 10，line 18 使用 Signal 的 temp1 必須為 9。而 Line 30 使用 Variable 的 temp2 卻須為 9。如圖 3-3 所示，原因是依據圖 3-1 的 Rule 3 Update，temp2 當 clk 在 rising edge 時立刻由 0 增加了 1。而 Temp1 卻須在 clk 完成整個 cycle 後才由 0 增加到 1。

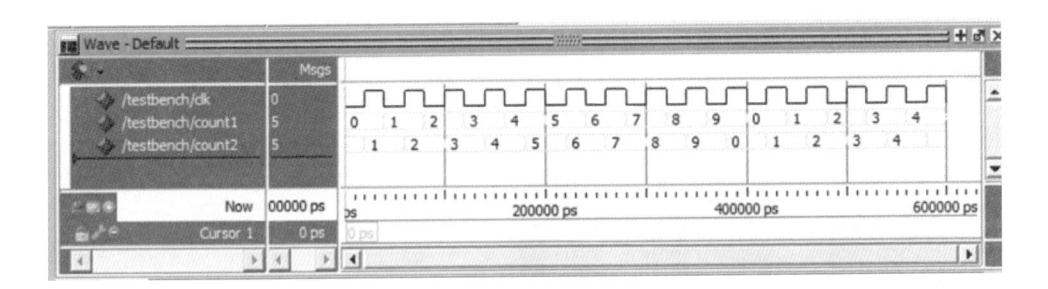

◦ 圖 3-3 ◦ 　選用 Signal 和 Variable 的 BCD counter 的不同結果

圖 3-2 的 *line 7* Signal count1 和 count2 依據圖 3-1 的 Rule 5：Multiple assignments 在整個 Program 中祇能被 assignment 一次，因而在 Process 中使用 temp1 和 temp2 來增加 1，最後在 *line 22* 和 line 34 祇用一次給了 count1 和 count2。

圖 3-1 的 Rule 6：被賦與 Signal 名稱的訊號，在遇到另一個訊號的過渡 transition 時 (如 clk'EVENT and ckl ='1')，隱含著該賦與 *Signal* 使用了 Register。所以圖 3-2 的 line 11 temp1 所造成的 counter 所用的 Registers，要比 line 26 temp2 所造成的 counter 多用了 4 個 Registers。(參考 "Circuit Design and Simulation with VHDL. 2nd" 習題 p.187 over register counter)。

3-2 頻率計的例子

頻率是單位時間內的**數目** (counts) 變化。因此構成**頻率計** (Frequency Meter) 的首要部分為產生一個單位時間窗口，如圖 3-4 右例的 $nT0 = 1$ s 計數用窗口。這個窗口可從電路系統中的 clk 經由一個特殊設計的 counter1 而獲得該計數用窗口。counter2 的作用為 rst > count > rst > count > ....。並且把計數所得的數目存到 register 內，以供讀出。

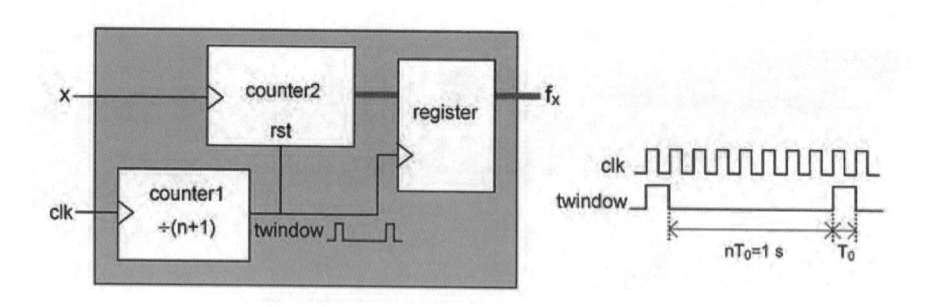

圖 3-4　頻率計的構成方塊圖

第二章的 slow_counter 電路是用 BCD_counter 和 BCD to SSG 二個 components 來組成如圖 2-7 所示。頻率計的構成將改用三個 Process 來構成，如圖 3-5 所示。頭一個 Process 在接受到 system clk 之後產生 twindow。第二個 Process 用 twindow 來計 x 信號的數量，並且暫存在 temp 的 register 中。第三個 Process 再將 temp 的 register 中的數量由 fx 來讀出。

```
-- Freq_meter.vhd ---------------------
LIBRARY ieee;
USE ieee.STD_LOGIC_1164.all;
USE ieee.std_logic_arith.all;
-----------------------------------------
ENTITY freq_meter IS
    GENERIC (fclk: INTEGER := 50;
             fxmax: INTEGER := 255);
    PORT (clk, x: IN STD_LOGIC;
          test: OUT STD_LOGIC;
          fx: OUT STD_LOGIC_VECTOR(7 downto 0));
END freq_meter;
-----------------------------------------
ARCHITECTURE behavioral OF freq_meter IS
    SIGNAL twindow: STD_LOGIC;
    SIGNAL temp: INTEGER RANGE 0 TO fxmax;
BEGIN
-- Generate Time window: -----------------
    PROCESS (clk)
      VARIABLE count1: INTEGER RANGE 0 TO fclk;
    BEGIN
        IF (clk'EVENT AND clk = '1') THEN
            count1 := count1 + 1;
            IF (count1 = fclk - 1) THEN
                twindow <= '1';
            ELSIF (count1 = fclk) THEN
                twindow <= '0';
                count1 := 0;
            END IF;
        END IF;
    END PROCESS;
-- Counts the pulses x: ------------------
    PROCESS (x, twindow)
      VARIABLE count2: INTEGER RANGE 0 TO 255;
    BEGIN
        IF (twindow = '1') THEN
            count2 := 0;
        ELSIF (x'EVENT AND x = '1') THEN
            count2 := count2 + 1;
        END IF;
        temp <= count2;
    END PROCESS;
-- Infers the Ouput Register: -----------
    PROCESS (twindow)
    BEGIN
        IF (twindow'EVENT AND twindow = '1') THEN
            fx <= CONV_STD_LOGIC_VECTOR(temp, 8);
        END IF;
    END PROCESS;
    test <= twindow;
END behavioral;
-----------------------------------------
```

圖 3-5　構成頻率計電路的 VHDL 檔

其中 fclk 所設定的數值跟 system clk 的關係為 time window 等於 system clk 除以 fclk 的數值。所以當圖 3-6 TestBench 的 fclk = 50，system clk 輸入必須為 20 Hz 時，twindow = 1 sec。圖 3-6 TestBench 的 x 訊號輸入為 4 MHz，system clk 為 10 MHz (100ns)，twindow = 50*100ns = 5000 ns。4 MHz 的週期為 250 ns 則 fx 的讀出當為 fx = 5000/250 = 20 − 1 =19，binary code 當為 "00010011"。

```
 1 -- TestBench.vhd -------------------------------
 2 LIBRARY ieee;
 3 USE ieee.STD_LOGIC_1164.all;
 4 ------------------------------------------------
 5 entity TestBench is
 6     GENERIC (fclk: INTEGER := 50;
 7              fxmax: INTEGER := 255);
 8 end TestBench;
 9 ------------------------------------------------
10 use work.all;
11 ------------------------------------------------
12 architecture stimulus of TestBench is
13     component freq_meter
14         PORT(clk, x: IN  STD_LOGIC;
15              test: OUT STD_LOGIC;
16              fx: OUT STD_LOGIC_VECTOR(7 downto 0));
17     end component;
18 ------------------------------------------------
19     SIGNAL  clk: STD_LOGIC := '0';
20     SIGNAL    x: STD_LOGIC := '0';
21     SIGNAL test: STD_LOGIC := '1';
22     SIGNAL   fx: STD_LOGIC_VECTOR(7 downto 0):="00000000";
23 begin
24     DUT: freq_meter port map ( clk, x, test, fx);
25
26     -- Concurrent Code for Periodical waveform
27     clk <= NOT clk AFTER 50 ns;
28       x <= NOT   x AFTER 125 ns;
29 end stimulus;
```

圖 3-6　測試頻率計電路的 TestBench 檔

圖 3-7 測試頻率計電路所得的結果也證實了以上的預測。

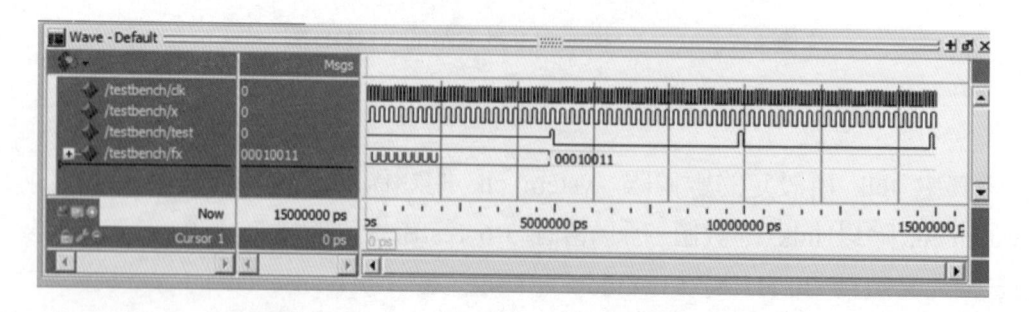

圖 3-7　測試頻率計電路所得的結果

模擬測試所需的時間，並不等於測試硬體所需的真實時間，而且往往高過測試硬體所需的真實時間的很多倍。

3-3 頻率計電路的合成例子

Quartus II 14.1 啟動之後點擊 "New Project Wizard" 進入第一階段 S1。如圖 3-8。

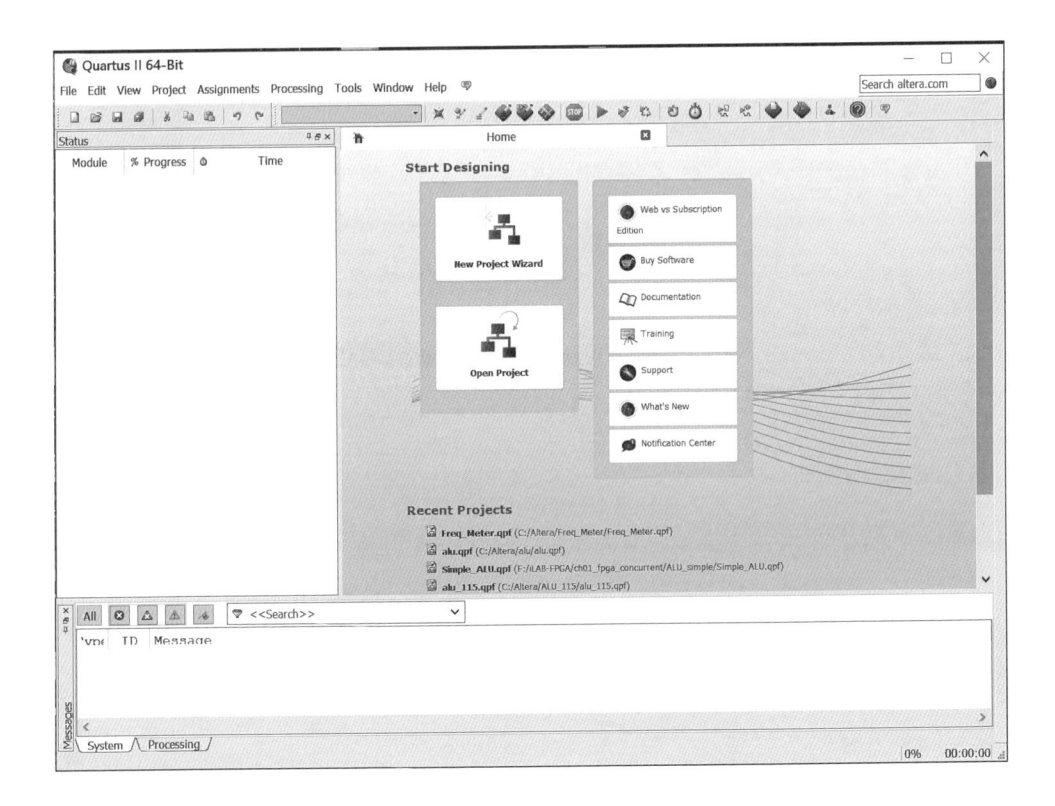

圖 3-8　　進入 New Project Wizard

New Project Wizard 視窗下有 Directory，Top-Level Entity，Project Type，和 Add Files 等視窗需要填寫和選用如圖 3-9 所示。

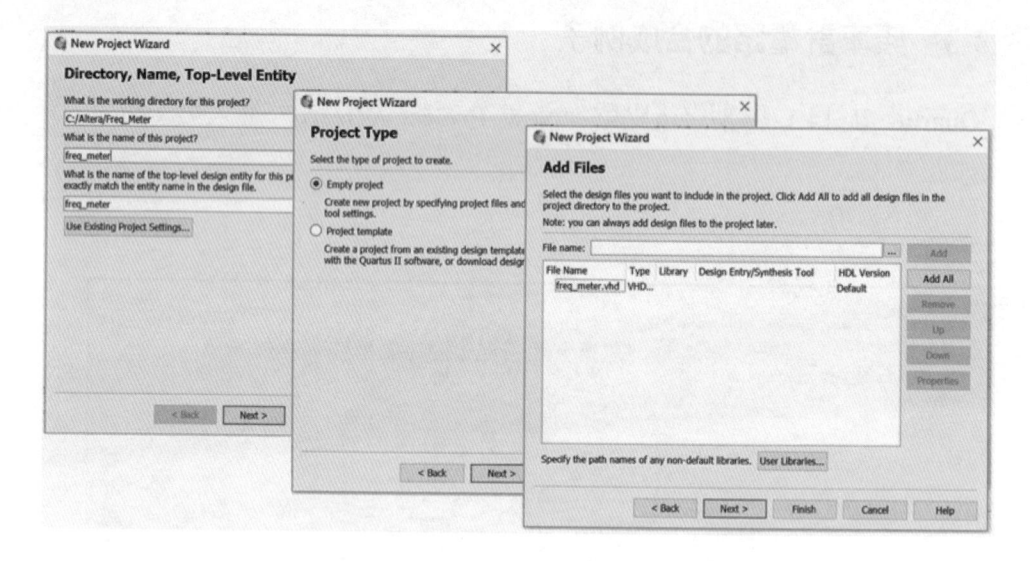

◎ 圖 3-9 ◎　　New Project Wizard 視窗下各視窗需要填寫和選用

第三章　常數、通用、信號與變數　77

最為重要的是 Family & Device Setting，DE2-115 用的是 Cyclone IV E Family 的 EP4CE115F29C7，如圖 3-10 所示。EDA Tool Setting 的 Simulation 的 Tool Name 為 VHDL。最後是 New Project Wizard 的 Summary。

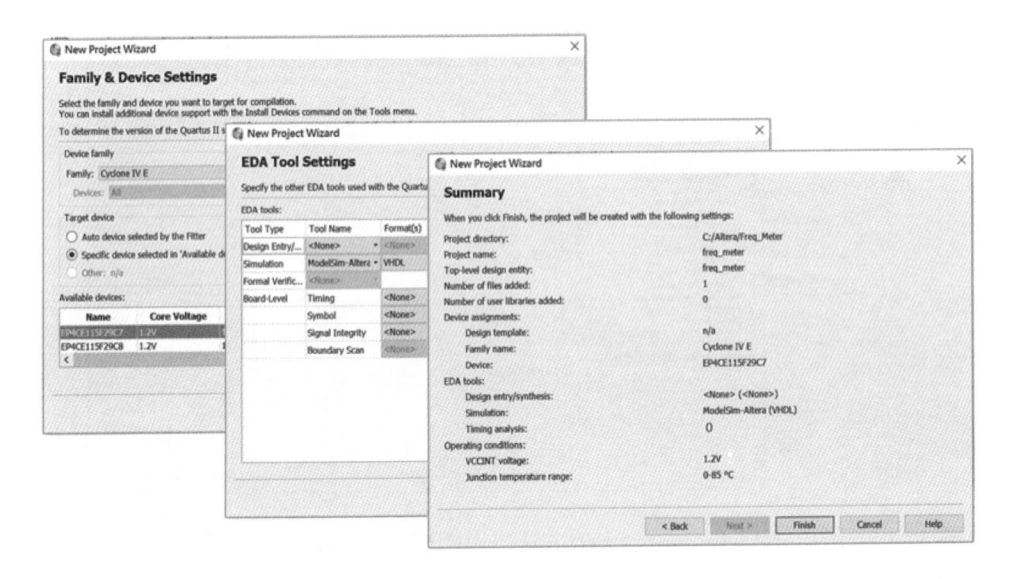

圖 3-10　DE2-115 Family & Device Setting 和 New Project Wizard 的 Summary

選用 **Project** 視窗中的 Assigments > Settings…可以得到如圖 3-11 以檢視 freq_meter.vhd 存在於 Project 中。

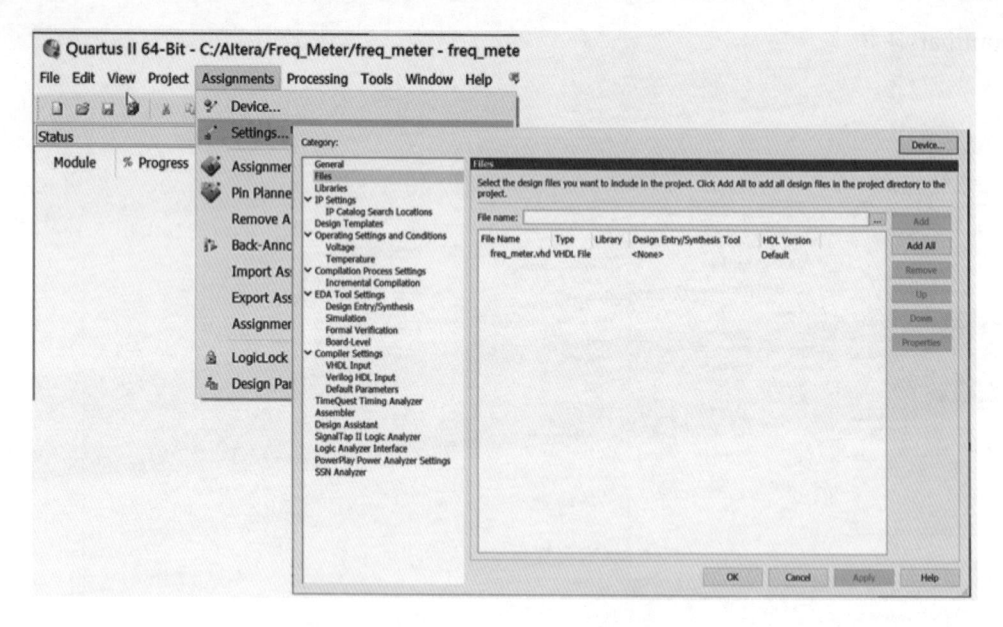

圖 3-11　Assigments > Settings… 以檢視 freq_meter.vhd 存在於 Project 中

第三章　常數、通用、信號與變數　79

最後還需要用 Processing > Start Compilation 來對這階段做檢驗，如圖 3-12 所示。Compilation 必須做到 0 errors。

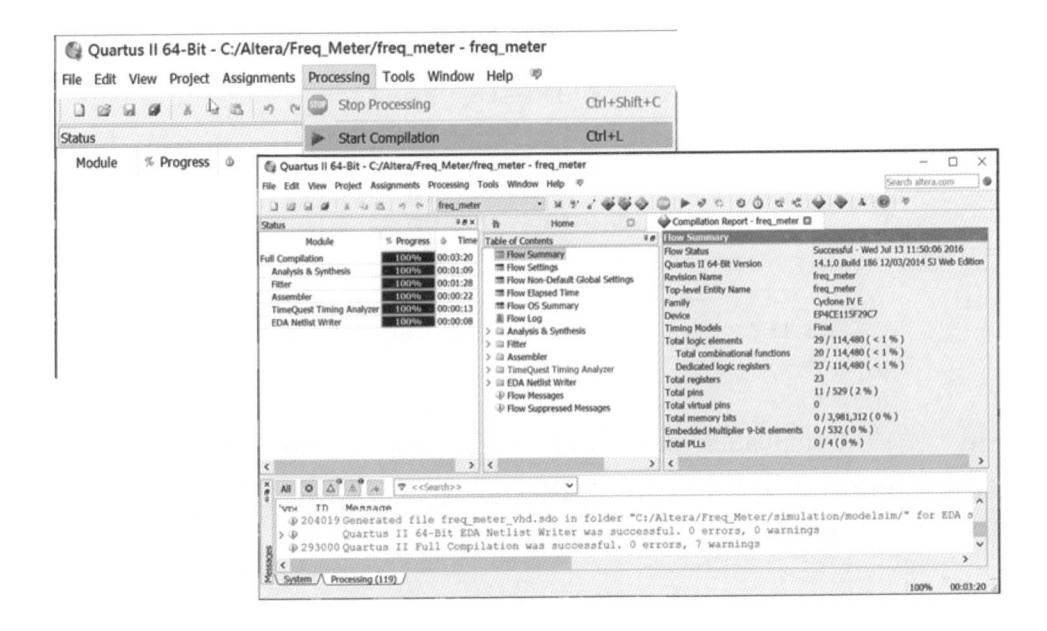

◦圖 3-12 ◦　用 Compilation 來對這個階段做檢驗，必須做到 0 errors

這個階段所產生的 Folder 和 Files 如圖 3-13 所示。

```
Freq_Meter                                              搜尋 Freq_Meter

名稱                          修改日期              類型            大小
  db                          2016/7/13 下午 12:...  檔案資料夾
  incremental_db              2016/7/13 上午 11:...  檔案資料夾
  output_files                2016/7/13 下午 12:...  檔案資料夾
  simulation                  2016/7/13 上午 11:...  檔案資料夾
  freq_meter                  2016/7/13 上午 11:...  QPF 檔案        2 KB
  freq_meter.qsf              2016/7/13 下午 12:...  QSF 檔案        4 KB
  freq_meter                  2016/7/12 下午 07:...  硬碟映像檔      2 KB
```

圖 3-13　　第一階段所產生的 Folder 和 Files

第一階段 Compilation pass 後就可以從 Assignments > Assignment Editor 進入第 2 階段選用 **FPGA/EP4CE115F29C7** 接腳的操作，如圖 3-14 所示。首先是單擊圖中左上角的 **<<new>>**。

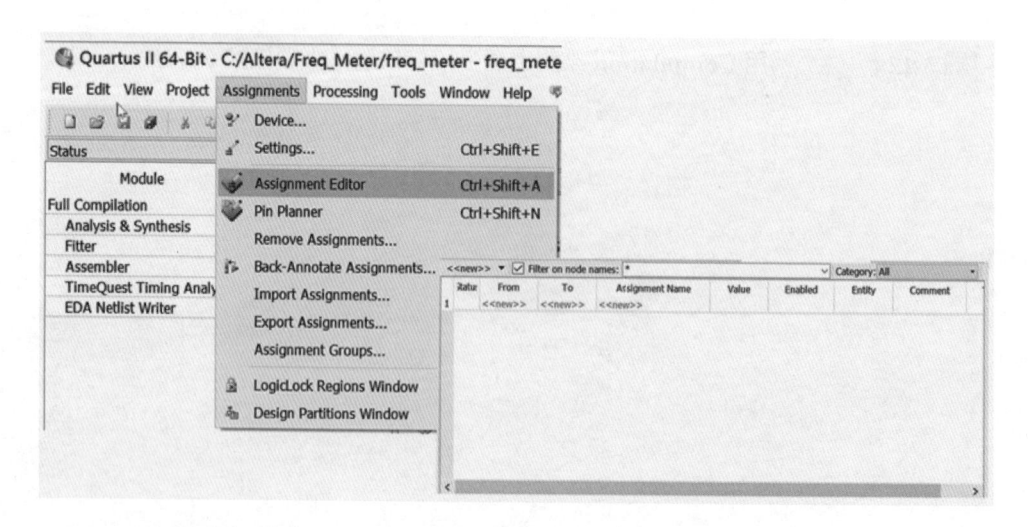

圖 3-14　　FPGA/EP4CE115F29C7 接腳的設定之一

然後是雙擊 Home 視窗 To 下面的望遠鏡。待 Node Finder 視窗出現時，再單擊圖中的 List。然後在圖中的 Nodes Found 內選取 clk，x，test，fx(0) ~ fx(7) 經 > 移入到 Selected Nodes 內，最後再按 OK。如圖 3-15 所示。

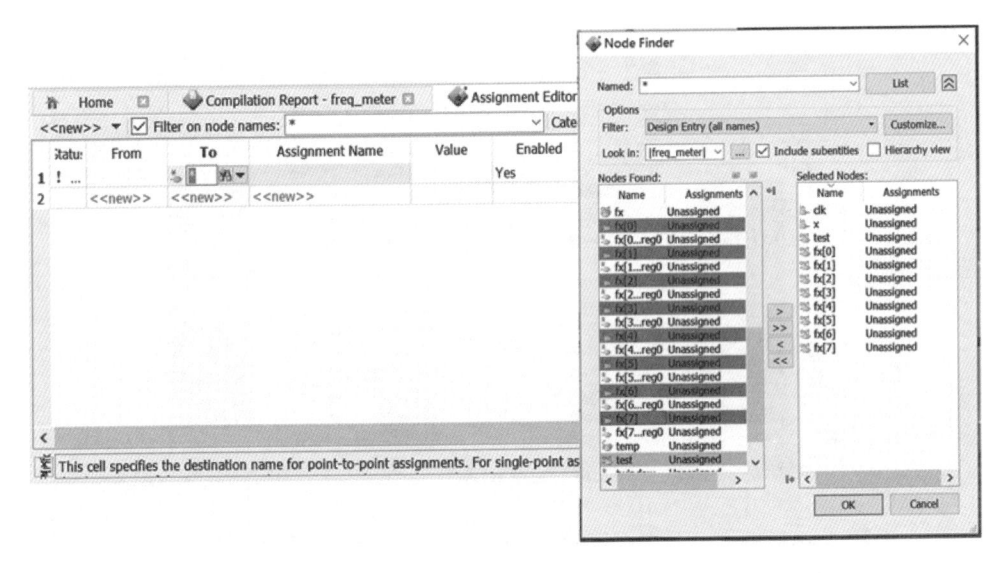

圖 3-15　FPGA/EP4CE115F29C7 接腳的設定之二

EP4CE115F29C7 的接腳的選用，關係到 freq_meter 電路的 I/O 和 DE2-115 Expansion Header JP5 間的測試連線。如圖 3-16 所示。

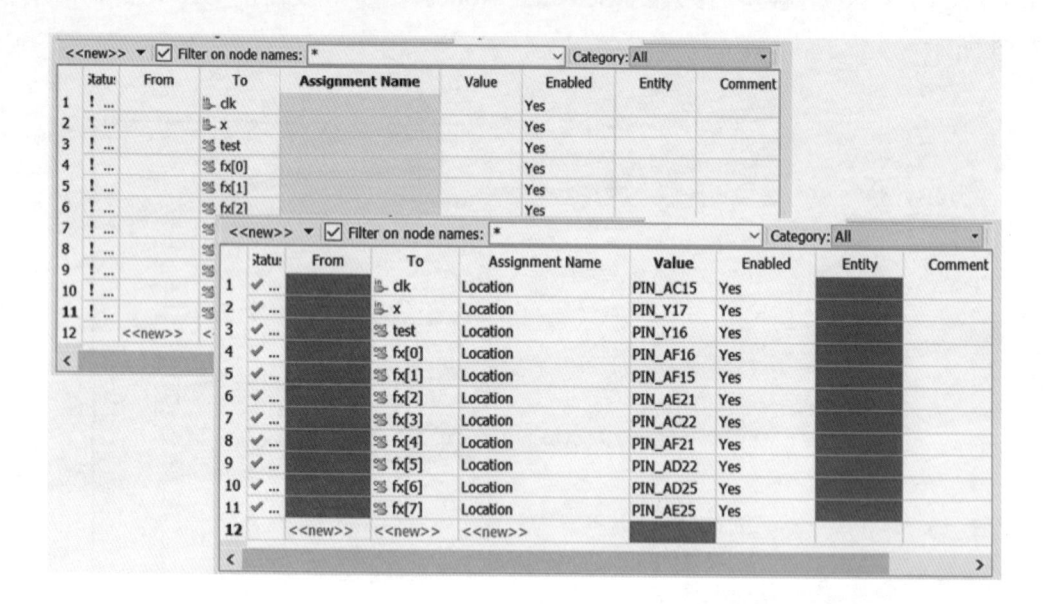

Status	From	To	Assignment Name	Value	Enabled
1		clk	Location	PIN_AC15	Yes
2		x	Location	PIN_Y17	Yes
3		test	Location	PIN_Y16	Yes
4		fx[0]	Location	PIN_AF16	Yes
5		fx[1]	Location	PIN_AF15	Yes
6		fx[2]	Location	PIN_AE21	Yes
7		fx[3]	Location	PIN_AC22	Yes
8		fx[4]	Location	PIN_AF21	Yes
9		fx[5]	Location	PIN_AD22	Yes
10		fx[6]	Location	PIN_AD25	Yes
11		fx[7]	Location	PIN_AE25	Yes

圖 3-16　FPGA/EP4CE115F29C7 接腳的設定之三

freq_meter 電路的 I/O 和 JP5 間的測試連線的 Assignment 過程中是否有錯，可以再一次使用 Processing > Start Compilation 如圖 3-17 加以測試，結果必須爲 Full Compilation was successful, 0 errors。

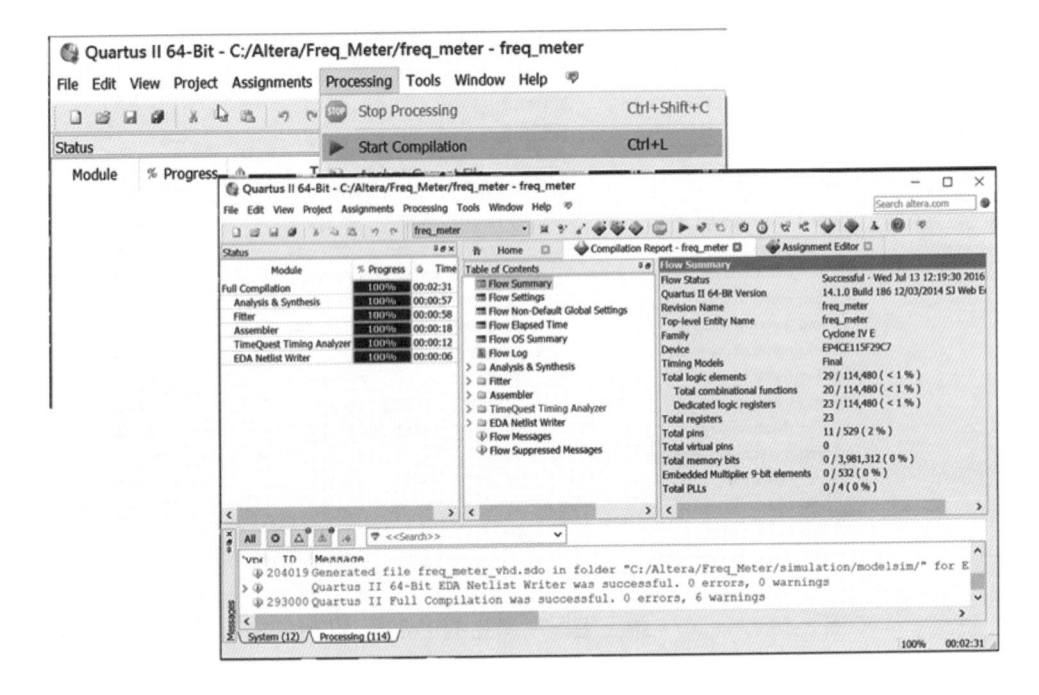

圖 3-17　freq_meter 電路的 I/O 和 JP5 間的連線，通過 Compilation 測試

第二階段所產生的 Folder 和 Files 如圖 3-18 所示。其中的 output_files Folder 裏面的 ALU_simple.sof 可用來直接 Programming FPGA/EP4CE115F29C7。

☉ 圖 3-18　　第二階段所產生的 Folder 和 Files

然後點選 **Project** 中的 **Tool > Programmer**，便有 Programmer 視窗的出現。如圖 3-19 所示。

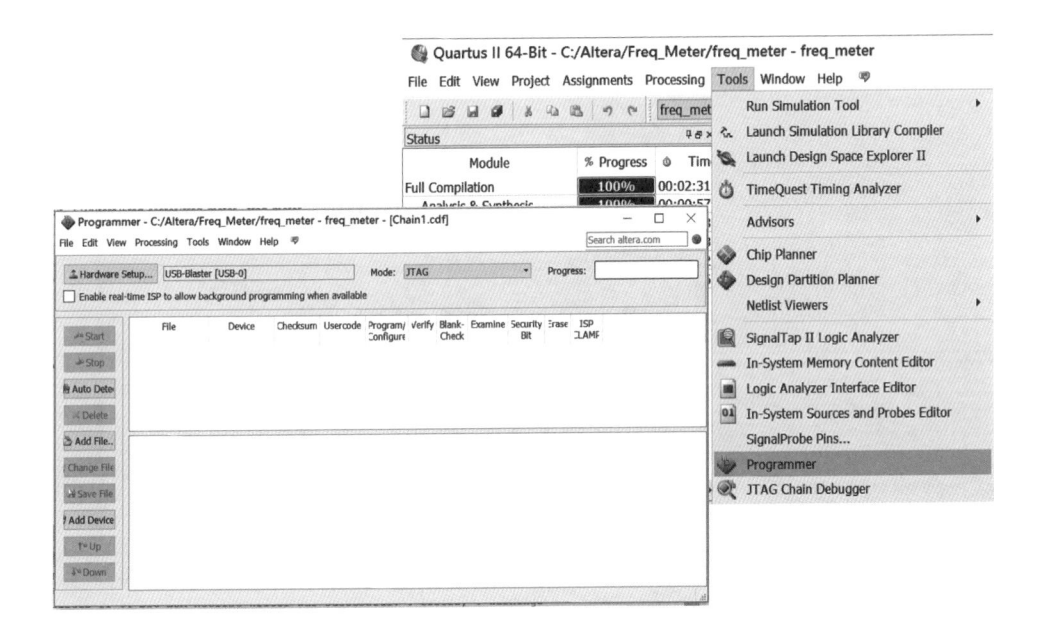

圖 3-19 Quartus II Tools 中的 Programmer 視窗

Programmer 視窗，它顯示的是 USB-Blaster(USB-0) Driver 已經設定。電源也已開啟，同時 Cable 已連接 DE2-115 到 PC 上，Program FPGA/EP4CE115F29 需要的 Mode 為 JTAG，點選 **Add File** 並從 output_files 的 folder 中選用 freq_meter.sof 如圖 3-20 所示。最後才點擊 **Start**，如果沒做錯當顯示 100% (Successful)。

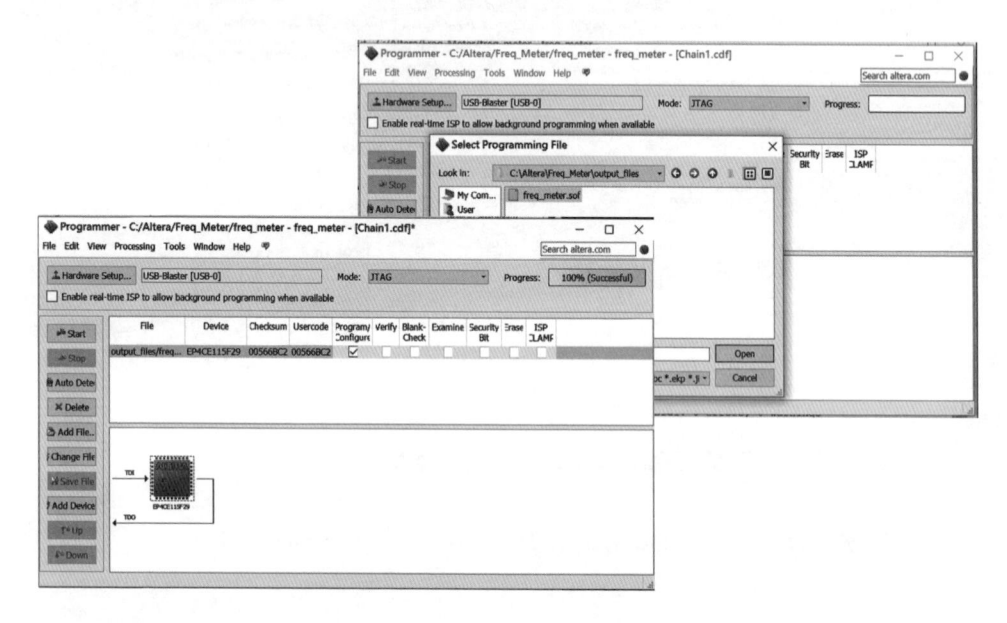

圖 3-20　　　**Program freq_meter 電路到 FPGA/EP4CE115F29**

◎ 3-4　頻率計電路的測試

依據圖 3-16，freq_meter 電路與 DE2-115 JP5 的接腳關係，使用 Analog Discovery 的 Pattern Generator 和 Logic Analyzer 的 DIO 來測試 freq_meter 電路，其連線當如圖 3-21 所示。

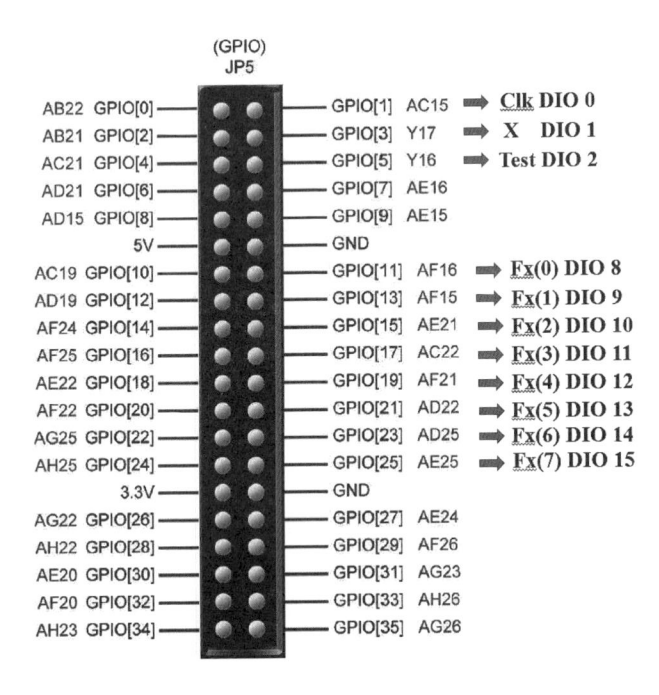

```
                (GPIO)
                 JP5
    AB22 GPIO[0]  ──  GPIO[1]  AC15  →  Clk DIO 0
    AB21 GPIO[2]  ──  GPIO[3]  Y17   →  X   DIO 1
    AC21 GPIO[4]  ──  GPIO[5]  Y16   →  Test DIO 2
    AD21 GPIO[6]  ──  GPIO[7]  AE16
    AD15 GPIO[8]  ──  GPIO[9]  AE15
         5V       ──  GND
    AC19 GPIO[10] ──  GPIO[11] AF16  →  Fx(0) DIO 8
    AD19 GPIO[12] ──  GPIO[13] AF15  →  Fx(1) DIO 9
    AF24 GPIO[14] ──  GPIO[15] AE21  →  Fx(2) DIO 10
    AF25 GPIO[16] ──  GPIO[17] AC22  →  Fx(3) DIO 11
    AE22 GPIO[18] ──  GPIO[19] AF21  →  Fx(4) DIO 12
    AF22 GPIO[20] ──  GPIO[21] AD22  →  Fx(5) DIO 13
    AG25 GPIO[22] ──  GPIO[23] AD25  →  Fx(6) DIO 14
    AH25 GPIO[24] ──  GPIO[25] AE25  →  Fx(7) DIO 15
         3.3V     ──  GND
    AG22 GPIO[26] ──  GPIO[27] AE24
    AH22 GPIO[28] ──  GPIO[29] AF26
    AE20 GPIO[30] ──  GPIO[31] AG23
    AF20 GPIO[32] ──  GPIO[33] AH26
    AH23 GPIO[34] ──  GPIO[35] AG26
```

◦ 圖 3-21 ◦　　Analog Discovery DIO 與 DE2-115 JP5 間的連線

Waveform 1 Program Digital Pattern Generator 對 freq_meter 電路輸入信號的設定，如圖 3-22 所示。

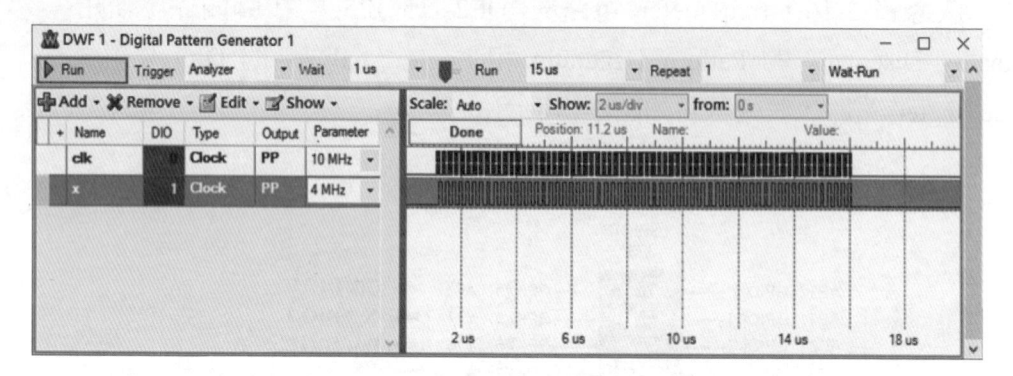

◦ 圖 3-22 ◦　　Waveform 1 Program Digital Pattern Generator 的設定

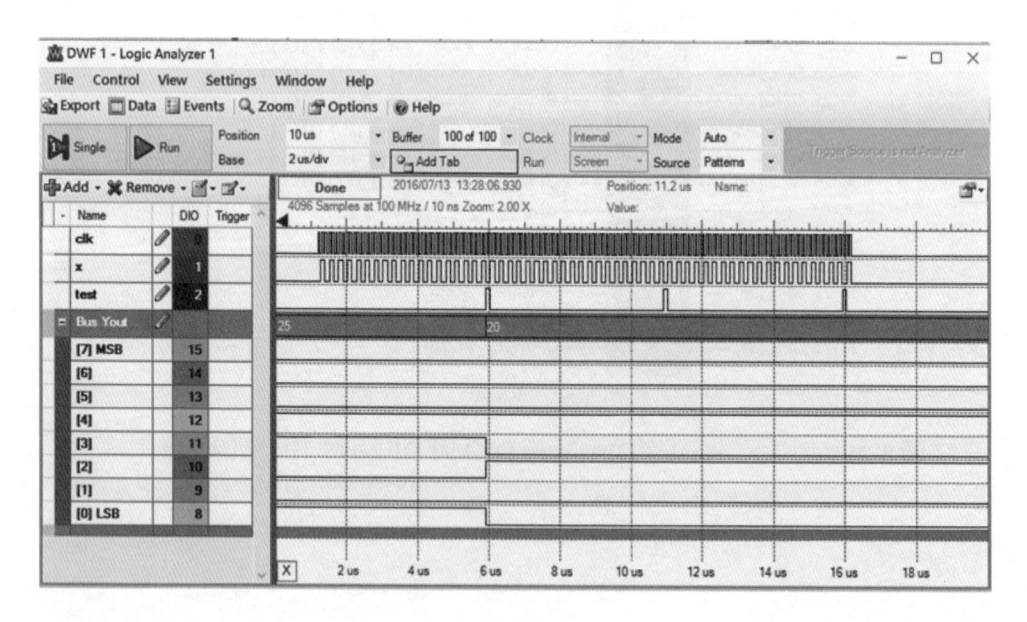

◦ 圖 3-23 ◦　　為 Waveform 1 Program Logic Analyzer 的設定與測試的結果

3-5 課外練習

(1) 圖 3-23 為 Waveform 1 Program Logic Analyzer 的設定，在 system clk 為 10 MHz，被測試頻率 x = 4 MHz 的結果為 20，這個數目與圖 3-7 ModelSim 測試的結果 19 不同，原因何在？(答案須包括代碼份的修正和 simulation 或實體測試)

(2) 試設計一計時器，準確度為 1 sec，最大計時為 60 sec。計時器有 start 和 stop 按鈕開關。請寫出其 VHDL device code 和 TestBench codes，並用 DE2-115 合成電路。再用 Analog Discovery 測試之。

(3) 試將 圖 3-5：構成頻率計電路的 VHDL 檔中的 process (x, twindow) 和 process (twindow) 改寫成二個獨立的 component，重新完成頻率計電路的 simulation 設計。

(4) 試將 圖 2-7：BCD_counter 和 BCD to SSG 組成的 slow_counter 電路，由 component 改寫成由 process 來完成的 simulation 設計。

第四章　測試平台和電路模擬測試

電路設計，為的是要能夠**合成** (Synthesis) 硬體電路。使該硬體電路不但在功能上達到設計的目的，同時還能達到設計所期望的速度。電路的**模擬測試** (Simulation)，對於合成硬體電路來講，它是要在做合成硬體電路之前，預先測試其所設計的電路，在合成硬體電路後，確能達到電路設計所期望的目的。

對於 VHDL 代碼方面來講，設計合成硬體電路所用的 VHDL，僅為模擬測試的 VHDL 中的一部分。也就是說模擬測試 TestBench 中的許多功能：如**波形的產生** (Stimulus Generation)，**時間的遲延** (Time Delay) 等，祇能用在模擬測試的環境裏。不能用在合成硬體的電路上。

4-1　測試平台 (TestBench) 的種類

電路設計除了**功能** (Functional)，還有速度也就是**時間** (Timing)，加上手動和自動因此可分成以下，如圖 4-1 等 4 大類。

Testbench type	Circuit's propagation delays	Output waveform analysis
I	Not included (functional analysis)	Visual inspection (manual analysis)
II	Included (timing analysis)	Visual inspection (manual analysis)
III	Not included (functional analysis)	With VHDL (automated analysis)
IV	Included (timing analysis)	With VHDL (automated analysis)

圖 4-1　TestBench 的分類

圖 4-1 四大類中的第 I 和 II 類在設計上比較簡單，結果的功能和時間等輸出，卻須要有經驗的技術人員來判讀，才知道其結果。第 III 和 VI 類在設計上雖然比較複雜，但結果卻為 "是與非" 的一種宣判。電路的模擬測試，大多是由電路設計人員在操作，並非用在生產線上。故以使用第 I 和 II 類較多。

◎ 4-2 測試波形的產生

書寫 TestBench.vhd 最重要的部分，是產生 Stimulus 輸入信號。歸納起來，輸入信號一共有以下的五種：

1. 對稱並且重複的波形，Clock 就是它的代表，它的寫法如下：

    ```
    Signal clock : bit := '1';
    clock <= NOT clock AFTER 30 ns;
    ```

    也可以寫成：

    ```
    Signal clock : bit := '1';
    Wait for 30 ns;  clock <= NOT clock;
    ```

2. 單一的波形，reset 就是它的代表，它的寫法如下：

    ```
    Signal reset : bit := '0';
    Wait for 30 ns;  reset <= '1';
    Wait for 30 ns;  reset <= '0';
    Wait;
    ```

3. 不規則，而且不重複，多個 bit 的波形。它的寫法如下：

    ```
    Constant wave: bit_vector( 1 to 8):= "10110100";
    Signal x: bit := '0';
    For i IN wave'RANGE LOOP
    ```

```
x <= wave(i); wait for 30 ns;
END LOOP;
Wait;
```

4. 不規則，但重複，多個 bit 的波形。它的寫法如下：

```
Constant wave: bit_vector( 1 to 8):="10110100";
Signal y: bit := '0';
For i IN wave'RANGE LOOP
y <= wave(i); wait for 30 ns;
END LOOP;
```

5. 數字型，多個 bit 的波形。它的寫法如下：

```
Signal z: INTEGER RANGE 0 to 255;
Z <= 0; wait for 120 ns;
Z <= 33; wait for 120 ns;
Z <= 255; wait for 60 ns;
Z <= 99; wait for 180 ns;
```

產生以上五種不同 VHDL test signals 的 VHDL TestBench，如圖 4-2 所示。

```
1  Library ieee;
2  Use ieee.std_logic_1164.all;
3  ------------------------------------------------
4  Entity test_testbench is
5  End test_testbench;
6  ------------------------------------------------
7  Architecture testbench of test_testbench is
8      Signal a: std_logic := '1';
9      Signal b: std_logic := '0';
10     constant wave: bit_vector(1 to 8):="10110100";
11     Signal x: bit := '0';
12     Signal y: bit := '0';
13     Signal z: integer range 0 to 255;
14 Begin
15     ----- Generate a: -----
16     Process
17     Begin
18         wait for 30 ns;
19         a <= not a;
20     End Process;
21     ----- Generate b: -----
22     Process
23     Begin
24         wait for 30 ns;
25         b <= '1';
26         wait for 30 ns;
27         b <= '0';
28         wait;
29     End Process;
30     ----- Generate c: -----
31     Process
32     Begin
33         For i in wave'range loop
34             x <= wave(i); wait for 30 ns;
35         end loop;
36         wait;
37     End Process;
38     ----- Generate d: -----
39     Process
40     Begin
41         For i in wave'range loop
42             y <= wave(i); wait for 30 ns;
43         end loop;
44     End Process;
45     ----- Generate e: -----
46     Process
47     Begin
48         z <= 0;    wait for 120 ns;
49         z <= 33;   wait for 120 ns;
50         z <= 255;  wait for  60 ns;
51         z <= 99;   wait for 180 ns;
52     End Process;
53 ------------------------------------------------
54 End testbench;
```

圖 4-2　產生五種代表波形的 VHDL 寫法

如果用 Modelsim 來做 simulation，可獲得的波形，如圖 4-3 所示。

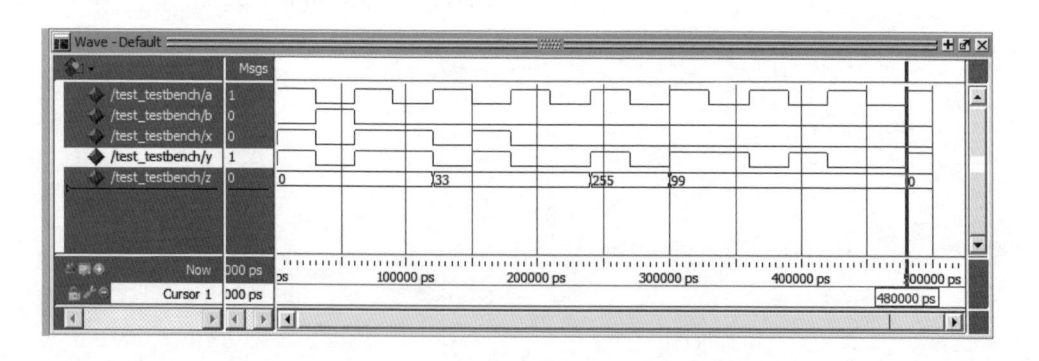

圖 4-3　Modelsim simulator 顯示的五種不同波形

4-3 第 I 類 TestBench 的寫法

圖 4-4 為第 I 類 TestBench 的標準寫法，這一類的測試以測試電路的功能為主，並不包含電路的**內部傳輸延遲** (internal propagation delay)，它的輸出是否 OK，要靠設計人員的判讀來決定。所以又稱 stimulus-only functional analysis。

圖中 Line 2～6 為 i 宣告 TestBench 用的是 ieee 的 std_logic_1164 Library。Line 8～17 為要測試的電路及 TestBench 測試訊號的名稱。Line 20 為被測試訊號與 TestBench 測試訊號的連接關係。Line 23～31 為 TestBench 測試訊號的產生。Line 33～37 為被測試的輸出**應有訊號** (Expected data)。

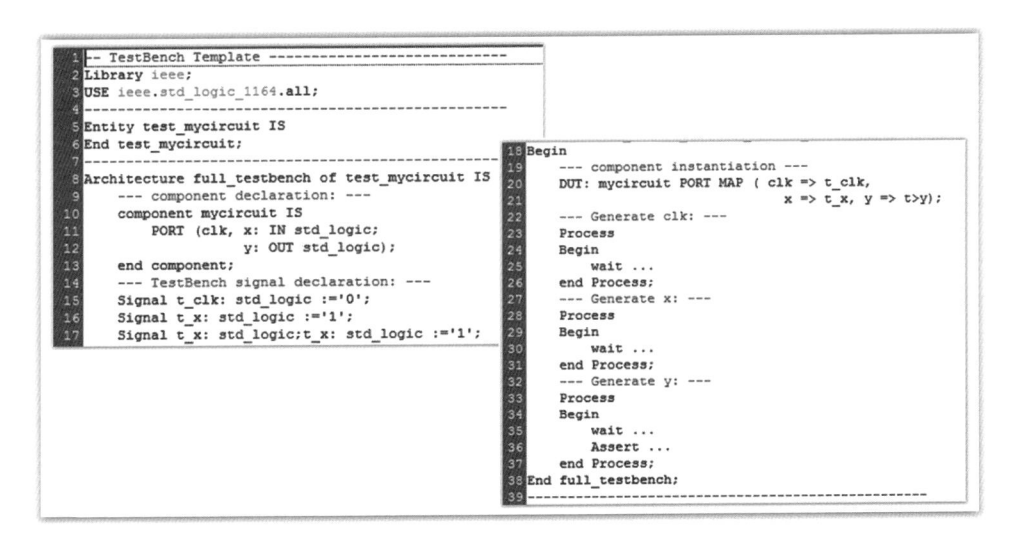

```
1  -- TestBench Template ---------------------
2  Library ieee;
3  USE ieee.std_logic_1164.all;
4  -------------------------------------------
5  Entity test_mycircuit IS
6  End test_mycircuit;
7  -------------------------------------------
8  Architecture full_testbench of test_mycircuit IS
9      --- component declaration: ---
10     component mycircuit IS
11         PORT (clk, x: IN std_logic;
12               y: OUT std_logic);
13     end component;
14     --- TestBench signal declaration: ---
15     Signal t_clk: std_logic :='0';
16     Signal t_x: std_logic :='1';
17     Signal t_x: std_logic;t_x: std_logic :='1';
18 Begin
19     --- component instantiation ---
20     DUT: mycircuit PORT MAP ( clk => t_clk,
21                               x => t_x, y => t>y);
22     --- Generate clk: ---
23     Process
24     Begin
25         wait ...
26     end Process;
27     --- Generate x: ---
28     Process
29     Begin
30         wait ...
31     end Process;
32     --- Generate y: ---
33     Process
34     Begin
35         wait ...
36         Assert ...
37     end Process;
38 End full_testbench;
39 -------------------------------------------
```

圖 4-4　stimulus-only functional analysis 的標準寫法

範例　**Stimulus-only functional analysis** 的例子

圖 4-5 為被測試的一個輸入頻率除 10 的電路。其中當 ena = '0' 時，output 保持該被除的數值。該電路的輸出不須測試**應有訊號** (Expected data) 的功能，所以適用 Stimulus-only functional analysis 的 TestBench 如圖 4-6 所示。

```vhdl
1  ---- Design file (clock_divider.vhd): ------------
2  ENTITY clock_divider IS
3      PORT (clk, ena: IN BIT;
4              output: OUT BIT);
5  END clock_divider;
6  ---------------------------------------------------
7  ARCHITECTURE clock_divider OF clock_divider IS
8      CONSTANT max: INTEGER := 5;
9  BEGIN
10     PROCESS(clk)
11         VARIABLE count: INTEGER RANGE 0 TO 7 := 0;
12         VARIABLE temp: BIT;
13     BEGIN
14         IF (clk'EVENT AND clk='1') THEN
15             IF (ena='1') THEN
16                 count := count + 1;
17                 IF (count=max) THEN
18                     temp := NOT temp;
19                     count := 0;
20                 END IF;
21             END IF;
22             output <= temp;
23         END IF;
24     END PROCESS;
25 END clock_divider;
26 ---------------------------------------------------
```

圖 4-5　第 I 類適用 Stimulus-only functional analysis 的電路

下圖為測試第 I 類 Stimulus-only functional analysis 電路的 TestBench。

```
 1 -- Test file (test_clock_divider.vhd):---------------------
 2 ENTITY test_clock_divider IS
 3 END test_clock_divider;
 4 ---------------------------------------------------------
 5 ARCHITECTURE test_clock_divider OF test_clock_divider IS
 6     ---- component declaration: ------------------
 7     COMPONENT clock_divider IS
 8         PORT (clk, ena: IN BIT;
 9                output: OUT BIT);
10 END COMPONENT;
11 ---- signal declarations: -----------
12     SIGNAL clk: BIT := '0';
13     SIGNAL ena: BIT := '0';
14     SIGNAL output: BIT;
15 BEGIN
16     ----- component instantiation: ---------------
17     dut: clock_divider PORT MAP (clk=>clk, ena=>ena,
18                                  output=>output);
19     ----- generate clock: ------------------------
20     PROCESS
21     BEGIN
22         WAIT FOR 30 ns;
23         clk <= NOT clk;
24     END PROCESS;
25     ----- generate enable: -----------------------
26     PROCESS
27     BEGIN
28         WAIT FOR 60 ns;
29         ena <= '1';
30         WAIT; --optional
31     END PROCESS;
32 END test_clock_divider;
33 ---------------------------------------------------------
```

☜ 圖 4-6 ☞　　測試第 I 類 Stimulus-only functional analysis 電路的 TestBench

　　圖 4-6 第 I 類 Stimulus-only functional analysis 電路測試的結果如圖 4-7 所示。

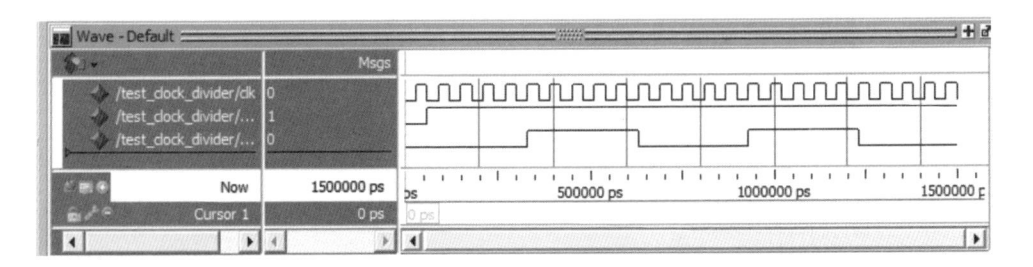

☜ 圖 4-7 ☞　　第 I 類 Stimulus-only functional analysis 電路測試的結果

4-4　第 II 類 TestBench 的寫法

　　第 II 類 TestBench 與第 I 類的不同點在於除了功能之外還需顧慮到電路的內部傳輸延遲。它對電路的輸出依然沒有自動測試的能力。所以有稱 stimulus-only timing analysis。這類 TestBench 的寫法，與第 I 類 TestBench 的寫法，完全相同。其不同點是在電路設計的檔案上，以圖 4-5 和圖 4-8 做一比較，圖 4-8 的 Line 16 加進了 8 ns 的 counter delay 和 Line 23 的 5 ns 的輸出訊號延遲。

```
1  ---- Design file (clock_divider.vhd): -------------
2  ENTITY clock_divider IS
3      PORT (clk, ena: IN BIT;
4              output: OUT BIT);
5  END clock_divider;
6  ---------------------------------------------------
7  ARCHITECTURE clock_divider OF clock_divider IS
8      CONSTANT max: INTEGER := 5;
9  BEGIN
10     PROCESS
11         VARIABLE count: INTEGER RANGE 0 TO 7 := 0;
12         VARIABLE temp: BIT;
13     BEGIN
14         WAIT UNTIL (clk'EVENT AND clk='1');
15         IF (ena='1') THEN
16             WAIT FOR 8 ns; --counter delay=8ns
17             count := count + 1;
18             IF (count=max) THEN
19                 temp := NOT temp;
20                 count := 0;
21             END IF;
22         END IF;
23         WAIT FOR 5 ns; --output delay=5ns
24         output <= temp;
25     END PROCESS;
26 END clock_divider;
27 ---------------------------------------------------
```

圖 4-8　　第 II 類電路的測試在於電路設計中加入了 I/O 的延遲

由於圖 4-8 電路設計中加入了 I/O 的延遲，結果引起輸出波形在 Timing 上的變化如圖 4-9 所示。

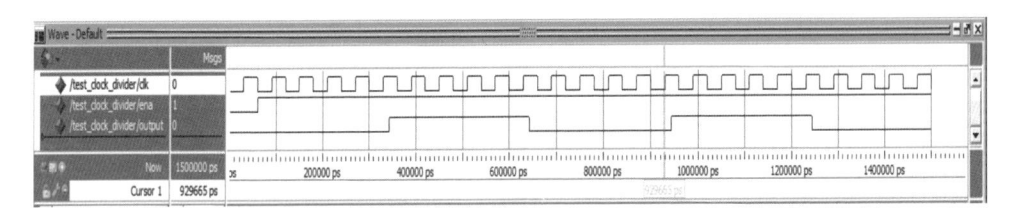

圖 4-9　電路設計中加入了 I/O 的延遲造成輸出波形在 Timing 上的變化

4-5　第 III 類 TestBench 的寫法

第 III 類 TestBench 被稱謂**自動功能測試** (automated functional analysis)，這類的測試，祇對被測試電路 (DUT) 的個別的輸出做自動檢測，而不對電路的內在傳輸延遲加以考慮。

用測試圖 4-5 的 clock_divider 檔為例，它的測試檔如圖 4-10 所示。這個測試檔 test_clock_divider 為了能夠做自動測試，增加了 line 31~39 的產生**期望** (Expect) data 的 generate template 部分和用來比較輸出 *data* 和期望 *data* 的 line 40~48 的 verify output 二個部分。

```
1 ---- TestBench file (test_clock_divider.vhd):-----------------
2 ENTITY test_clock_divider IS
3 END test_clock_divider;
4 --------------------------------------------------------------
5 ARCHITECTURE test_clock_divider OF test_clock_divider IS
6   ---- component declaration: --------
7   COMPONENT clock_divider IS
8       PORT (clk, ena: IN BIT;
9             output: OUT BIT);
10  END COMPONENT;
11  ---- signal declarations: ----------
12  SIGNAL clk: BIT := '0';
13  SIGNAL ena: BIT := '0';
14  SIGNAL output: BIT;
15  SIGNAL template: BIT := '0'; --for output verification
16 BEGIN
17     ----- Component instantiation: ----------
18     DUT: clock_divider PORT MAP (clk=>clk, ena=>ena, output=>output);
19     ----- generate clock: --------------------
20     PROCESS
21     BEGIN
22         WAIT FOR 30 ns;
23         clk <= NOT clk;
24     END PROCESS;
25     ----- generate enable: ------------------
26     PROCESS
27     BEGIN
28         WAIT FOR 60 ns;
29         ena <= '1';
30     END PROCESS;
31     ----- generate template: -----------------
32     PROCESS
33     BEGIN
34         WAIT FOR 330 ns;
35         WHILE ena='1' LOOP
36             template <= NOT template;
37             WAIT FOR 300 ns;
38         END LOOP;
39     END PROCESS;
40     ----- verify output: --------------------
41     PROCESS
42     BEGIN
43         WAIT FOR 1 ns;
44         ASSERT (output=template)
45             REPORT "Output differs from template!"
46             SEVERITY FAILURE;
47             --SEVERITY WARNING;
48     END PROCESS;
49 END test_clock_divider;
50 --------------------------------------------------------------
```

圖 4-10 第 III 類自動功能測試 TestBench 的組成

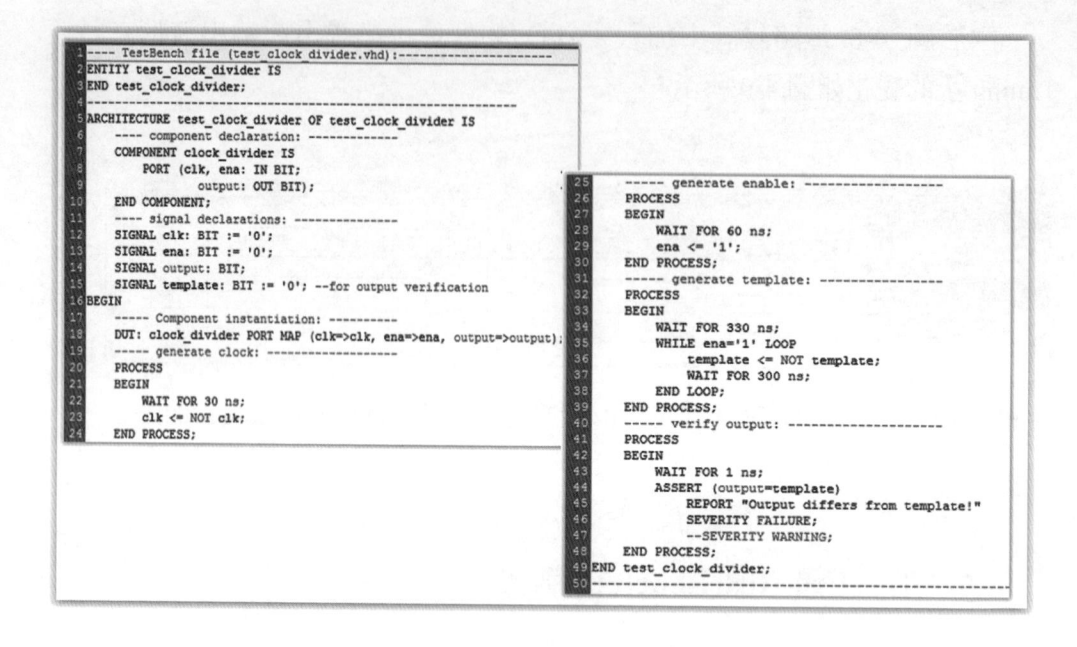

第三類自動功能測試 TestBench 的 generate template 可從圖 4-7：第 I 類 Stimulus-only functional analysis 電路測試的輸出訊號轉變成文字得。Verify output 部分的主要構成是 line 44 的 Assert statement，它的結構如下：

ASSERT (Boolean_expresson) [REPORT "message"] [SEVERITY severity_level] Severity level 有 NOTE、WARNING、ERROR、和 FAILURE 等四種。

當 (Boolean_expression) 為錯誤時，[REPORT "message"]，同時有以下的顯示：

NOTE：to pass information from compiler/simulator
WARNING：to inform that something unusual has occurred

```
ERROR：to inform that aserious unusual condition has been found
FAILURE：a complete unacceptable condition has occurred
```

在 NOTE 和 WARNING 的情況下 simulation 會繼續進行。ERROR 和 FAILURE 時 simulation 則停止運作。

圖 4-11 為自動測試 TestBench 對圖 4-7 電路測試的結果。

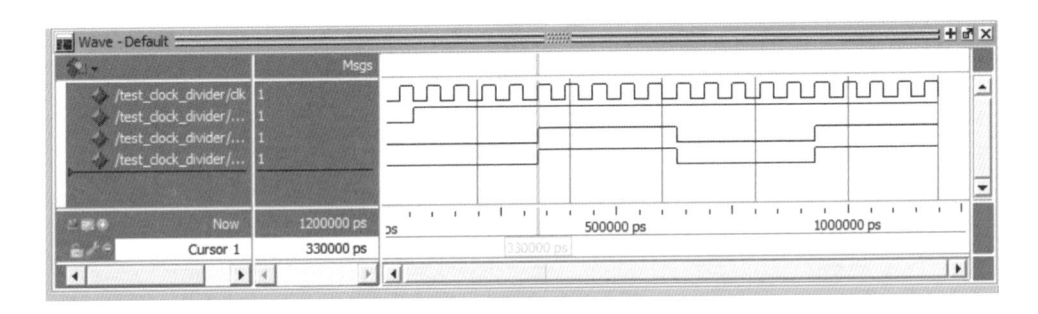

圖 4-11　　　　圖 4-10 自動測試對圖 4-7 電路測試的結果

4-6　第 IV 類 TestBench 的寫法

第 IV 類 TestBench 被稱謂**全自動時序測試** (full-bench or automated timing analysis)，這類的測試，祇對被測試電路 (DUT) 的個別的輸出功能和內在傳輸延遲做完整的自動檢測。

用測試圖 4-8 設計中加入了 I/O 延遲的電路為例，它的測試檔如圖 4-12 所示。這個測試檔 test_clock_divider 為了能夠做自動測試，增加了 line 31~39 的產生**期望** (Expect) data 的 generate template 部分和用來比較輸出 *data* 和期望 *data* 的 line 40~48 的 verify output 二個部分。

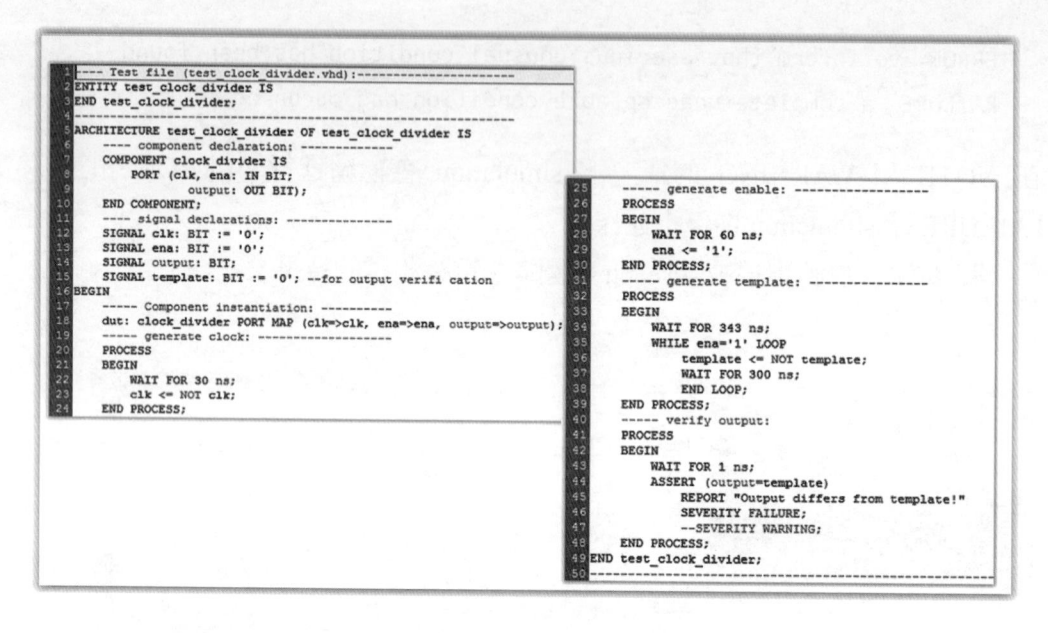

```
1 ---- Test file (test_clock_divider.vhd):---------------------
2 ENTITY test_clock_divider IS
3 END test_clock_divider;
4
5 ARCHITECTURE test_clock_divider OF test_clock_divider IS
6     ---- component declaration: ---------------
7     COMPONENT clock_divider IS
8         PORT (clk, ena: IN BIT;
9               output: OUT BIT);
10    END COMPONENT;
11    ---- signal declarations: ---------------
12    SIGNAL clk: BIT := '0';
13    SIGNAL ena: BIT := '0';
14    SIGNAL output: BIT;
15    SIGNAL template: BIT := '0'; --for output verification
16 BEGIN
17    ---- Component instantiation: ---------
18    dut: clock_divider PORT MAP (clk=>clk, ena=>ena, output=>output);
19    ---- generate clock: -------------------
20    PROCESS
21    BEGIN
22        WAIT FOR 30 ns;
23        clk <= NOT clk;
24    END PROCESS;
25    ----- generate enable: -----------------
26    PROCESS
27    BEGIN
28        WAIT FOR 60 ns;
29        ena <= '1';
30    END PROCESS;
31    ----- generate template: ---------------
32    PROCESS
33    BEGIN
34        WAIT FOR 343 ns;
35        WHILE ena='1' LOOP
36            template <= NOT template;
37            WAIT FOR 300 ns;
38        END LOOP;
39    END PROCESS;
40    ----- verify output:
41    PROCESS
42    BEGIN
43        WAIT FOR 1 ns;
44        ASSERT (output=template)
45            REPORT "Output differs from template!"
46            SEVERITY FAILURE;
47            --SEVERITY WARNING;
48    END PROCESS;
49 END test_clock_divider;
50 ------------------------------------------
```

◎ 圖 4-12　　第 IV 類自動功能及時序測試 TestBench 的組成

　　圖 4-12 的 TestBench 與圖 4-10 的不同點，是在它們的 generate template 部分的 line 34：WAIT FOR 343 ns。這 343 ns 的由來是圖 4-10 的 300 ns 加上圖 4-8 的 counter delay = 5 ns 和 output delay = 8 ns 的結果。

　　圖 4-13 為其測試的結果，從圖中可以看到 clock 的 rising edge 與 output 訊號的 rising edge 並不同步，它們之間相差為 counter delay + output delay = 13 ns。

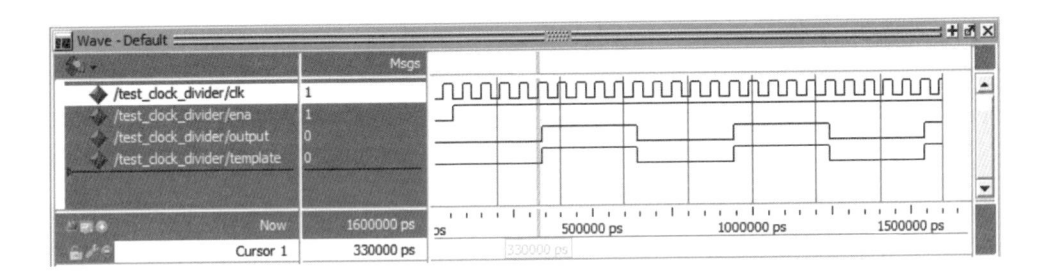

圖 4-13　圖 4-12 自動測試對圖 4-8 電路測試的結果

4-7 課外練習

(1) 試用 Analog Discovery 的 Pattern Generator，產生如下圖的波形，T = 1 uS。

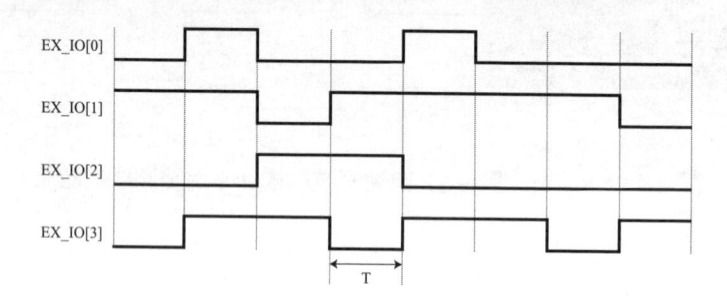

```
EX_IO[0] ____|‾‾|____|‾‾‾‾|____
EX_IO[1] _____|‾‾‾‾‾‾|_____|‾
EX_IO[2] _____|‾‾‾‾|_____
EX_IO[3] __|‾|_____|‾‾|__|‾|__
              |←T→|
```

(2) 試述 VHDL code 在 Simulation 和 Synthesis 中，有何不同點？testbench 為何只能用在 Synthesis 之前？第 IV 類自動測試功能，對設計者意義不大？

(3) Analog Discovery 的 Logic Analyzer 是用來做什麼的？它須要像 Pattern Generator 一樣地設定嗎？原因何在？

(4) 試述 Analog Discovery 的硬體有那幾種？軟體又有那幾種？其共用性與非共用性？

第五章　有限狀態機

有限狀態機 (Finite State Machine，簡稱 FSM)，是順序邏輯系統的另一種設計，其操作可以通過包含所有可能的系統狀態，以及一個不太長的 I/O 定義，從一個狀態到另一個狀態，可以列表來描述。系統必須在每個狀態下產生輸出值。這種類型的設計將在以下用 VHDL 進一步說明。

5-1　FSM 的模式和它的狀態轉移圖

圖 5-1 是 FSM 的模式，其中 (a) 為其簡化的硬體模式，(b) 為 Mealy machine，(c) 為 Moore machine。

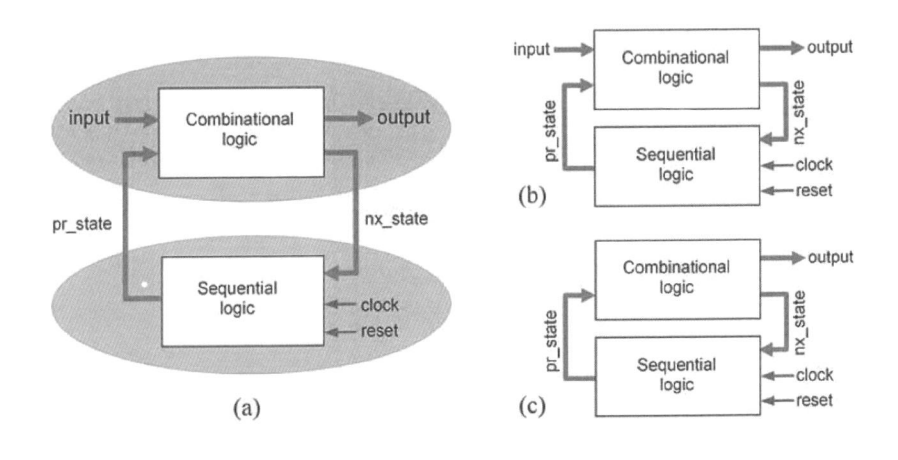

　　　　　　　圖 5-1　　有限狀態機 FSM 的模式

Mealy machine 和 Moore machine 的不同點是 Moore machine 沒有外界訊號的輸入，counter 電路就是一個例子。Mealy machine 則有外界訊號的輸入。其它如 clock、reset、output、present state、next state、二者均有之。

順序系統的電路規格可以用圖 5-2 的狀態轉變圖來表示，

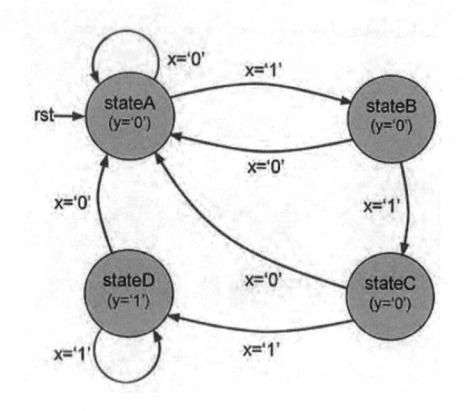

◦ 圖 5-2 ◦　　順序系統的電路規格可以用狀態轉變圖來表示

圖 5-2 表示出該順序系統的電路有四個狀態，它們是 StateA、StateB、StateC、和 StateD。當 stateA 在的時候，它的輸出 y = '0'，當外來訊號 x = '0' 時 stateA 不變，但若外來訊號 x = '1' 則 state 由 stateA 改變成 stateB。又當輸入為 reset 時，不論現在為那種 state，都將改變成 stateA。其它的依此類推。

5-2　有限狀態機 FSM 的 VHDL 模型

圖 5-3 為圖 5-1 有限狀態機 FSM 的 VHDL 模型。這個模型把圖 5-1(a) 的 Sequential logic 部分，從它在圖中的位置稱之為 Lower section。而在它上方的 combinational logic 部分，稱之為 Upper section。

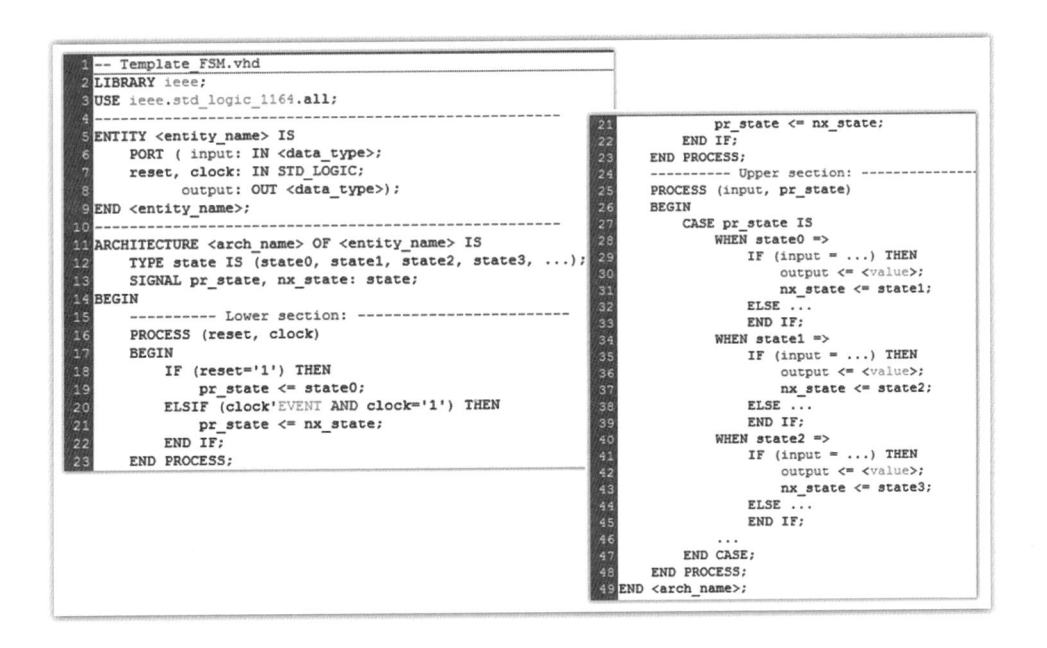

```vhdl
1  -- Template_FSM.vhd
2  LIBRARY ieee;
3  USE ieee.std_logic_1164.all;
4  -----------------------------------------
5  ENTITY <entity_name> IS
6      PORT ( input: IN <data_type>;
7      reset, clock: IN STD_LOGIC;
8             output: OUT <data_type>);
9  END <entity_name>;
10 
11 ARCHITECTURE <arch_name> OF <entity_name> IS
12     TYPE state IS (state0, state1, state2, state3, ...);
13     SIGNAL pr_state, nx_state: state;
14 BEGIN
15     --------- Lower section: -----------------
16     PROCESS (reset, clock)
17     BEGIN
18         IF (reset='1') THEN
19             pr_state <= state0;
20         ELSIF (clock'EVENT AND clock='1') THEN
21             pr_state <= nx_state;
22         END IF;
23     END PROCESS;
24     --------- Upper section: -----------------
25     PROCESS (input, pr_state)
26     BEGIN
27         CASE pr_state IS
28             WHEN state0 =>
29                 IF (input = ...) THEN
30                     output <= <value>;
31                     nx_state <= state1;
32                 ELSE ...
33                 END IF;
34             WHEN state1 =>
35                 IF (input = ...) THEN
36                     output <= <value>;
37                     nx_state <= state2;
38                 ELSE ...
39                 END IF;
40             WHEN state2 =>
41                 IF (input = ...) THEN
42                     output <= <value>;
43                     nx_state <= state3;
44                 ELSE ...
45                 END IF;
46             ...
47         END CASE;
48     END PROCESS;
49 END <arch_name>;
```

圖 5-3　圖 5-1 有限狀態機 FSM 的 VHDL 模型。

Lower section 部分跟 clock 和 reset 有關。Upper section 則跟 input 和 output 有關係。這樣的區分，簡化了 FSM 的 VHDL 檔的組成。

5-3　Moore machine 的一個例子

　　Moore machine 電路的特點是沒有外界訊號的輸入，BCD_counter 電路就是一個最好的例子，如圖 5-4 所示。圖中 zero ~ nine 代表 BCD_counter 電路的 10 個 state。而 → 代表每當 clock 發生時。

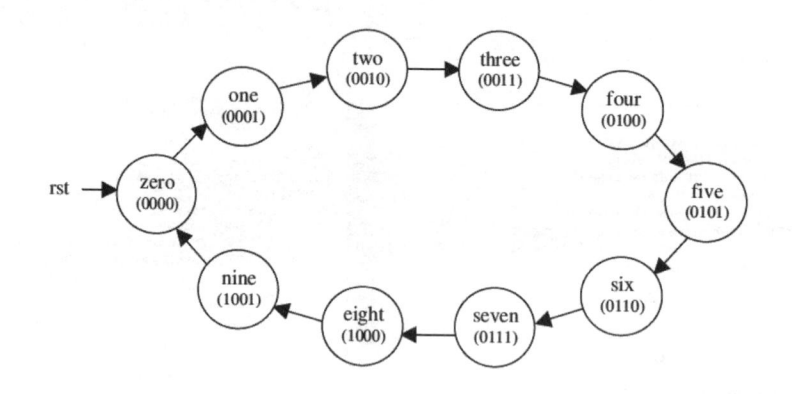

◎ 圖 5-4　Moore machine BCD_counter 電路的狀態轉變圖

圖 5-5 是圖 5-4 的 VHDL 的 FSM 檔。在 ARCHITECTURE 之下，首先要宣告的是 zero ~ nine 等 10 個 TYPE 為 state 的狀態。接下來是二個也屬於 state 的訊號 pr_state 和 nx_state。

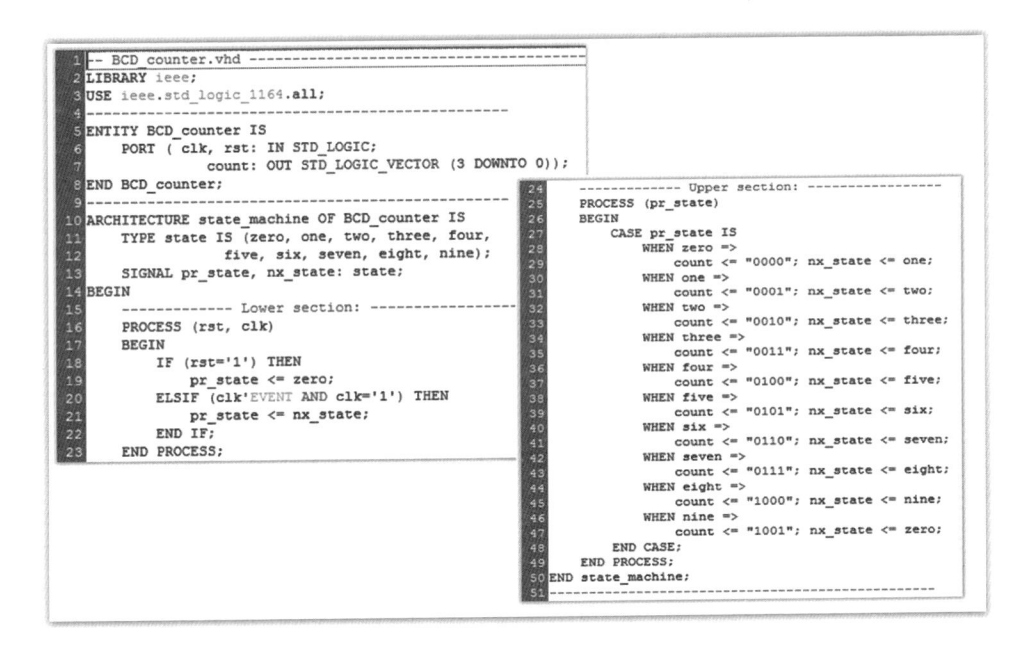

```
1 -- BCD_counter.vhd ------------
2 LIBRARY ieee;
3 USE ieee.std_logic_1164.all;
4 ------------------------------
5 ENTITY BCD_counter IS
6   PORT ( clk, rst: IN STD_LOGIC;
7          count: OUT STD_LOGIC_VECTOR (3 DOWNTO 0));
8 END BCD_counter;
9 ------------------------------
10 ARCHITECTURE state_machine OF BCD_counter IS
11   TYPE state IS (zero, one, two, three, four,
12                  five, six, seven, eight, nine);
13   SIGNAL pr_state, nx_state: state;
14 BEGIN
15 ------------ Lower section: ------------
16   PROCESS (rst, clk)
17   BEGIN
18     IF (rst='1') THEN
19       pr_state <= zero;
20     ELSIF (clk'EVENT AND clk='1') THEN
21       pr_state <= nx_state;
22     END IF;
23   END PROCESS;
24 ------------ Upper section: ------------
25   PROCESS (pr_state)
26   BEGIN
27     CASE pr_state IS
28       WHEN zero =>
29         count <= "0000"; nx_state <= one;
30       WHEN one =>
31         count <= "0001"; nx_state <= two;
32       WHEN two =>
33         count <= "0010"; nx_state <= three;
34       WHEN three =>
35         count <= "0011"; nx_state <= four;
36       WHEN four =>
37         count <= "0100"; nx_state <= five;
38       WHEN five =>
39         count <= "0101"; nx_state <= six;
40       WHEN six =>
41         count <= "0110"; nx_state <= seven;
42       WHEN seven =>
43         count <= "0111"; nx_state <= eight;
44       WHEN eight =>
45         count <= "1000"; nx_state <= nine;
46       WHEN nine =>
47         count <= "1001"; nx_state <= zero;
48     END CASE;
49   END PROCESS;
50 END state_machine;
51 ------------------------------
```

圖 5-5　BCD_counter 的 FSM 檔

Lower section 提供 reset 和 clock，每當 clock 的 rising edge 來到時，pr_state 為 nx_state 所取代。Upper section 宣告在 pr_state 時各 10 個不同 state 輸出的內容，同時進階到下一個 state (如 zero 到 one，one 到 two，…….)。

BCD_counter 的 TestBench 檔如圖 5-6 所示。TestBench 的測試 SIGNAL 如果採用被測試 component 的 SIGNAL 相同，可以使 port map 的內容更為簡單，如圖 5-6 的 Line 23 所示。

```
 1 --TestBench.vhd --------------------------
 2 Library IEEE;
 3 use IEEE.std_logic_1164.all;
 4 --------------------------------------------
 5 entity TestBench is
 6 end TestBench;
 7 --------------------------------------------
 8 use work.all;
 9 --------------------------------------------
10 architecture stimulus of TestBench is
11 --First, declare lower-level entity that to be test
12     component BCD_counter
13         PORT ( clk, rst: IN STD_LOGIC;
14             count: OUT STD_LOGIC_VECTOR (3 DOWNTO 0));
15 end component;
16 --------------------------------------------
17 -- Next, declare TstBench's SINGNALs ------
18 SIGNAL   clk: std_logic := '0';
19 SIGNAL   rst: std_logic := '0';
20 SIGNAL   count : std_logic_vector(3 downto 0):= "0000";
21 --------------------------------------------
22 begin
23     DUT:  BCD_counter port map (clk, rst, count);
24     --- Concurrent Code for Periodical waveform ----
25     clk <= NOT clk AFTER 50 ns;
26     rst <= '1' AFTER 100 ns, '0' AFTER 200 ns;
27 end stimulus;
```

圖 5-6　　BCD_counter 的測試檔 TestBench

測試的結果如圖 5-7 所示。由於 clock 的週期為 100 ns，10 個 clock 週期加上 reset，和另外 3 個 clock，全部加起來 run 1500 ns，即可觀其全部。

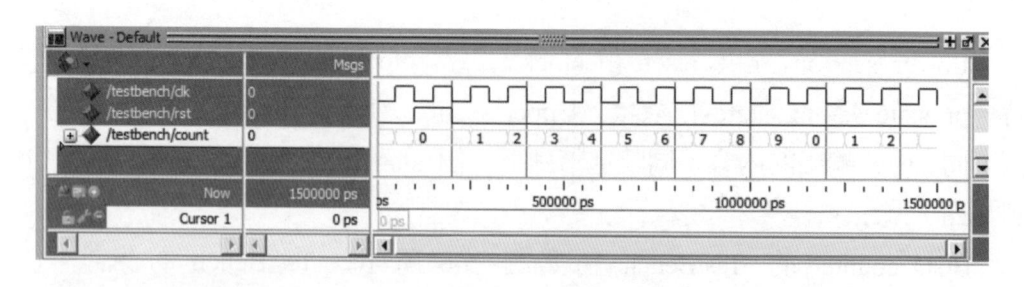

圖 5-7　　run 1500 ns 即可完整測試 BCD_counter

5-4　Mealy machine 的一個例子

　　Mealy machine 和 Moore machine 的不同點是 Mealy machine 有外界訊號的輸入，用 string_detector 來做一個例子，它的電路方塊圖和狀態轉變圖如圖 5-8 所示。這個 string_detector 專門檢驗 ascii_in 是否輸入為連續的 "mp3" 三個字，如果是則 string_detected 的輸出為 '1'，否則為 '0'。電路中當然還有 clk 和 rst。依據以上的輸入和輸出的條件，產生了狀態轉變圖。

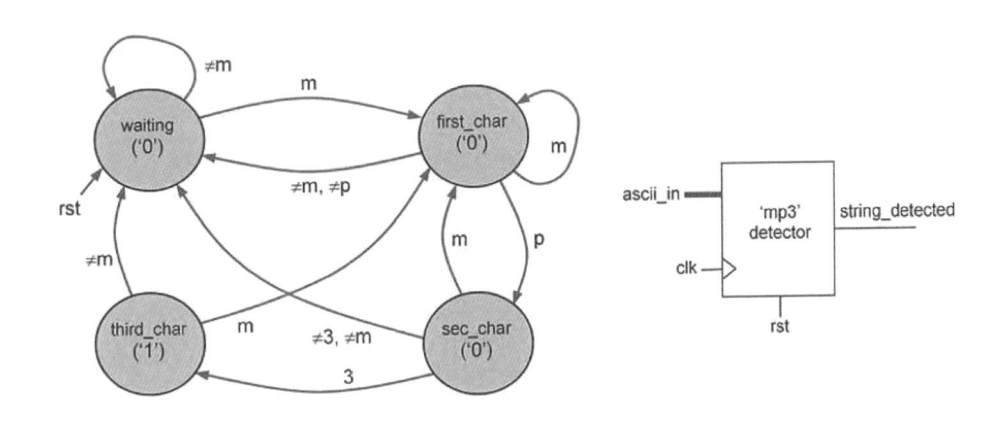

◎ 圖 5-8　string_detector 電路的方塊圖和狀態轉變圖

　　電路的 VHDL 如圖 5-9 所示。它的 lower section 由於 clk 和 rst 的條件和圖 5-5 完全一樣，每當 clk 的 rising edge 來到時 nx_state 取代了 pr_state。所以跟圖 5-5 一樣不需要改變。

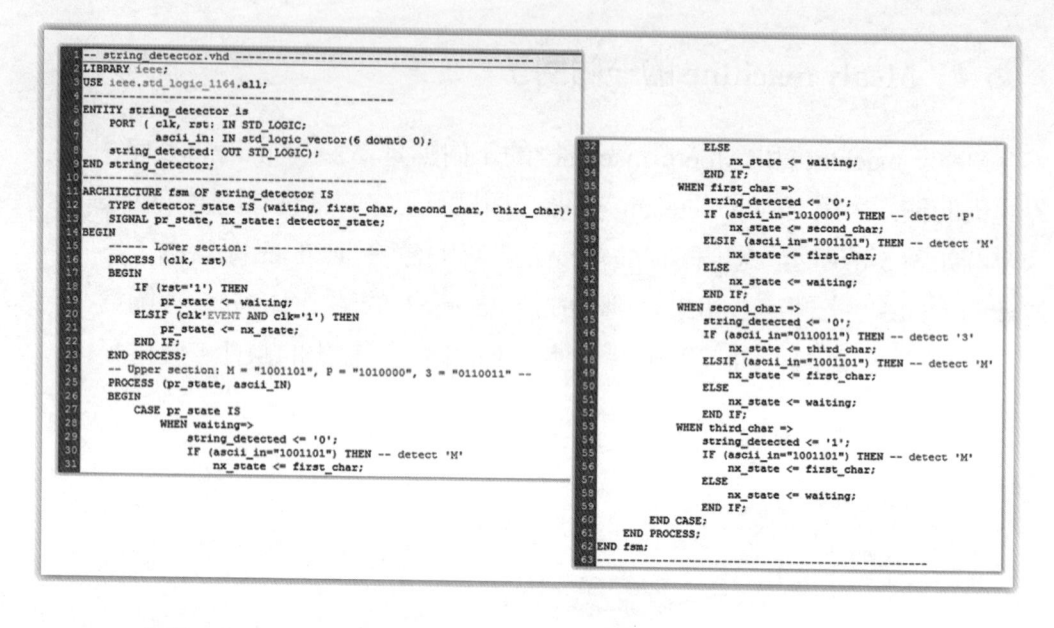

```
-- string_detector.vhd ------------------------
LIBRARY ieee;
USE ieee.std_logic_1164.all;
------------------------------------------------
ENTITY string_detector is
    PORT ( clk, rst: IN STD_LOGIC;
           ascii_in: IN std_logic_vector(6 downto 0);
    string_detected: OUT STD_LOGIC);
END string_detector;
------------------------------------------------
ARCHITECTURE fsm OF string_detector IS
    TYPE detector_state IS (waiting, first_char, second_char, third_char);
    SIGNAL pr_state, nx_state: detector_state;
BEGIN
    ------ Lower section: --------------------
    PROCESS (clk, rst)
    BEGIN
        IF (rst='1') THEN
            pr_state <= waiting;
        ELSIF (clk'EVENT AND clk='1') THEN
            pr_state <= nx_state;
        END IF;
    END PROCESS;
    -- Upper section: M = "1001101", P = "1010000", 3 = "0110011" --
    PROCESS (pr_state, ascii_IN)
    BEGIN
        CASE pr_state IS
            WHEN waiting=>
                string_detected <= '0';
                IF (ascii_in="1001101") THEN -- detect 'M'
                    nx_state <= first_char;
                ELSE
                    nx_state <= waiting;
                END IF;
            WHEN first_char =>
                string_detected <= '0';
                IF (ascii_in="1010000") THEN -- detect 'P'
                    nx_state <= second_char;
                ELSIF (ascii_in="1001101") THEN -- detect 'M'
                    nx_state <= first_char;
                ELSE
                    nx_state <= waiting;
                END IF;
            WHEN second_char =>
                string_detected <= '0';
                IF (ascii_in="0110011") THEN -- detect '3'
                    nx_state <= third_char;
                ELSIF (ascii_in="1001101") THEN -- detect 'M'
                    nx_state <= first_char;
                ELSE
                    nx_state <= waiting;
                END IF;
            WHEN third_char =>
                string_detected <= '1';
                IF (ascii_in="1001101") THEN -- detect 'M'
                    nx_state <= first_char;
                ELSE
                    nx_state <= waiting;
                END IF;
        END CASE;
    END PROCESS;
END fsm;
------------------------------------------------
```

圖 5-9　由 string_detector 的狀態轉變圖寫成的 VHDL 檔

Upper section 的動作如下：

1. 當 pr_state 為 waiting 開始，首先是令輸出 string_detected='0'，然後如果輸入 ascii_in 為 'm'，則 nx_state 為 first_char，否則 nx_state 不變，依舊為 waiting。

2. 當 pr_state 為 first_char 時，首先是令輸出 string_detected='0'，然後如果輸入 ascii_in 為 'p'，則 nx_state 為 second_char，否則如果 ascii_in 為 'm' 則 nx_state 變為 first_char，要不然 nx_state 變為 waiting。

3. 當 pr_state 為 second_char 時，首先是令輸出 string_detected='0'，然後如果輸入 ascii_in 為 '3'，則 nx_state 為 third_char，否則如果 ascii_in 為 'm' 則 nx_state 變為 first_char，要不然 nx_state 變為 waiting。

4. 當 pr_state 為 third_char 時，首先是令輸出 string_detected = '1'，然

後如果輸入 ascii_in 為 'm'，則 nx_state 為 first_char，否則 nx_state 變為 waiting。

仔細觀察 Upper section 的四個動作，在模式上可以說是相似。這是 FSM 容易寫的地方，但 state 不能太多，以圖 5-5 BCD_counter 的 FSM 檔為例，需要 50 條 lines 來構成，已嫌太長了一點。

測試圖 5-9 string_detector.vhd 的 TestBench 如圖 5-10 所示。TestBench 的測試 SIGNAL 也採用與被測試 component 的 SIGNAL 相同。

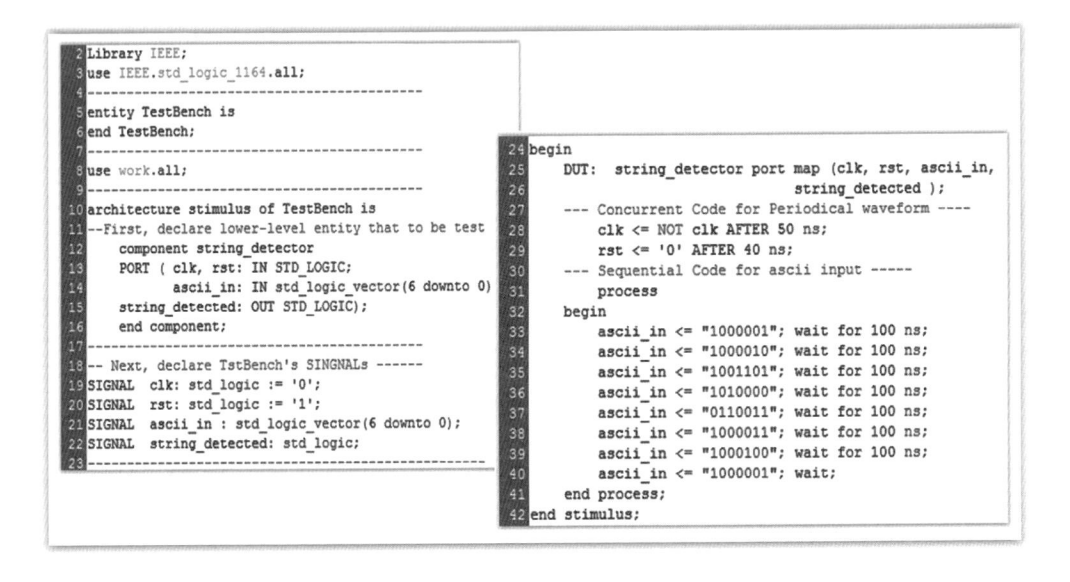

```
Library IEEE;
use IEEE.std_logic_1164.all;
----------------------------------
entity TestBench is
end TestBench;
----------------------------------
use work.all;
----------------------------------
architecture stimulus of TestBench is
--First, declare lower-level entity that to be test
  component string_detector
  PORT ( clk, rst: IN STD_LOGIC;
         ascii_in: IN std_logic_vector(6 downto 0);
         string_detected: OUT STD_LOGIC);
  end component;
----------------------------------
-- Next, declare TstBench's SINGNALs ------
SIGNAL clk: std_logic := '0';
SIGNAL rst: std_logic := '1';
SIGNAL ascii_in : std_logic_vector(6 downto 0);
SIGNAL string_detected: std_logic;
----------------------------------
begin
  DUT:  string_detector port map (clk, rst, ascii_in,
                                  string_detected );
  --- Concurrent Code for Periodical waveform ----
  clk <= NOT clk AFTER 50 ns;
  rst <= '0' AFTER 40 ns;
  --- Sequential Code for ascii input -----
  process
  begin
    ascii_in <= "1000001"; wait for 100 ns;
    ascii_in <= "1000010"; wait for 100 ns;
    ascii_in <= "1001101"; wait for 100 ns;
    ascii_in <= "1010000"; wait for 100 ns;
    ascii_in <= "0110011"; wait for 100 ns;
    ascii_in <= "1000011"; wait for 100 ns;
    ascii_in <= "1000100"; wait for 100 ns;
    ascii_in <= "1000001"; wait;
  end process;
end stimulus;
```

圖 5-10　string_detector.vhd 的測試檔 TestBench.vhd

測試的結果如圖 5-11 所示。由於 clock 的週期為 100 ns，測試的過程中 ascii_in 的輸入為 'A'、'B'、'M'、'P'、'3'、'C'、'D'、'A'. 每個字佔時 100 ns，共 run 800 ns。

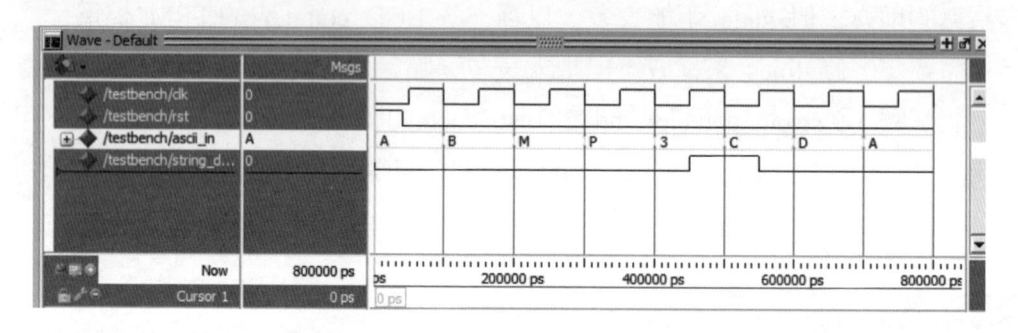

圖 5-11　string_detector.vhd 的測試檔 TestBench run 800 ns 的結果

5-5　液晶顯示器的例子

DE2-115 中的 HD44780U **液晶顯示器** (LCD)，它是一個能顯示二行，每行 16 個字的液晶顯示器。這種顯示器的 LCD 控制器，通常安裝在顯示器的背面，負責驅動該顯示器的**黑點** (Dot)。這一節將介紹如何用 FSM 來控制液晶顯示器的**寫入** (write) 和**讀出** (read) 並將文字顯示出來。

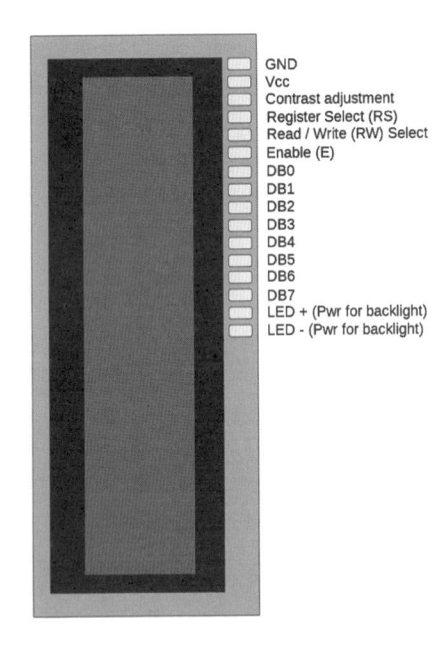

圖 5-12　HD44780U 液晶顯示器（LCD）

使用 LCD，第一步是要了解 LCD 控制器。圖 5-12 為 LCD 接腳及其名稱。

我們觀察到，除了供電，下面的四個信號必須被發送到控制器：

- **E** (啓用)：必須在高脈衝時才能寫東西到控制器的寄存器 (實際寫作發生在 E 的下降沿)。對於 E 簡化的時序圖如圖 5.13 所示。

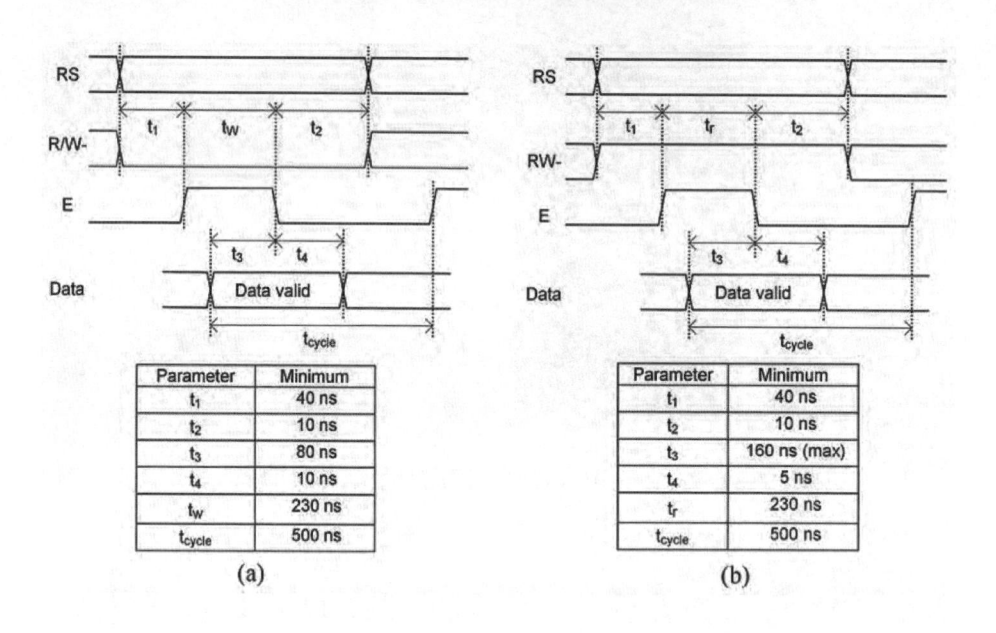

Parameter	Minimum
t_1	40 ns
t_2	10 ns
t_3	80 ns
t_4	10 ns
t_w	230 ns
t_{cycle}	500 ns

(a)

Parameter	Minimum
t_1	40 ns
t_2	10 ns
t_3	160 ns (max)
t_4	5 ns
t_r	230 ns
t_{cycle}	500 ns

(b)

圖 5-13　E 簡化的時序圖：(a) 為 write，(b) 為 read

- **RS** (寄存器選擇)：'0' 選擇控制器的 Instruction Register，而 '1' 選擇其 Data Register (後者是用於在液晶顯示的字符)。
- **R/W-** (被讀/寫)：如果 '0'，下一個 E (使能) 脈衝將導致本指令或數據被寫入到由 RS 所選控制器的寄存器，而 '1' 會從該讀出的數據控制器的寄存器。
- **DB** (數據群線)：8 位群線 (Bus)，其內容 (數據或指令) 在下一脈衝若 R/ W- = '0' 時被寫入控制器的寄存器。而當 R / W- = '1' 時，將數據從控制器的寄存器內讀出。

- **忙標誌 (BusyFlag)**：該信號控制器提供，通過位 DB(7) 的數據群線，如爲 '1'，表示控制器正忙。在實踐中，通常都避免對這個信號的使用，而用在所述指令的時間間隔，比用於完成指令所需的最大長度爲長之時。如表圖 5.14 所示。如果是這樣，R / W- 可以保持永久 low。

Instruction	RS	RW-	DB7 ... DB0	Description	Max. exec. time (*)
1) Clear Display	0	0	00000001	Clears display and sets DD RAM address to zero.	1.52 ms
2) Return Home	0	0	0000001X (X=don't care)	Returns display to origin and sets DD RAM address to zero.	1.52 ms
3) Entry Mode Set	0	0	000001 I/D S	Sets cursor direction and display shift during read and write. I/D=1 increment DD RAM address, =0 decrement S=1 shift display, =0 do not shift	37 us
4) Display ON/OFF Control	0	0	00001 D C B	D=1 display on, =0 off C=1 cursor on, =0 off B=1 blink char., =0 do not blink	37 us
5) Cursor or Display Shift	0	0	0001 S/C R/L X X	Moves cursor or display without changing DD RAM contents. S/C=1 shift display, =0 shift cursor R/L=1 shift to right, =0 shift to left	37 us
6) Function Set	0	0	001 DL N F X X	Sets bus size, number of lines, and digit size (font). DL=1 8-bit bus, =0 4-bit bus N=0 1-line operation, =1 2-line F=0 5x8 dots, =1 5x10 dots	37 us
7) Set CG RAM Address	0	0	01 A A A A A A	Sets CG RAM address to AAAAAA	37 us
8) Set DD RAM Address	0	0	1 A A A A A A A	Sets DD RAM address to AAAAAAA	37 us
9) Read Busy Flag and Address	0	1	BF A A A A A A A	Reads busy flag and address counter	0 us
10) Write Data to CG or DD RAM	1	0	D D D D D D D D	Writes data into DD RAM or CG RAM (defined by last DD or CG RAM address set)	41 us
11) Read Data from CG or DD RAM	1	1	D D D D D D D D	Reads data from DD RAM or CG RAM (defined by last DD or CG RAM address set)	41 us

(*) For 270 kHz internal oscillator; for other frequencies (100 to 500 kHz), multiply time given by 270 kHz/foscillator.

表圖 5-13　HD44780U LCD 控制器的指令表

該控制器指令集與解釋，如表圖 5.14，它們的主要特點總結如下：

- Reset 存儲器地址 (Clear Display)，除了 (Return Home) 之外，還能夠清除顯示。
- 增量或減量顯示的位置 (Entry Mode Set)，再加上單獨 cursor 或顯示移位的偏移或不偏移 (Cursor or Display Shift)。
- 顯示器，cursor 的單獨選擇，和閃爍 ON/OFF 模式 (Display ON/OFF Control)。
- 4- 或 8 位群的線操作，5×8 位或 5×10 點的字符一行，或 5×8 點的字符兩行的顯示 (Function Set)。
- 顯示字符為 7 bits 地址 (Set DD RAM Address)，使 LCD 可以有 128 位置，分成兩行每何為 64 個字符。第一個行中的第一個字符的地址是 0，第二行中的第一個字符的地址為 64，不管 LCD 字符的實際數量。

表圖 5-14 還列出了用於所有指令所須的時間，它是控制器內部的振盪器在 270 KHz 時的最大執行時間。這個頻率是由 75 KΩ，外部電阻設置的 (VDD=3 V) 和 91 KΩ (當 VDD=5V) 所得的頻率。如果採用其它電阻值而得到不同的頻率 (覆蓋的範圍大約為 100 千赫-500 千赫)，則執行時間必須乘以 270 KHz/ 該振盪器頻率。

一個重要的設計考慮是該控制器的初始化過程，該過程可以用兩種方式來完成：在加上電時或由指令自動完成。由指令自動完成的組成如下：

1. 打開電源。
2. 等待 >15 us 當 VDD 上升到 4.5 V 後 (或 >40 us 後，當 VDD 上升到 2.7 V 後)。
3. 執行指令 "Function set"(37uS) 使用 RW='0'，RS='0'，DB ="0011× xxx"。

4. 等待 > 4.1 毫秒。
5. 執行指令 "Function set"(37us) 使用 RW='0'，RS='0'，DB ="0011× xxx"。
6. 等待 > 100us 的。
7. 執行指令 "Function set" (37us) 與 RW='0'，RS='0'，DB ="0011 xxxx"。
8. 執行指令 "Function set" (37us) 與 RW='0'，RS='0'，DB ="0011 NFxx" (選擇 N 和 F)。
9. 執行指令 "Display on/off control" 與 RW='0'，RS='0'，DB ="00001000" (37us)。
10. RW='0'，RS='0'，DB ="00000001" 執行指令 "Clear display" (1.52us)。
11. 執行指令 "Entry mode set" (37us) ）與 RW='0'，RS='0'，DB ="000001 I D S" (選擇 I / D 和 S)。

設計實例：

使用 LCD 為 HD44780U，當模式為 2-line/8-bit Bus 時，使其顯示 "VHDL" 寫在第一行的四個位置如圖 5.15 所示。

圖 5-15 　　　HD44780U LCD 設計實例

圖 5-16 為這個設計的流程圖。圖的左測為 LCD 的四個 FunctionSet1~ 4 送到控制器以設定 2-line/8-bit Bus/5x8-dot 的字符。接下來的 ClearDisplay 清除了顯示器的舊有字符，並使 MemoryAddress 歸零。DisplayControl 將顯示器開啓，cursor 和 blink 則關閉。最後的

EntryMode，使 RAM 的 address 設置為增加的模式。注意：這七個 states 都是用來設置 Instruction Register，所以 RS 設定為 '0'。圖的右測為 LCD 的四個 WriteData，RS 設定為 '1'，選用的是 Data Register，將 'V' = "01010110"，'H' = "01001000"，'D' = "01000100"，和 'L' = "01001100"，寫了進去。最後的 state 是 ReturnHome，它將 Memory Address 歸零，使得 cursor 重新回到 line 1。全部的過程中，訊號 E 至為重要。當 E 的 negative edge 時，Data 被寫入 Data Register。

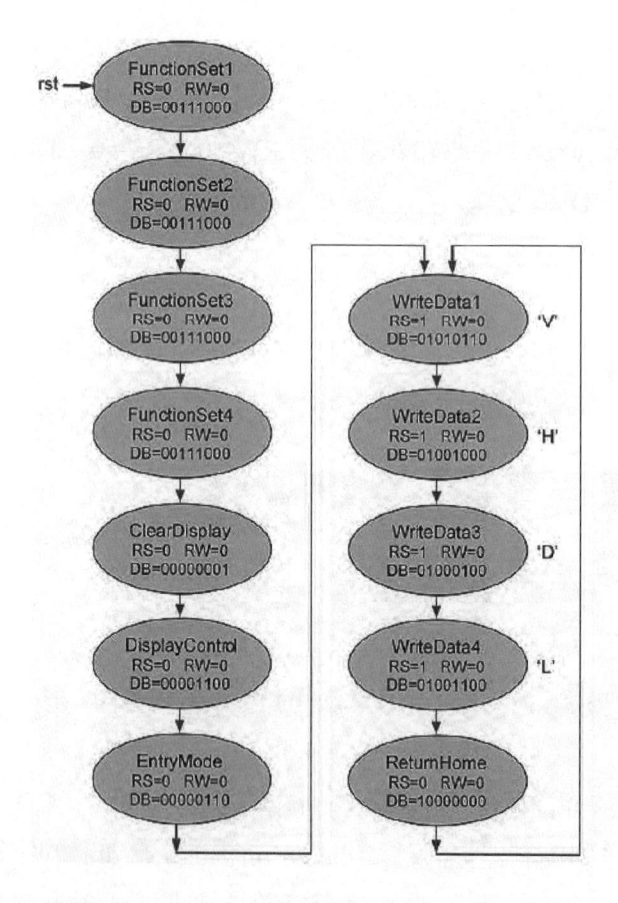

圖 5-16　LCD 顯示 "VHDL" 設計的流程圖

第五章　有限狀態機　121

　　表圖 5.14 的 HD44780U LCD 控制器的指令表中的 Max exec time 所標示的是每一個 states 處理至少應給的時間。其中佔時最長為 1.25 ms，因此為了簡單起見採用 500 Hz，佔時為 2 ms 來做 E 訊號，當能滿足表圖 5.14 的各項處理工作。

　　圖 5-17A 的 LCD_DRIVER.VHD 的 Clock Generator 便是由此而來。在它的 Lower section process (E) 跟上面二個例子的 Lower section，除了沒有 reset 之外，幾乎完全一樣。

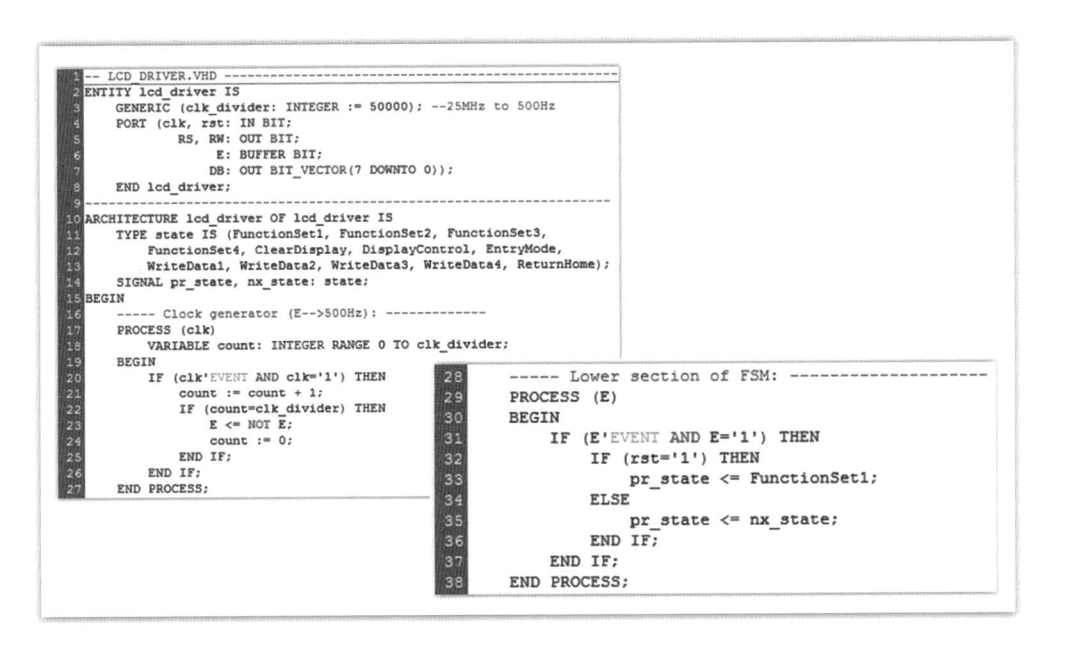

```vhdl
-- LCD_DRIVER.VHD ------------------------------
ENTITY lcd_driver IS
    GENERIC (clk_divider: INTEGER := 50000); --25MHz to 500Hz
    PORT (clk, rst: IN BIT;
          RS, RW: OUT BIT;
          E: BUFFER BIT;
          DB: OUT BIT_VECTOR(7 DOWNTO 0));
END lcd_driver;
-----------------------------------------------
ARCHITECTURE lcd_driver OF lcd_driver IS
    TYPE state IS (FunctionSet1, FunctionSet2, FunctionSet3,
         FunctionSet4, ClearDisplay, DisplayControl, EntryMode,
         WriteData1, WriteData2, WriteData3, WriteData4, ReturnHome);
    SIGNAL pr_state, nx_state: state;
BEGIN
----- Clock generator (E-->500Hz): -------------
PROCESS (clk)
    VARIABLE count: INTEGER RANGE 0 TO clk_divider;
BEGIN
    IF (clk'EVENT AND clk='1') THEN
        count := count + 1;
        IF (count=clk_divider) THEN
            E <= NOT E;
            count := 0;
        END IF;
    END IF;
END PROCESS;

----- Lower section of FSM: --------------------
PROCESS (E)
BEGIN
    IF (E'EVENT AND E='1') THEN
        IF (rst='1') THEN
            pr_state <= FunctionSet1;
        ELSE
            pr_state <= nx_state;
        END IF;
    END IF;
END PROCESS;
```

圖 5-17A　　LCD_DRIVER.VHD 的 E 訊號和 Lower section 部分

　　圖 5-17B 為 LCD_DRIVER.VHD 的 Upper section 部分。左側的 7 個 states 是用來設置 LCD 的 Instruction Register。左側的 5 個 states 是用來寫字符到 Data Register 中。LCD_DRIVER 的字符顯示，是在字符進到 Data Register 後的下一個 E 訊號來到時發生。

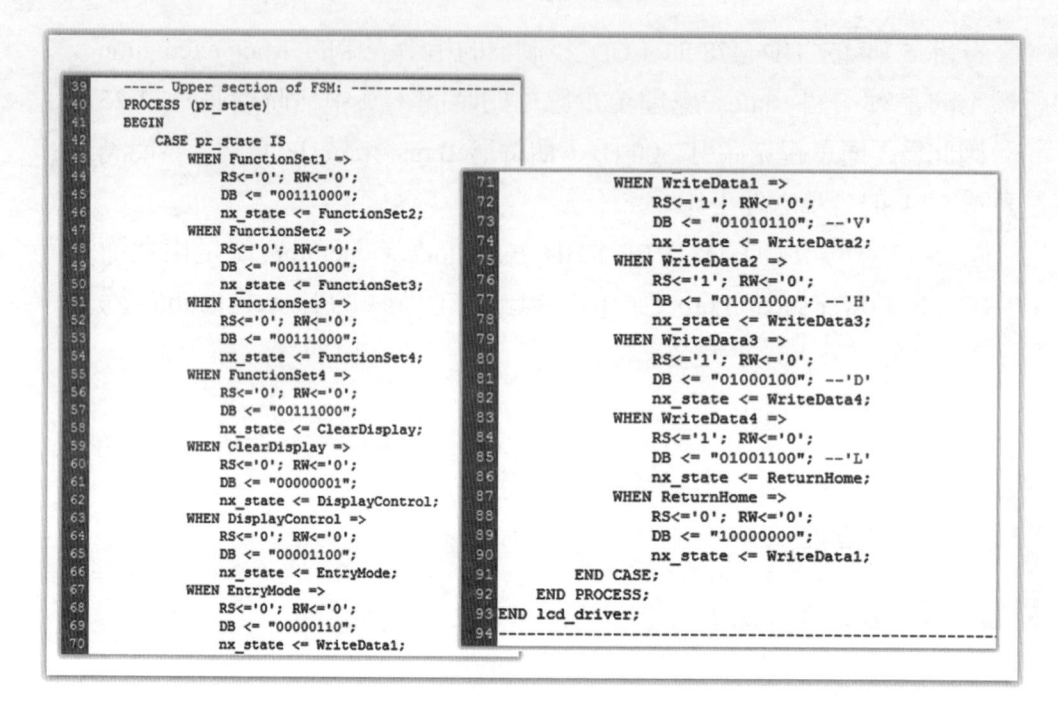

```
----- Upper section of FSM: ------------------
PROCESS (pr_state)
BEGIN
    CASE pr_state IS
        WHEN FunctionSet1 =>
            RS<='0'; RW<='0';
            DB <= "00111000";
            nx_state <= FunctionSet2;
        WHEN FunctionSet2 =>
            RS<='0'; RW<='0';
            DB <= "00111000";
            nx_state <= FunctionSet3;
        WHEN FunctionSet3 =>
            RS<='0'; RW<='0';
            DB <= "00111000";
            nx_state <= FunctionSet4;
        WHEN FunctionSet4 =>
            RS<='0'; RW<='0';
            DB <= "00111000";
            nx_state <= ClearDisplay;
        WHEN ClearDisplay =>
            RS<='0'; RW<='0';
            DB <= "00000001";
            nx_state <= DisplayControl;
        WHEN DisplayControl =>
            RS<='0'; RW<='0';
            DB <= "00001100";
            nx_state <= EntryMode;
        WHEN EntryMode =>
            RS<='0'; RW<='0';
            DB <= "00000110";
            nx_state <= WriteData1;
        WHEN WriteData1 =>
            RS<='1'; RW<='0';
            DB <= "01010110";  --'V'
            nx_state <= WriteData2;
        WHEN WriteData2 =>
            RS<='1'; RW<='0';
            DB <= "01001000";  --'H'
            nx_state <= WriteData3;
        WHEN WriteData3 =>
            RS<='1'; RW<='0';
            DB <= "01000100";  --'D'
            nx_state <= WriteData4;
        WHEN WriteData4 =>
            RS<='1'; RW<='0';
            DB <= "01001100";  --'L'
            nx_state <= ReturnHome;
        WHEN ReturnHome =>
            RS<='0'; RW<='0';
            DB <= "10000000";
            nx_state <= WriteData1;
    END CASE;
END PROCESS;
END lcd_driver;
```

圖 5-17B　LCD_DRIVER.VHD 的 Upper section 部分

5-6　DE2-115 液晶顯示器實作

　　圖 5-17A/B 的 VHDL 電路是依據 HD44780U LCD 控制器和 DE2-115 規格來設計，由於 HD44780U LCD 控制器的模式並不存在，所以無法寫一個 TestBench 來做完整軟體的測試。Modelsim 祇能做到對 lcd_driver.vhd 的 syntax 是否有錯而已。實質上祇能接上 DE2-115 的 LCD 來做電路的合成來測試。其中最重要的是 HD44780U 與 DE2-115 間的接腳連線，它們間的關係，如圖 5-18 所示。

```
-- LCD Driver to DE2-115 PINs
clk -> PIN_Y2;
rst -> PIN_M23(KEY[0]);
 RS -> PIN_M2;
 RW -> PIN_M1
  E -> PIN_L4;
DB0 -> PIN_L3;
DB1 -> PIN_L1;
DB2 -> PIN_L2;
DB3 -> PIN_K7;
DB4 -> PIN_K1;
DB5 -> PIN_K2;
DB6 -> PIN_M3;
DB7 -> PIN_M5;
```

圖 5-18　HD44780U 與 DE2-115 間的接腳連線

這個 HD44780U 與 DE2-115 間的接腳連線，將在做合戰的第 2 階段 S2 的 Assignment Editor 選用 FPGA/EP4CE115F29C7 的接腳被用到。

使用 Altera/Quantus II 軟體來合成 LCD_driver 的 S1 和 S2 階段的步驟，請參照第一章 1-6 節 Software Synthesis ALU_simple 電路的例子。它們間的第一階段 S1 除了 Project 的名稱不同外，其它的完全相同。第二階段的 S2，接腳 PIN 的關係如圖 5-19 所示。

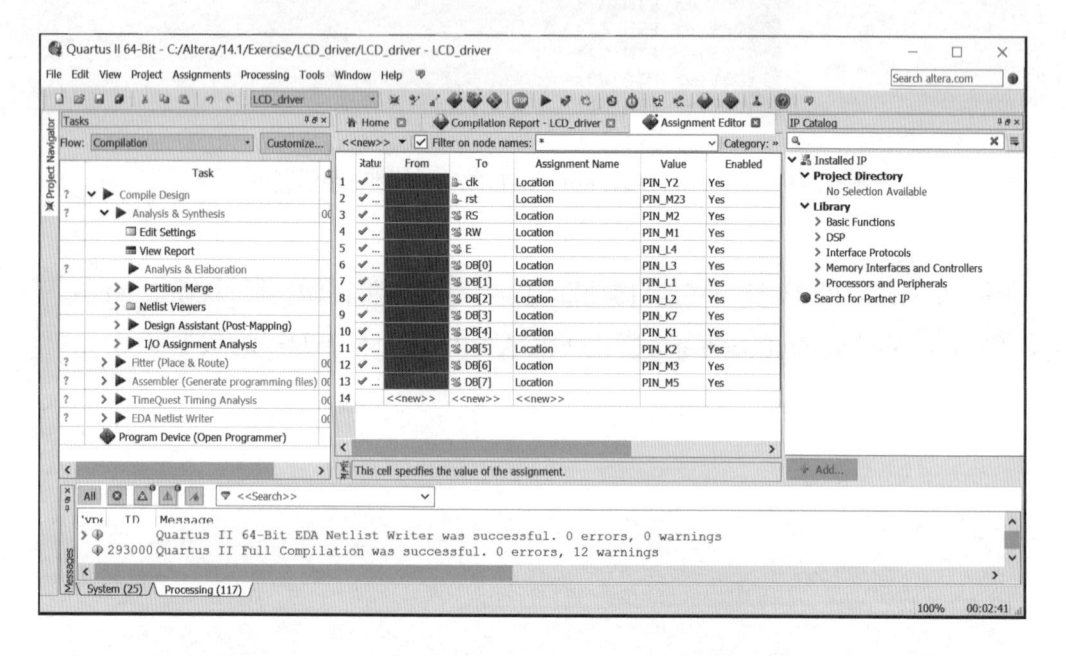

圖 5-19　LCD_driver 在 DE2-115 的接腳 PIN assignment

第五章　有限狀態機　　125

　　S2 的 compile 結果如圖 5-20 所示。Flow Summary 顯示共用了 48/114,400 個 combinational function，29/114,400 個 Dedicated logic registers 和 13/529 個 PIN。全部僅為 FPGA 總量的 1% 左右。

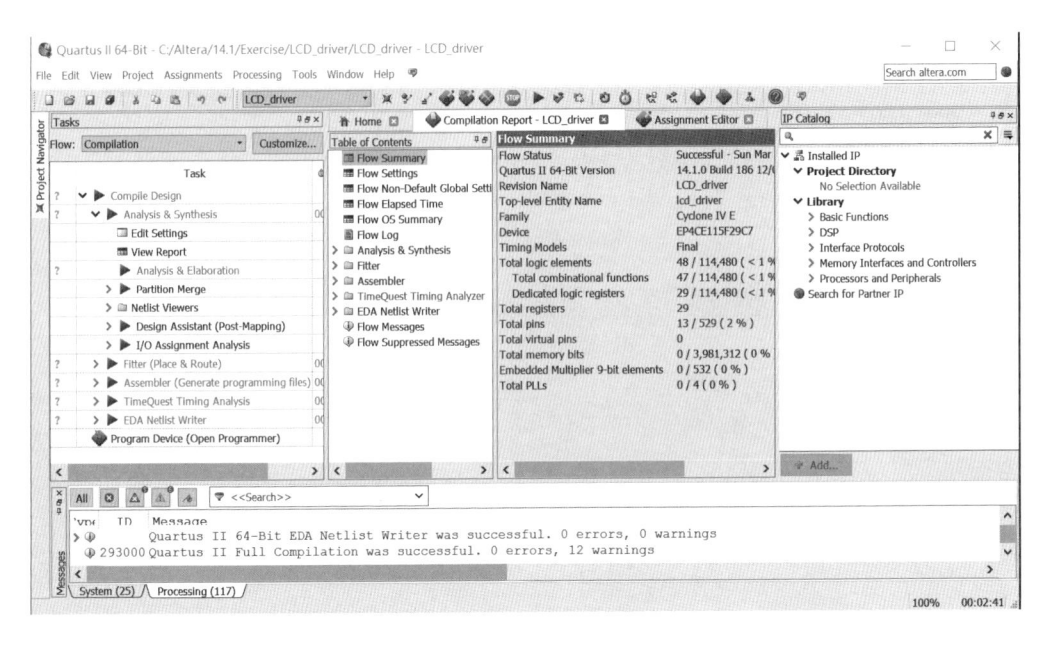

圖 5-20　S2 的 compile 結果，全部僅為 FPGA 總量的 1% 左右

連接 DE2-115 板子到 PC，並且接上電源，Programmer LCD_driver 的結果如圖 5-21 所示。

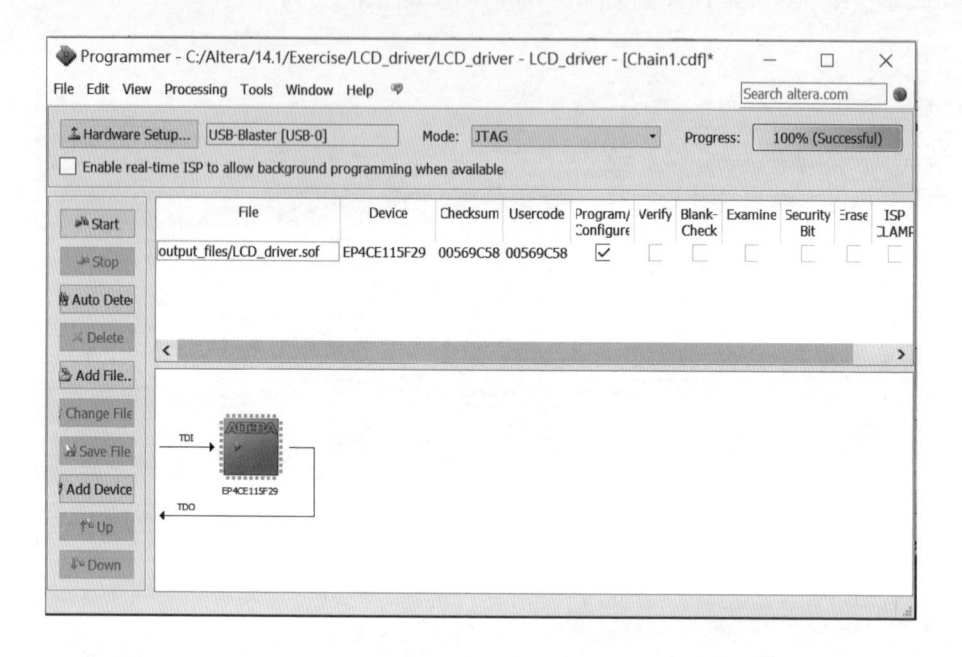

◎ 圖 5-21 ◎　　S2 的 compile 結果，全部僅為 FPGA 總量的 1%左右 Programmer LCD_driver 的結果

圖 5-22 為 Programmer LCD_driver 成功後的照相，"VHDL" 四個字顯示在 LCD 的第一行上。

圖 5-22 Programmer LCD_driver 成功後，"VHDL" 四個字顯示在 LCD 的第一行上

5-7 課外練習

(1) FSM 是數位系統設計中的一種方法。試述在眾多的數位系統設計中，最適用於哪類的數位系統設計中？

(2) FSM 數位系統設計中，state type 的選用會直接影響到系統中使用 flip-flop 數量的多寡，和速度的快慢。試列表說明之。

(3) 試用 Moore machine 來設計一個直接推動 SSD 的 BCD Counter。並寫出測試該 Counter 的 Test Bench。並用 ModelSim 來測試之。

第六章　Intel / Altera 方塊圖電路的設計

Altera 在提供 Quartus II 的同時還提供了電路的**電路庫** (Libraries)，其中不但有最基本的 Gates、Flip Flops、Adders 等。在 Mega-function Library 中還有 PLL、Multiplex，CPU、Memory、UART 等。除了提供其 VHDL/Verilog files 之外，還提供相對的**線路圖** (Schematic Diagrams)。所以用 FPGA 來做電路的設計，幾乎可說除非是特殊情況，否則祇要能熟悉所提供的電路庫，便可以事半功倍，完成電路的設計。本章將從線路圖入手。

◎ 6-1　VHDL 檔轉變成電路圖形

　　VHDL 代碼，基本上是同時進行，不分先後的代碼。而純並發邏輯聲明僅限於 WHEN、SELECT、和 GENERATE，這類的聲明祇能夠存在於**順序碼** (sequential code) 之外。並發碼祇適用於**組合電路** (combinational circuits) 的設計中。第一章的圖 1-2 為將 WHEN 和 SELECT 用在**複用器** (Multiplexer) 電路的設計上。

　　線路圖的設計，是從 Quartus II 所提供的電路 Libraries 中的電路**圖形** (Symbol) 拉到線路圖的 Project 內。然後連線、I/O 的命名後再編輯而成。但是電路的電路庫中如果沒有設計所需的電路圖形，而設計者有 VHDL 或 Verilog 檔，便可以經由 Quartus II 轉變成其對應的電路圖形。

VHDL 檔轉變成電路圖形的頭一步，是從 Quartus II 的 New Project Wizard 開始，如圖 6-1A 和 6-1B 所示。

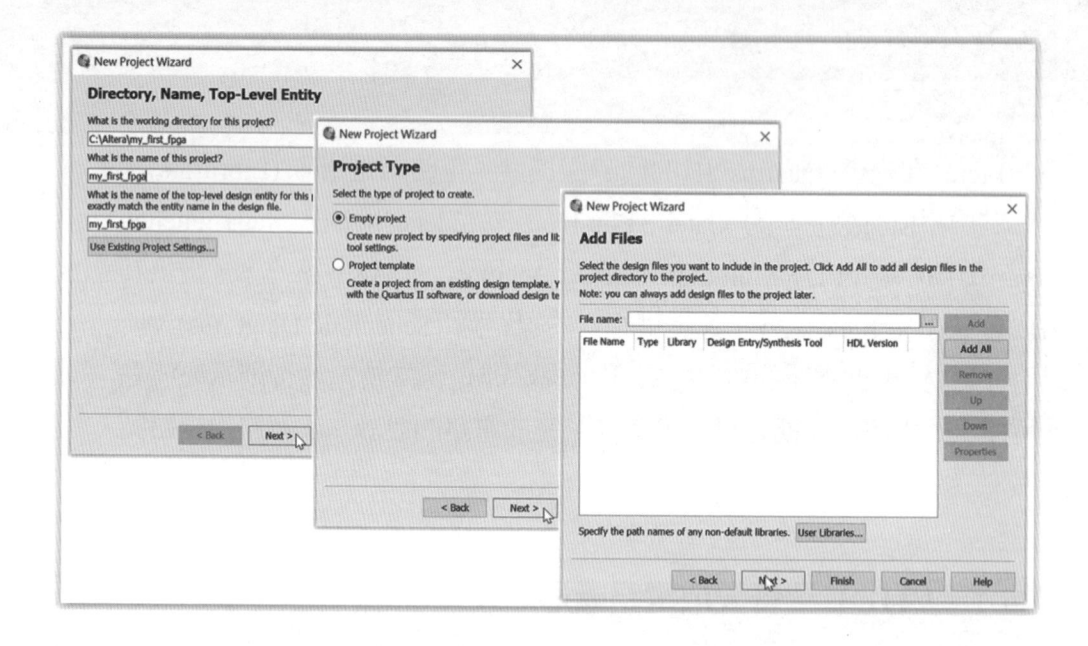

圖 6-1A　Quartus II 的 Assign The Device 之一

第六章　Intel / Altera 方塊圖電路的設計　131

　　圖 6-1B 為 FPGA 的選用，DE2-115 用的是 Cyclone IV E 的 EP4CE115F29C7。最後的 Summary 列出了 Project directory、Project 的名字、使用的 FPGA、Simulation 使用的 Simulator 和 FPGA 內部 Vccint 的電壓等。

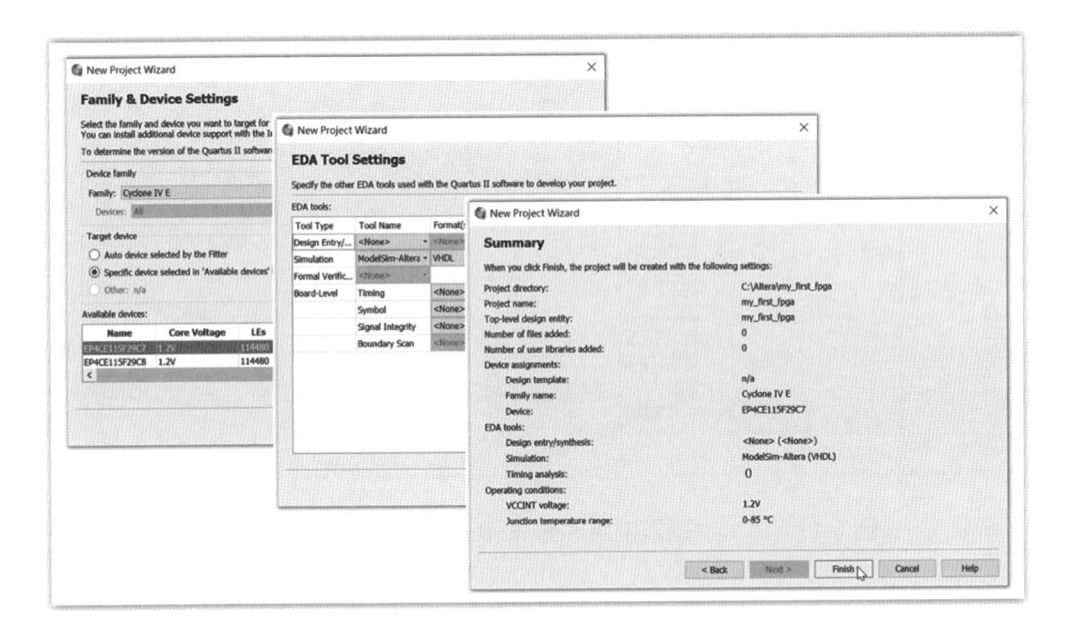

圖 6-1B　　　Quartus II 的 Assign The Device 之二

接下來選用 **File > New > Block Diagram / Schematic File** 如圖 6-2 所示。並選用 **File > Save As**，File Name 為 my_first_fpga.bdf。

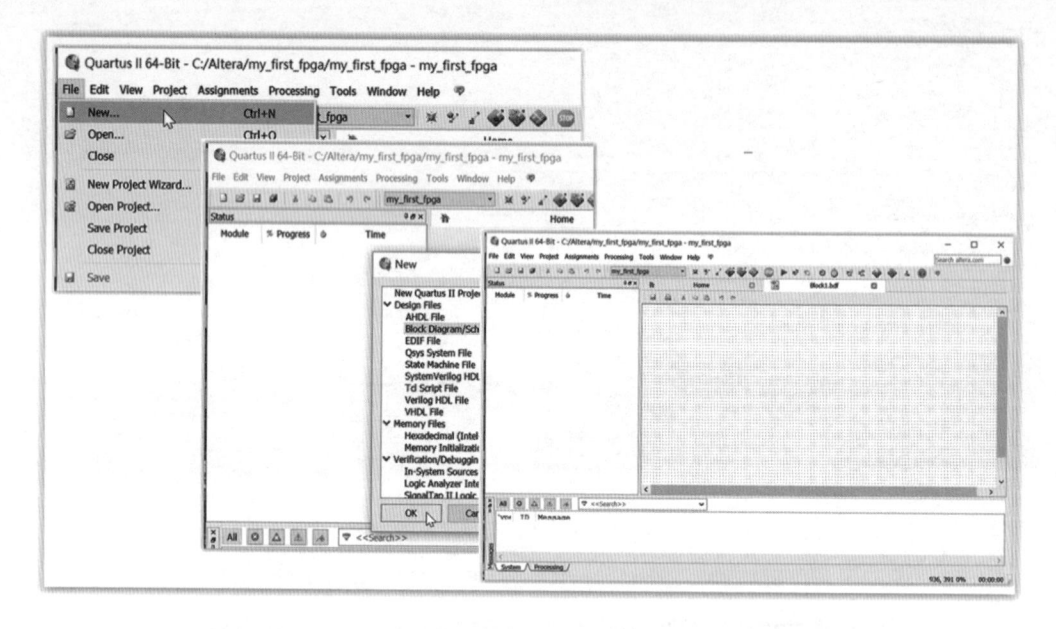

◦ 圖 6-2 ◦　　進入選用 Schematic Diagram 設計之一

第六章　Intel / Altera 方塊圖電路的設計

再選用 File > New >VHDL File 將要產生 Symbol 的 VHDL 檔加入到空白的 block diagram 中，如圖 6-3 所示。

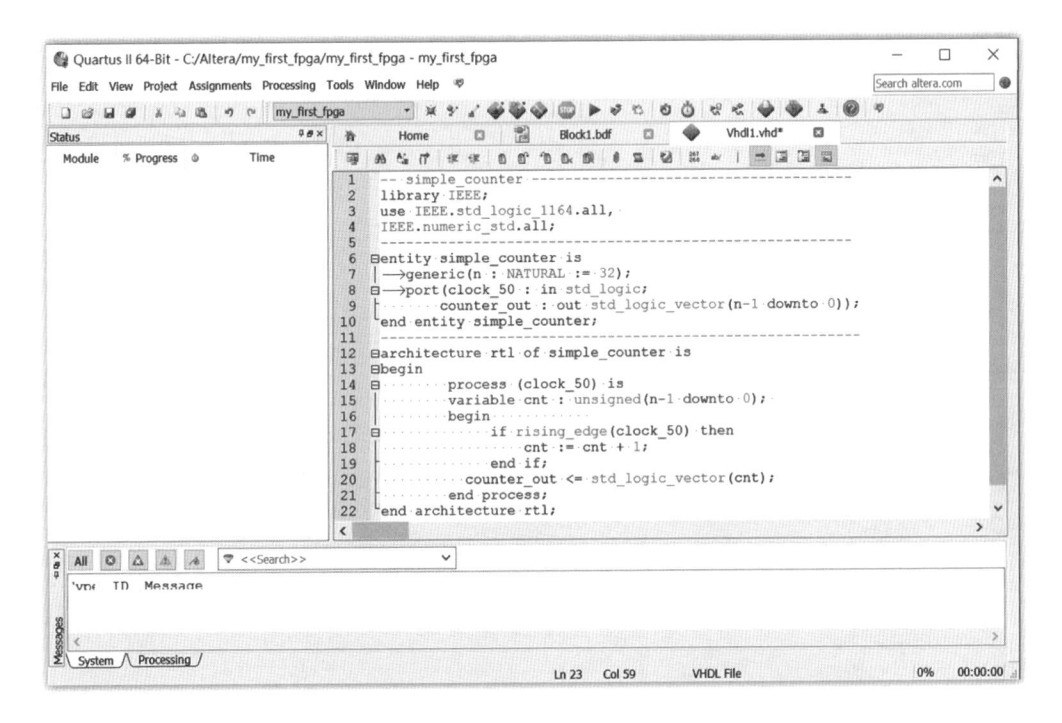

◉ 圖 6-3 ◦―― 將要產生 Symbol 的 VHDL 檔加入到空白的 block diagram 中

再選用 **File > Save As** 檔名為 simple_counter.vhd，如圖 6-4 所示。在 QuartusⅡ 開啟 simple_counter.vhd 檔案。然後選用 **File > Create/Create Symbol** 使 simple_counter.vhd 轉變為方塊圖式的 simple_counter.sym 檔。這樣便能夠使 VHDL 代碼檔用在 BDF 方塊圖上。

Quartus II 產生了 simple_counter.sym 檔的訊息，如圖 6-4 所示。

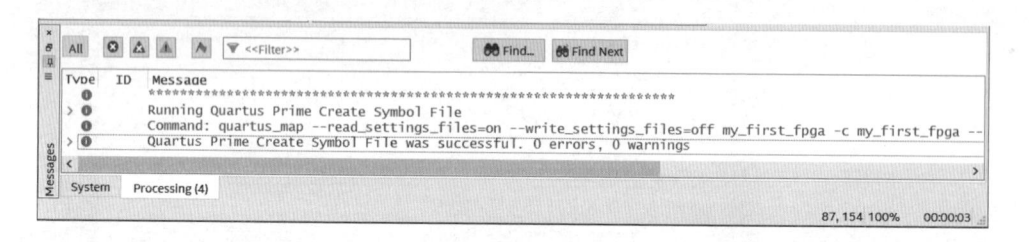

```
Type   ID    Message
       0     ****************************************************************
       0     Running Quartus Prime Create Symbol File
       0     Command: quartus_map --read_settings_files=on --write_settings_files=off my_first_fpga -c my_first_fpga --
       0     Quartus Prime Create Symbol File was successful. 0 errors, 0 warnings
```

圖 6-4　　Quartus II 產生了 simple_counter.sym 檔的訊息

第六章　Intel / Altera 方塊圖電路的設計　135

為了將 simple_counter.vhd 的 Symbol 加入到設計的 top-level，雙擊 project 視窗中的 my_first_fpga.bdf。當 Symbol 視窗出現時，展開 Libraries 的 project，選取其中的 simple_counter，再單擊 **OK**，使 simple_counter 的 Symbol 加入到 project 視窗上。如圖 6-5 所示。

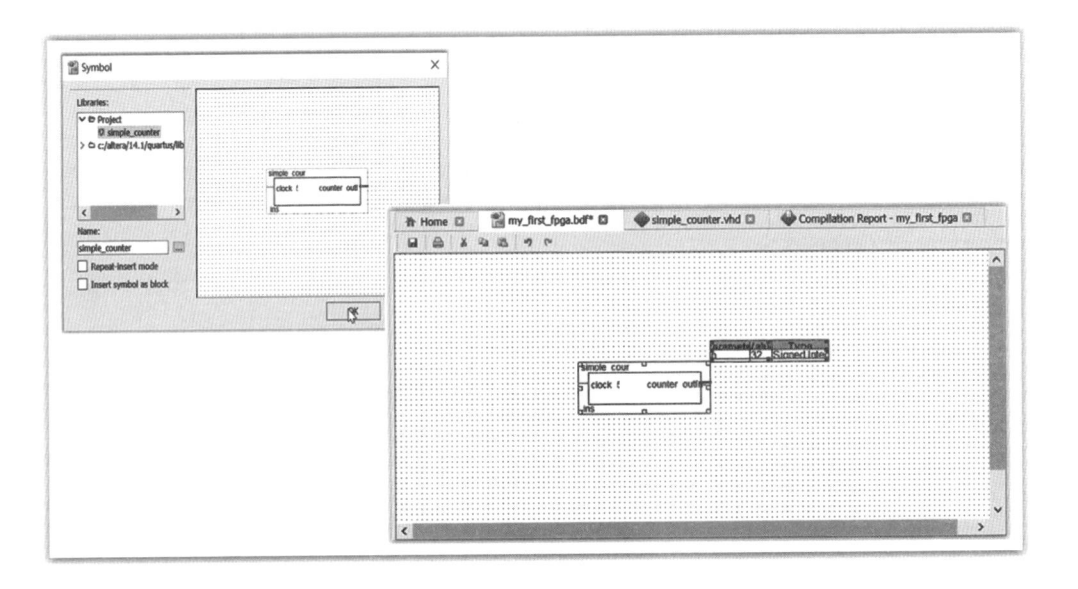

圖 6-5　simple_counter.vhd 的 Symbol 加入到設計的 top-level 上

6-2　選用電路庫中的電路 Symbols

欲將 Quartus II 14.1 的 Libraries 中的 PLL 和 MUX 等 Symbols 加入到 project 視窗，可以在 Symbol 視窗的最右側，展開 Installed IP 的 Library、Basic Function、Clocks、選取 PLL 中的 ALTPLL，再單擊 +Add。當 Save IP Variation 小視窗出現時在 my_first_fpga 的後面，加入 mypll。然後再單擊 Ok，如圖 6-6 所示。

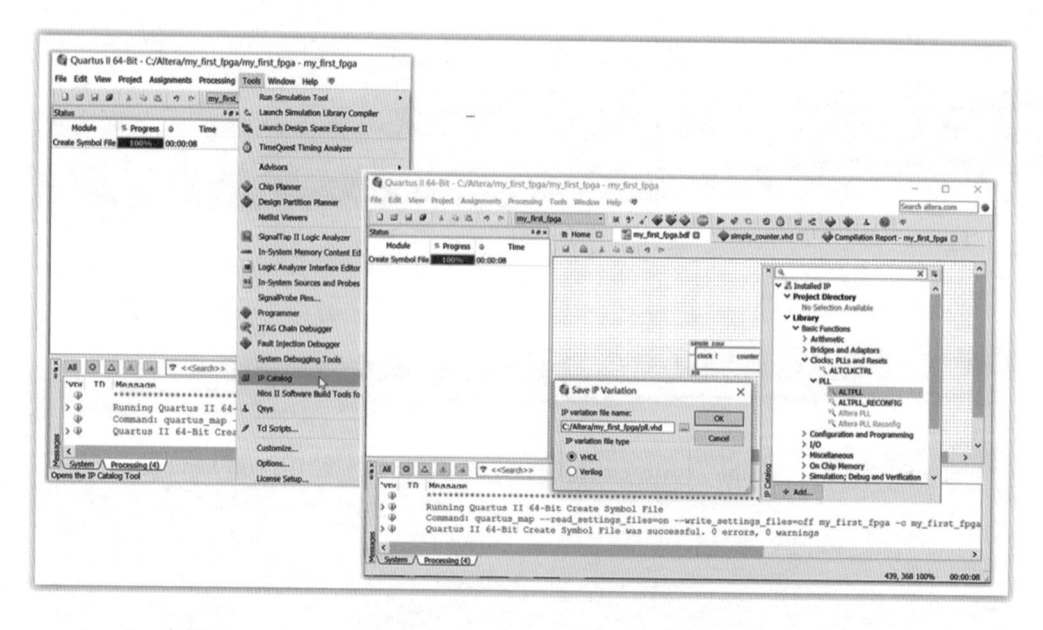

圖 6-6　選用 Quartus II 14.1 Libraries 中的 ALTPLL

第六章　Intel / Altera 方塊圖電路的設計　137

ALTPLL 的設定項目共 12 項，有 Parameter Settings、PLL Reconfiguration、Output Clock、EDA 和 Summary 等五類。

1. Parameter Settings 又有四頁，其第一頁與 DE2-115 的 System Clock 50 MHz 有關。如圖 6-7 所示。

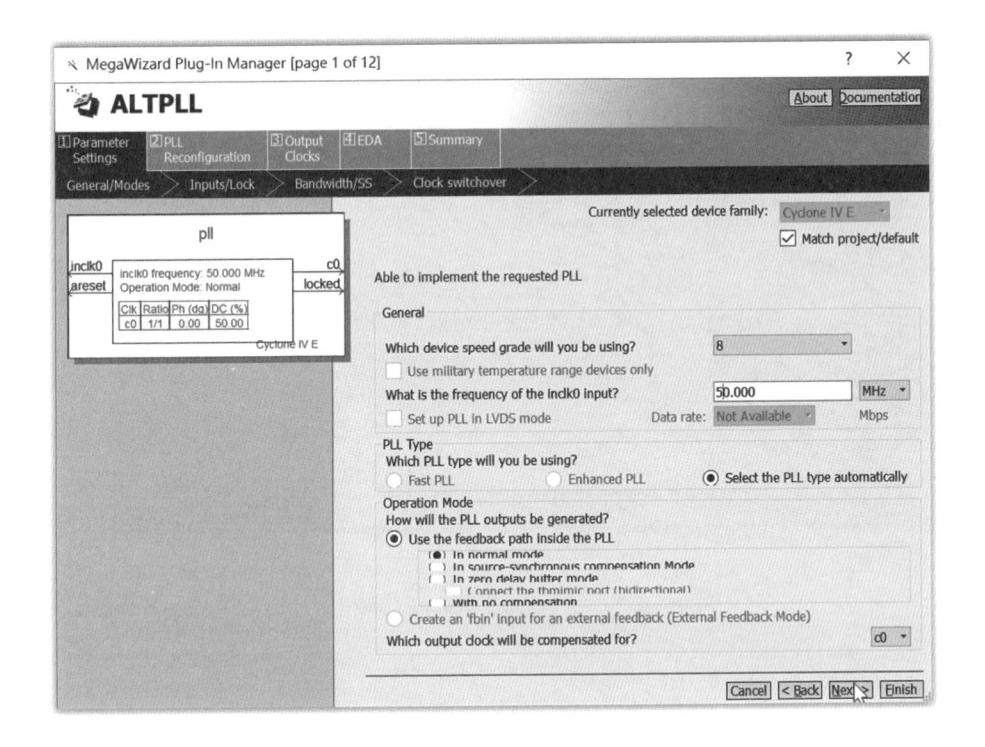

◦ 圖 6-7 ◦　　inclk0 由 DE2-115 的 System Clock 50 MHz 所提供

第 2 頁選用 areset 為輸入，取消 locked。如圖 6-8 所示。

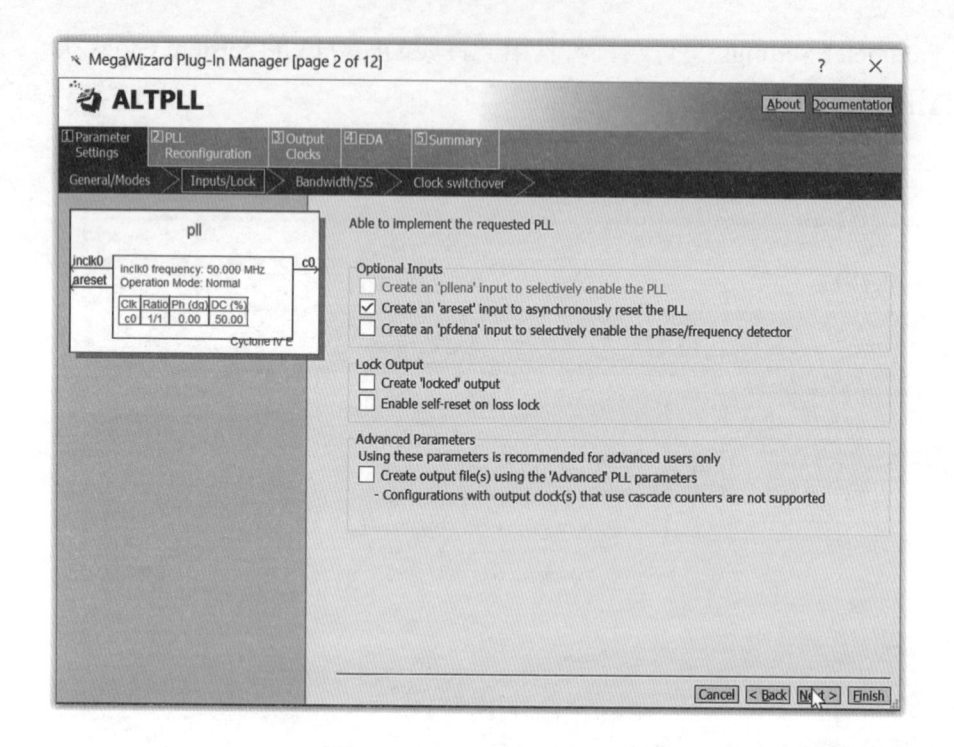

◦ 圖 6-8 ◦　　選用 areset 為輸入

其它 3、4、5 頁，在本例子中可以忽略按 next。

第 6 頁為 c0-Core/External Output Clock 的設定，如圖 6-9 所示。

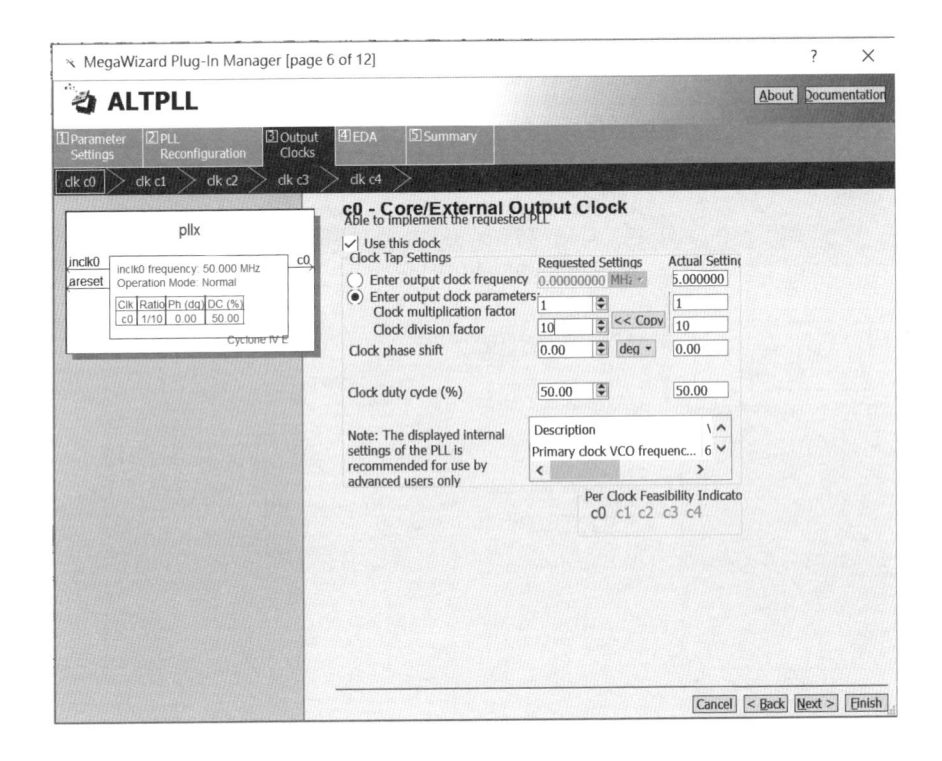

圖 6-9　c0-Core/External Output Clock 的設定

第 7、8、9、10、和 11 頁，在本例子中可以忽略按 **next**。設定的總結，與其應產生的檔案如圖 6-10。

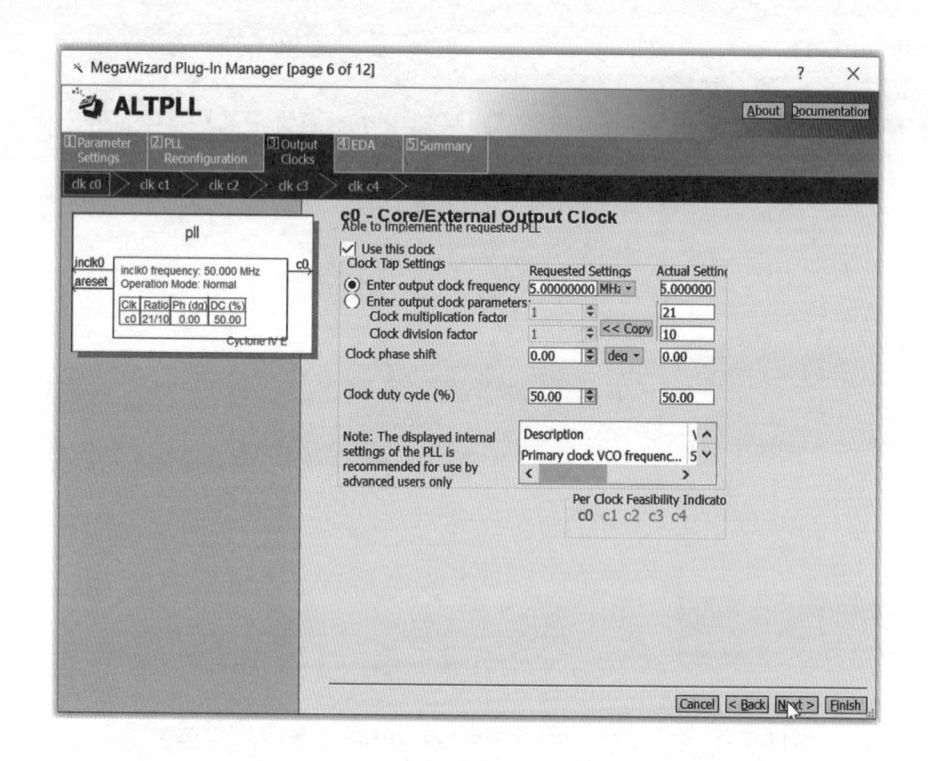

　　　　圖 6-10　　　ALTPLL 設定的總結，與其應產生的檔案

當 Symbol 視窗出現時，在視窗最右側的頂端輸入 LPM_MUX。再將 Library 展開，選取 LPM_MUX。然後點擊視窗最右側的末端 **+Add**。

第六章　Intel / Altera 方塊圖電路的設計　141

又當 Save IP Variation 視窗出出現時，在 my_first_fpga/ 的後面寫入 myMUX，及單擊 **OK**，如圖 6-11 所示。

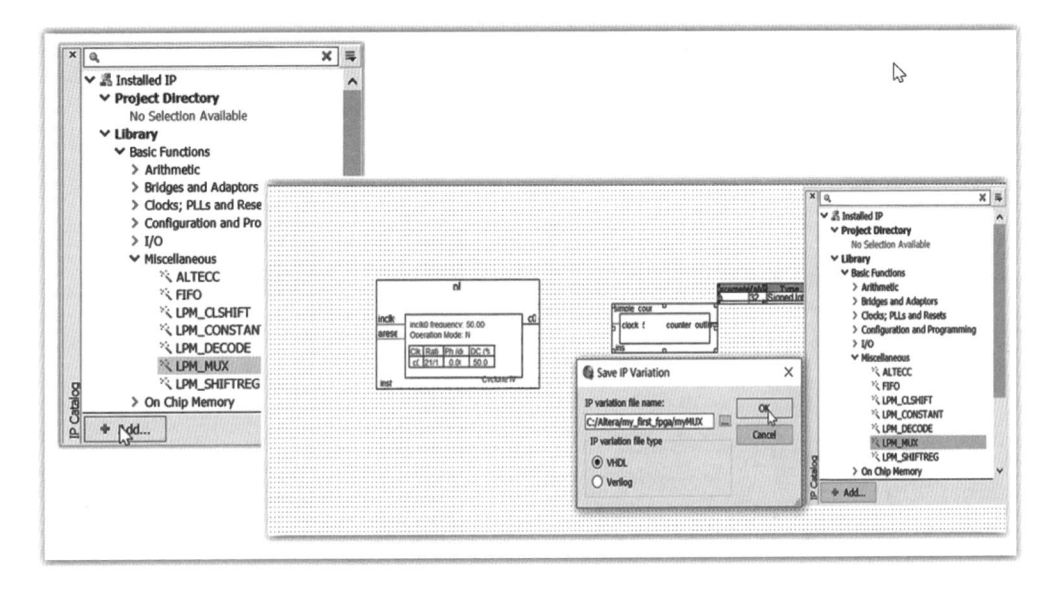

⚙ **圖 6-11**　　　加入 LPM_MUX 的步驟

當 MegaWizard Plug-In Manger 視窗出現 [page 1 of 3] 時,在 How many 'data' inputs do you want? 選取 **2**。How wide should the 'data' input and the 'result' outout buses be? 選取 **4**。Do you want to pipeline the multiplexer? 選 **No**。如圖 6-12A 所示。

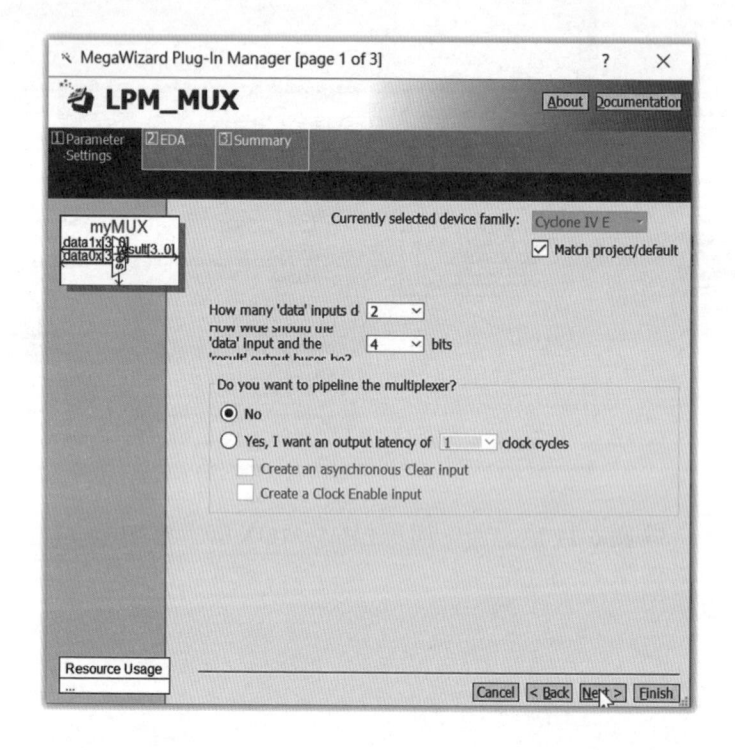

圖 6-12A　　LPM_MUX 設定之一

第六章　Intel / Altera 方塊圖電路的設計　143

　　MegaWizard Plug-In Manger 視窗的 [page 2 of 3]，可以忽略。[page 3 of 3] 應勾選 myMUX.cmp 和 myMUX.bsf，然後單擊 Finish，如圖 6-12B 所示。

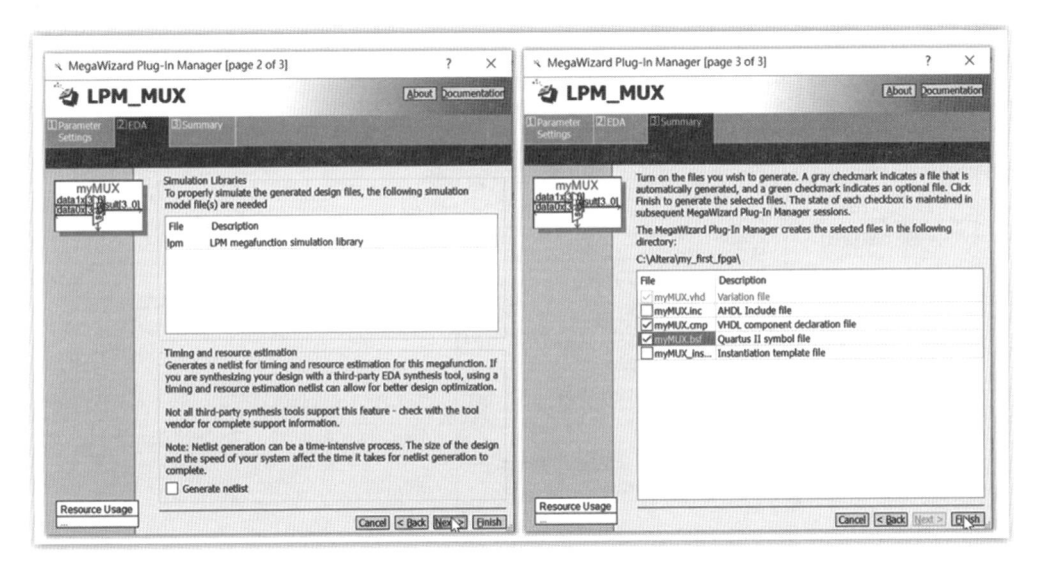

圖 6-12B　　LPM_MUX 設定之二

當 Symbol 視窗再度出現，將 myMUX 拉進到 project BDF 視窗的下方，如圖 6-13A 所示。並且在 Symbol 視窗的 Libraries 的 primitives 中選用 pin，Name 其為 input 或 output，再單擊 OK。同理在 logic 中選用 not gate，Name 其為 input，再單擊 OK。

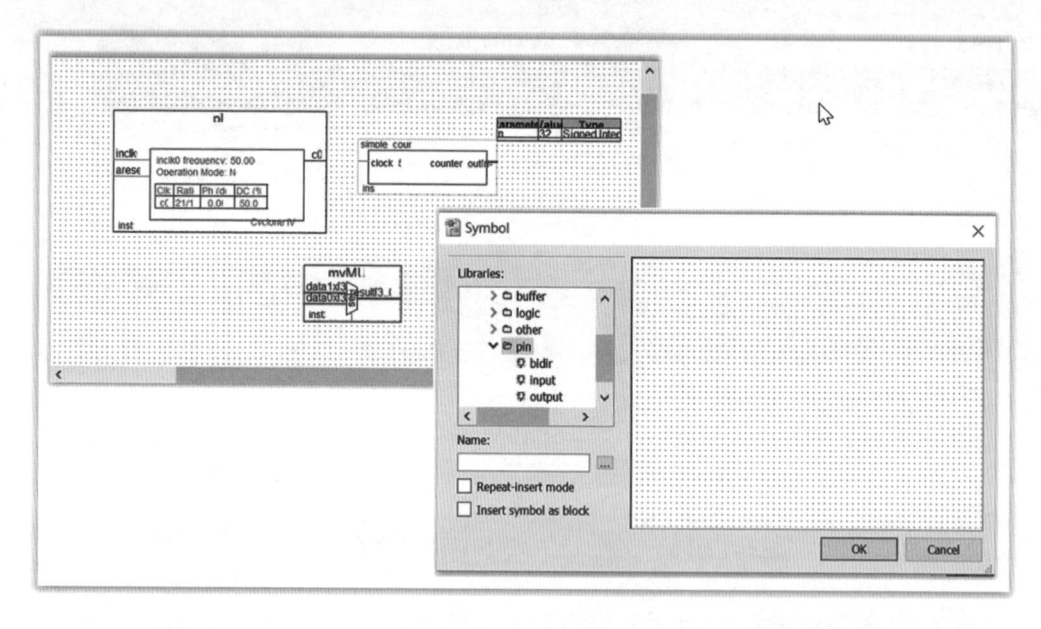

圖 6-13A　Project BDF 視窗內 I/O pin、及附件的設定

輸入或輸出接點名稱的設定和修改可從右擊目標，再點擊 **Properties** 入手，如圖 6-13B 所示。

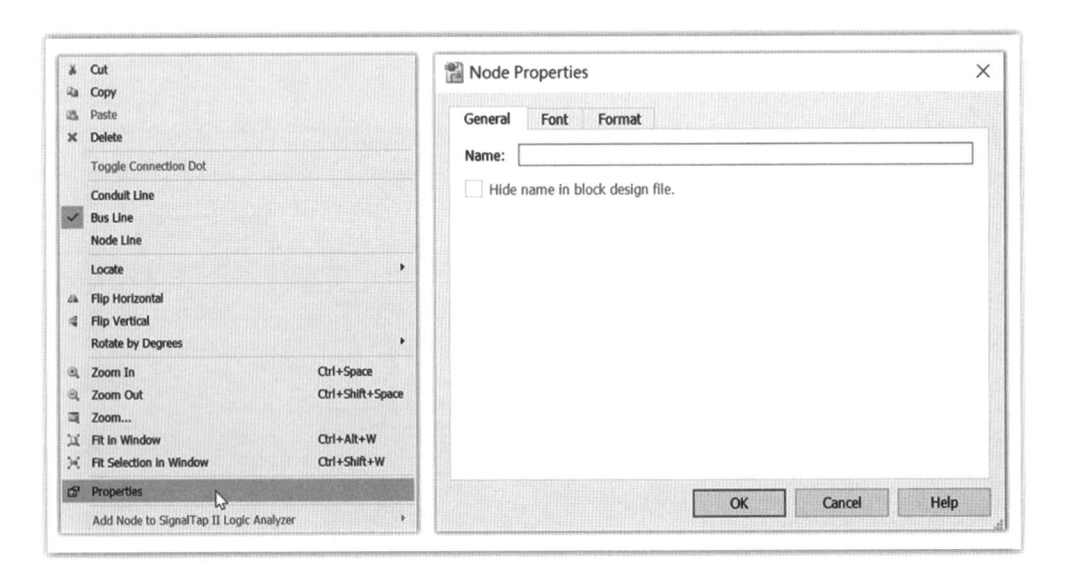

圖 6-13B　輸入或輸出接點名稱的設定和修改

146 iLAB FPGA 數位系統設計、模擬測試與實體除錯

接下來是在 Project BDF 視窗中方塊間的連線及 I/O ports 的命名，如圖 6-14 所示。Windows 10 對於圖 6-13B 名稱的設定效果很不理想，clock_50 只看到 clock_5，KEY[1] 只看到 KEY[1，counter[31..0] 只看到 counter[31 等。

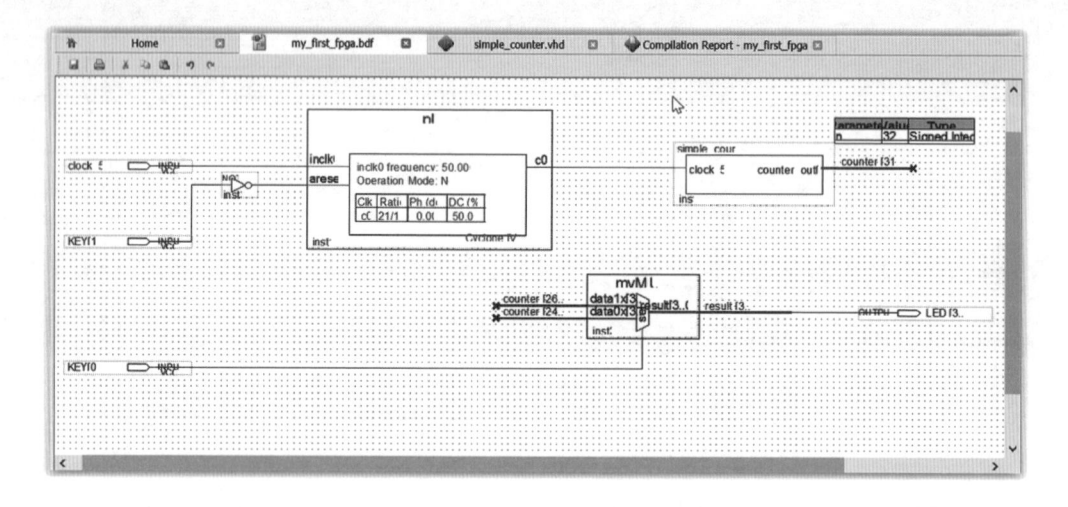

◎ 圖 6-14 　 Project BDF 視窗中方塊間的連線及 I/O ports 的命名

6-3 電路方塊圖完成後的設計工作

電路方塊圖完成後必須再次 Processing > Start compilation 以確認沒有錯誤。圖 6-15 compilation 便是這樣的結果，0 errors。

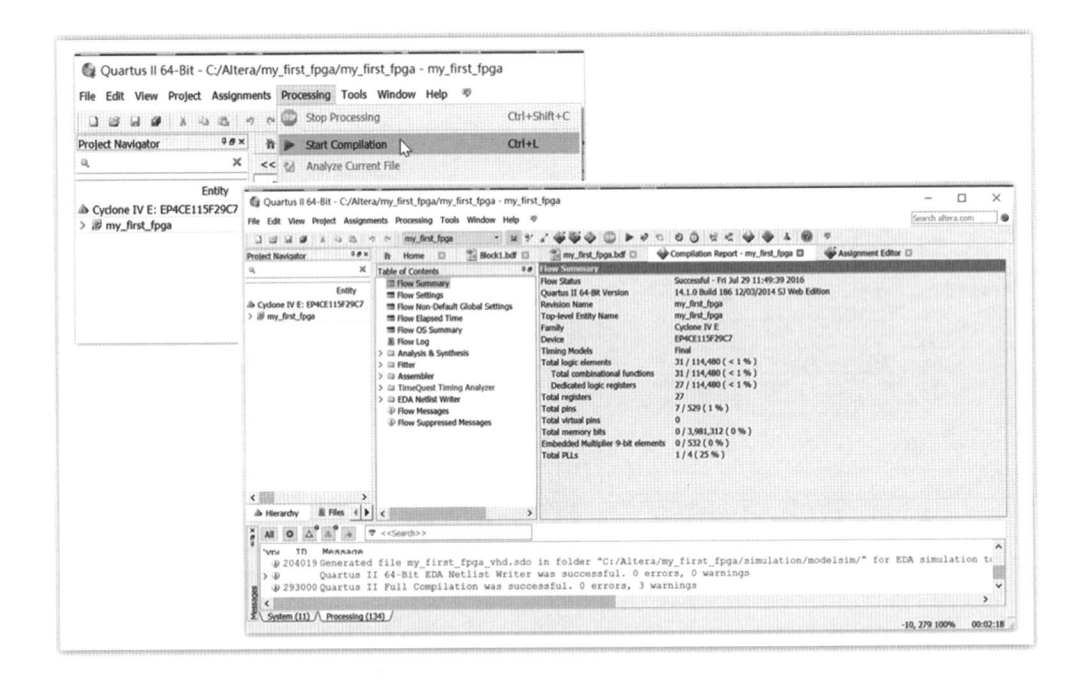

圖 6-15　Compilation 的結果確認沒有錯誤，0 errors。

148　iLAB FPGA 數位系統設計、模擬測試與實體除錯

　　完成 Project BDF 視窗中方塊間的連線及 I/O ports 的命名後，為了能配合 DE2-115 的 FPGA 的 I/O pins 的連接，要在 Project window 的工具欄上選用 Assignments > Assignment Editor，然後點擊左上角的 <<new>>，再點擊 To 下面的 Node Finder，當 Node Finder 視窗出現時點擊右上方的 List，再在 Nodes Found 欄選取並 > 到 Selected Nodes 欄內。如圖 6-16A 所示。

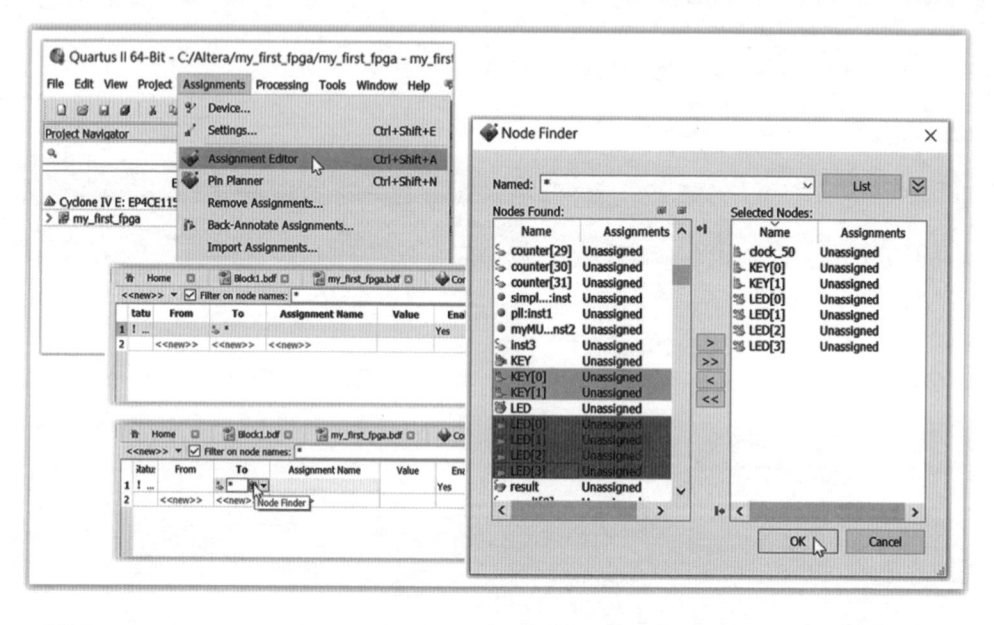

圖 6-16A　my_first_fpga project 與 DE2-115 pin assignment 步驟之一

第六章　Intel / Altera 方塊圖電路的設計　149

　　Clock_50、KEY[0]、KEY[1]、LED[0]~LED[3] 的 pins location 都詳細登錄在 DE2-115 使用者手冊中，如圖 6-16B 所示。

Status	From	To	Assignment Name	Value	Enabled	Entity	Comment	Tag
1 ✓ ...		clock_50	Location	PIN_Y2	Yes			
2 ✓ ...		KEY[0]	Location	PIN_M23	Yes			
3 ✓ ...		KEY[1]	Location	PIN_M21	Yes			
4 ✓ ...		LED[0]	Location	PIN_G19	Yes			
5 ✓ ...		LED[1]	Location	PIN_F19	Yes			
6 ✓ ...		LED[2]	Location	PIN_E19	Yes			
7 ✓ ...		LED[3]	Location	PIN_F21	Yes			
8		<<new>>	<<new>>	<<new>>				

圖 6-16B　my_first_fpga project 與 DE2-115 pin assignment 步驟之二

　　圖 6-16B pin assignment 填寫完畢之後，不要忘記 File > Save。然後再做一次如圖 6-15 般的 Processing > Start compilation 以確認 pin assignment 沒有錯誤，0 errors。

Compilation 確認沒有錯誤的結果，就可以在 Altera > my_first_fpga 的 output_files 中獲得 my_first_fpga.sof。如圖 6-17 所示。

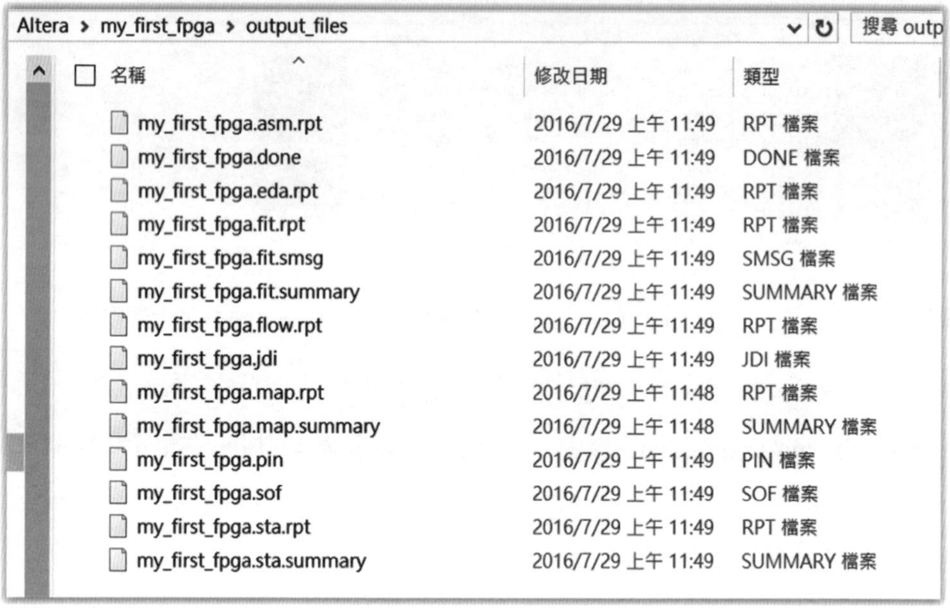

圖 6-17　compilation 無誤的結果產生的 my_first_fpga.sof

第六章　Intel / Altera 方塊圖電路的設計　151

將 DE2-115 的電源 ON，並且將 USB cable 連接到 PC 上，選用 **Project window** 工具欄中的 **Tools > Programmer**，如圖 6-18 所示。點擊 **Add File** 選用其中的 my_first_fpga.sof，再單擊 Start，當 Progress 欄顯示 100% (Successful) 表示 Project 成功。

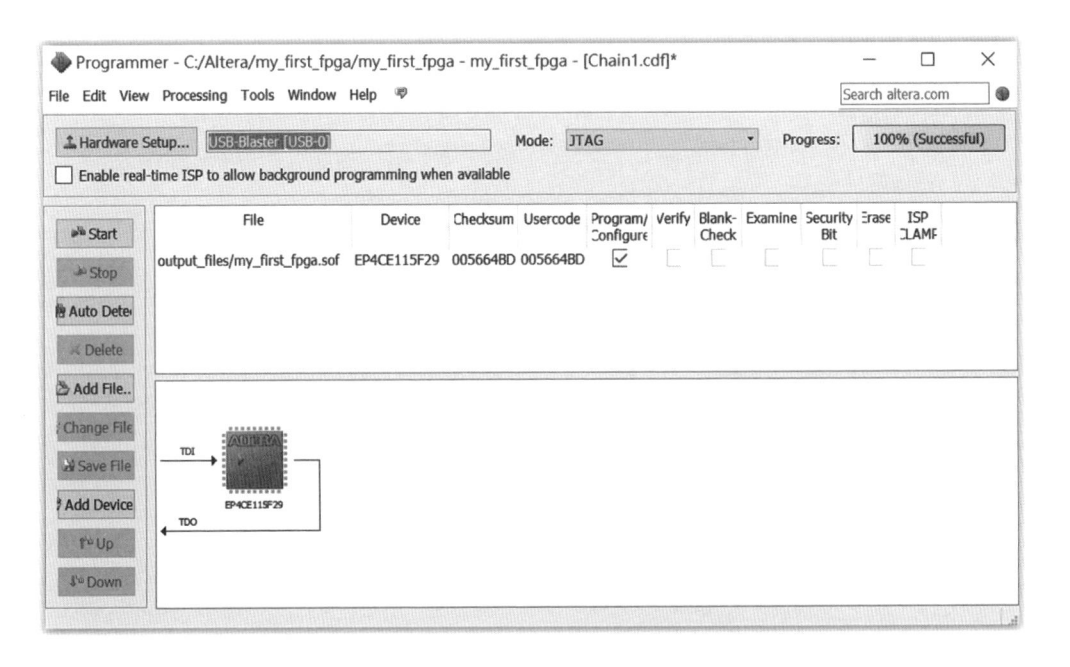

圖 6-18　Programmer 視窗顯示 100% 成功

6-4 課外練習

(1) 試述如何將 VHDL 檔為 simple_counter 所產生的 Symbol 添加到一個屬於設計者的 MyLib 中？

(2) Altera 方塊圖電路的設計，雖然擺脫了電路零件的設計，但對於其所提供電路零件組參數的選用，並不全然是容易的事。因此對於是否選用方塊圖電路來做電路的設計，應該在具備那些條件下進行？試詳細分析之。

第七章　串行平行乘法器

◎ 7-1　串行平行乘法器的設計

乘法器在數位濾波器、神經網絡等電路擔負著不可缺的任務。乘法器的設計，種類很多，串行平行乘法器的設計，假定 a()="1100"，b()="1101" 它的基本數學演算規則如圖 7-0 所示。

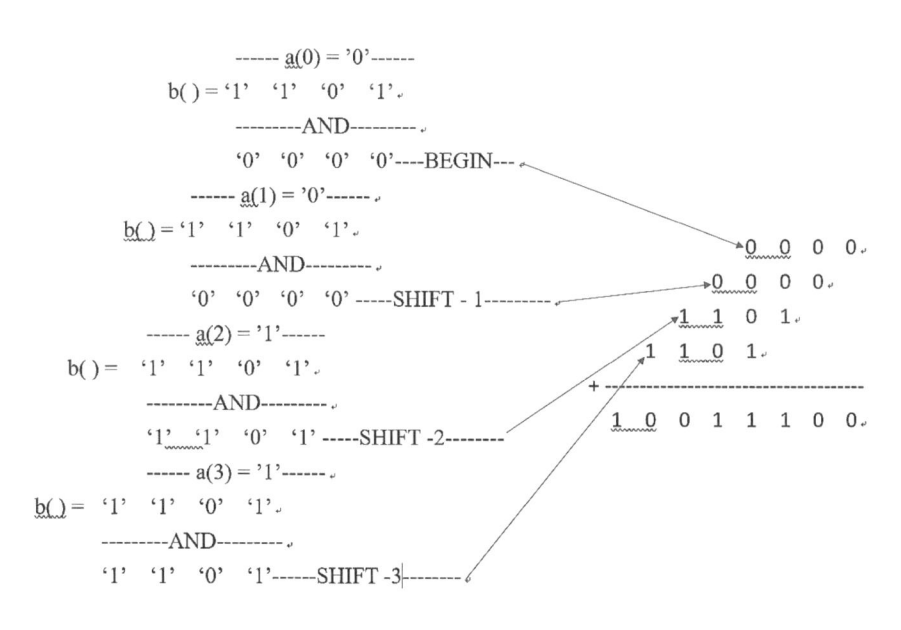

```
                ------ a(0) = '0'------
            b( ) = '1'  '1'  '0'  '1'
                ---------AND---------
                '0'  '0'  '0'  '0'----BEGIN---
                                                    0 0 0 0
                ------ a(1) = '0'------
            b( ) = '1'  '1'  '0'  '1'
                ---------AND---------              0 0 0 0
                '0'  '0'  '0'  '0' -----SHIFT - 1---------
                                                  1 1 0 1
                ------ a(2) = '1'------
            b( ) = '1'  '1'  '0'  '1'           1 1 0 1
                ---------AND---------         + ---------------------
                '1'  '1'  '0'  '1' -----SHIFT -2---------
                                              1 0 0 1 1 1 0 0
                ------ a(3) = '1'------
            b( ) = '1'  '1'  '0'  '1'
                ---------AND---------
                '1'  '1'  '0'  '1'-----SHIFT -3---------
```

圖 7-0　乘法的基本數學演算規則

153

圖 7-1 所示為 RTL 型串行平行組成的乘法器。它的一個訊號 a，是以串行每次以一個 Bit 的方式，由 LSB 為首輸入到電路中。另外一個訊號 b( )，則以所有的 Bits 同時並行輸入到電路裏。如果訊號 a 為 M bits，訊號 b( ) 為 N bits，則乘法器的 prod 輸出當為 M+N bits。因此當訊號 a 的 M bits 輸入到電路之後，必須再加入 N 個 '0'，為 prod 的輸出預留位置。

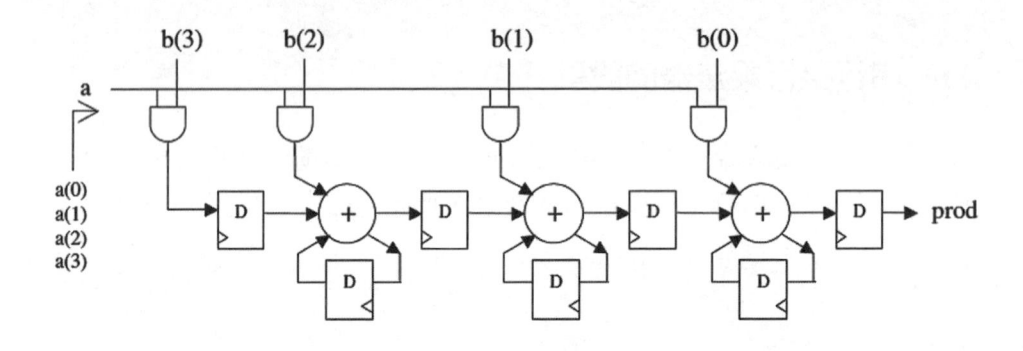

◦ 圖 7-1 ◦　　串行平行組成的乘法器電路圖

從圖 7-1 可以看到該系統使用**流水線化** (Pipelined) 的設計，並且使用 AND gate，和 full-adder 來建構。流水線的每個單元（除了最左邊的那個單元）都需要用一個加法器和兩個 Dff，再外加一個 AND gate 來組成，如圖 7-5 所示。

下面是組成每個單元的零件代碼，以及聲明包含所有零件的包裹 (PACKAGE)，最後是乘法器的主代碼 multiplier.vhd。這種結構型的設計，是多階實例化的一個例子。先由零件組成了 pipe，而 pipe 又成為組成 multiplier 的零件。

圖 7-7 所示，為 2 input AND gate 的代碼。圖 7-3 所示，為 DFF 的代碼。圖 7-4 所示，為 1 bit full adder 的代碼。它們都是組成 串行平行的乘法器電路圖的基本零件。

第七章　串行平行乘法器

```vhdl
------ and_2.vhd (component): ---------
LIBRARY ieee;
USE ieee.std_logic_1164.all;
----------------------------------------
ENTITY and_2 IS
    PORT ( a, b: IN STD_LOGIC;
           y: OUT STD_LOGIC);
END and_2;
----------------------------------------
ARCHITECTURE and_2 OF and_2 IS
BEGIN
    y <= a AND b;
END and_2;
----------------------------------------
```

圖 7-2　2 input AND gate 零件 and_2.vhd 的代碼

圖 7-1 串行並行組成的乘法器電路圖中的 reg，每個 DFF 都省略了 rst 和 clk 輸入訊號的標示。實際上所有的 rst 必須連接在一起，而且所有的 clk 也要連接在一起。

```vhdl
------ reg.vhd (component): -----------
LIBRARY ieee;
USE ieee.std_logic_1164.all;
----------------------------------------
ENTITY reg IS
    PORT ( d, clk, rst: IN STD_LOGIC;
           q: OUT STD_LOGIC);
END reg;
----------------------------------------
ARCHITECTURE reg OF reg IS
BEGIN
    PROCESS (clk, rst)
    BEGIN
        IF (rst='1') THEN q<='0';
        ELSIF (clk'EVENT AND clk='1') THEN q<=d;
        END IF;
    END PROCESS;
END reg;
----------------------------------------
```

圖 7-3　DFF 零件 reg.vhd 的代碼

```vhdl
1 ------ fau.vhd (component): ------------
2 LIBRARY ieee;
3 USE ieee.std_logic_1164.all;
4 ----------------------------------------
5 ENTITY fau IS
6     PORT ( a, b, cin: IN STD_LOGIC;
7            s, cout: OUT STD_LOGIC);
8 END fau;
9 ----------------------------------------
10 ARCHITECTURE fau OF fau IS
11 BEGIN
12     s <= a XOR b XOR cin;
13     cout <= (a AND b) OR (a AND cin) OR (b AND cin);
14 END fau;
15 ----------------------------------------
```

圖 7-4 1 bit full adder 零件 fau.vhd 的代碼

```vhdl
1 ------ pipe.vhd (component): -----------
2 LIBRARY ieee;
3 USE ieee.std_logic_1164.all;
4 USE work.my_components.all;
5 ----------------------------------------
6 ENTITY pipe IS
7     PORT ( a, b, clk, rst: IN STD_LOGIC;
8            q: OUT STD_LOGIC);
9 END pipe;
10 ----------------------------------------
11 ARCHITECTURE structural OF pipe IS
12     SIGNAL s, cin, cout: STD_LOGIC;
13 BEGIN
14     U1: COMPONENT fau PORT MAP (a, b, cin, s, cout);
15     U2: COMPONENT reg PORT MAP (cout, clk, rst, cin);
16     U3: COMPONENT reg PORT MAP (s, clk, rst, q);
17 END structural;
18 ----------------------------------------
```

圖 7-5 由 fau 和 reg 構成的 pipe.vhd 代碼

```
1 ----- my_components.vhd (package):-----
2 LIBRARY ieee;
3 USE ieee.std_logic_1164.all;
4 ------------------------
5 PACKAGE my_components IS
6 ------------------------
7 COMPONENT and_2 IS
8     PORT (a, b: IN STD_LOGIC; y: OUT STD_LOGIC);
9 END COMPONENT;
10 ------------------------
11 COMPONENT fau IS
12     PORT (a, b, cin: IN STD_LOGIC; s, cout: OUT STD_LOGIC);
13 END COMPONENT;
14 ------------------------
15 COMPONENT reg IS
16     PORT (d, clk, rst: IN STD_LOGIC; q: OUT STD_LOGIC);
17 END COMPONENT;
18 ------------------------
19 COMPONENT pipe IS
20     PORT (a, b, clk, rst: IN STD_LOGIC; q: OUT STD_LOGIC);
21 END COMPONENT;
22 ------------------------
23 END my_components;
24 ------------------------
```

圖 7-6　設計項目所用零件 my_components.vhd 的代碼

```
1 ----- multiplier.vhd (project): -------
2 LIBRARY ieee;
3 USE ieee.std_logic_1164.all;
4 USE work.my_components.all;
5 ------------------------
6 ENTITY multiplier IS
7     PORT ( a, clk, rst: IN STD_LOGIC;
8            b: IN STD_LOGIC_VECTOR (3 DOWNTO 0);
9            prod: OUT STD_LOGIC);
10 END multiplier;
11 ------------------------
12 ARCHITECTURE structural OF multiplier IS
13     SIGNAL and_out, reg_out: STD_LOGIC_VECTOR (3 DOWNTO 0);
14 BEGIN
15 U1: COMPONENT and_2 PORT MAP (a, b(3), and_out(3));
16 U2: COMPONENT and_2 PORT MAP (a, b(2), and_out(2));
17 U3: COMPONENT and_2 PORT MAP (a, b(1), and_out(1));
18 U4: COMPONENT and_2 PORT MAP (a, b(0), and_out(0));
19 U5: COMPONENT reg PORT MAP (and_out(3), clk, rst, reg_out(3));
20 U6: COMPONENT pipe PORT MAP (and_out(2), reg_out(3), clk, rst, reg_out(2));
21 U7: COMPONENT pipe PORT MAP (and_out(1), reg_out(2), clk, rst, reg_out(1));
22 U8: COMPONENT pipe PORT MAP (and_out(0), reg_out(1), clk, rst, reg_out(0));
23     prod <= reg_out(0);
24 END structural;
25 ------------------------
```

圖 7-7　設計項目 multiplier.vhd 的代碼

7-2 串行平行乘法器的模擬測試

模擬測試檔 test_bench.vhd 如圖 7-8 所示。測試結果的邏輯時序如圖 7-9 所示。

```
1 ----- test_bench_for_multiplier.vhd --------
2 LIBRARY ieee;
3 USE ieee.std_logic_1164.all;
4 USE work.all;
5 -----------------------------------------
6 Entity test_bench is
7 End test_bench;
8 -----------------------------------------
9 ARCHITECTURE stimulus OF test_bench IS
10 component multiplier
11     PORT ( a, clk, rst: IN STD_LOGIC;
12            b: IN STD_LOGIC_VECTOR (3 DOWNTO 0);
13            prod: OUT STD_LOGIC);
14 END component;
15 -----------------------------------------
16 Signal  rst: std_logic := '1';
17 Signal  clk: std_logic := '0';
18 Signal  a:   std_logic := '0';
19 Signal  b:   std_logic_vector(3 DOWNTO 0):= "1101";
20 Signal prod: std_logic := '0';
21 begin
22     DUT: multiplier port map( a, clk, rst, b, prod );
23     -- generate signal rst: --------
24     process
25     begin
26         wait for 100 ns; rst <= '0';
27     end process;
28     -- generate signal clk: --------
29     process
30     begin
31         wait for 50 ns; clk <= not clk;
32     end process;
33     -- generate signal a: --------
34     process
35     begin
36         wait for 300 ns; a <= '1';
37         wait for 200 ns; a <= '0';
38         wait;
39     end process;
40 end stimulus;
```

圖 7-8　模擬測試 test_bench.vhd 的代碼

邏輯時序圖的 rst 和 clk 作用明確。輸入訊號 a 是從 LSB 開始，為 "0011" 再加上 "0000"。輸入訊號 b( )，始終為 13。Prod 的輸出訊號也是從 LSB 開始，12 × 13 = 156，故其串行輸出為 "00111001"。

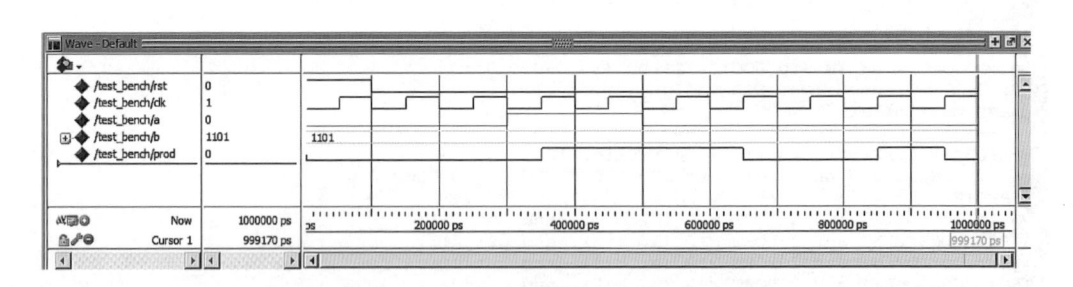

圖 7-9　模擬測試結果的 I/O 波形

7-3　串行平行乘法器電路合成的軟體部份

合成的軟體的第 1 部份從進入 new project wizard 開始,它要求填寫 directory,加入 project 所需的 files 和實驗板子 FPGA 的選用。

因此首先要新增一個名為 Multiplier Synthesis 的 folder,然後將通過 Simulation 的 and_2、fau、reg、pipe、my_components、和 multiplier 等 6 個 vhd files,copy 到 Multiplier Synthesis 的 folder 中。

在啟動 Quartus Prime Lite 16.1 後,選用 **File > New Project Wizard**,如圖 7S1-1 所示。

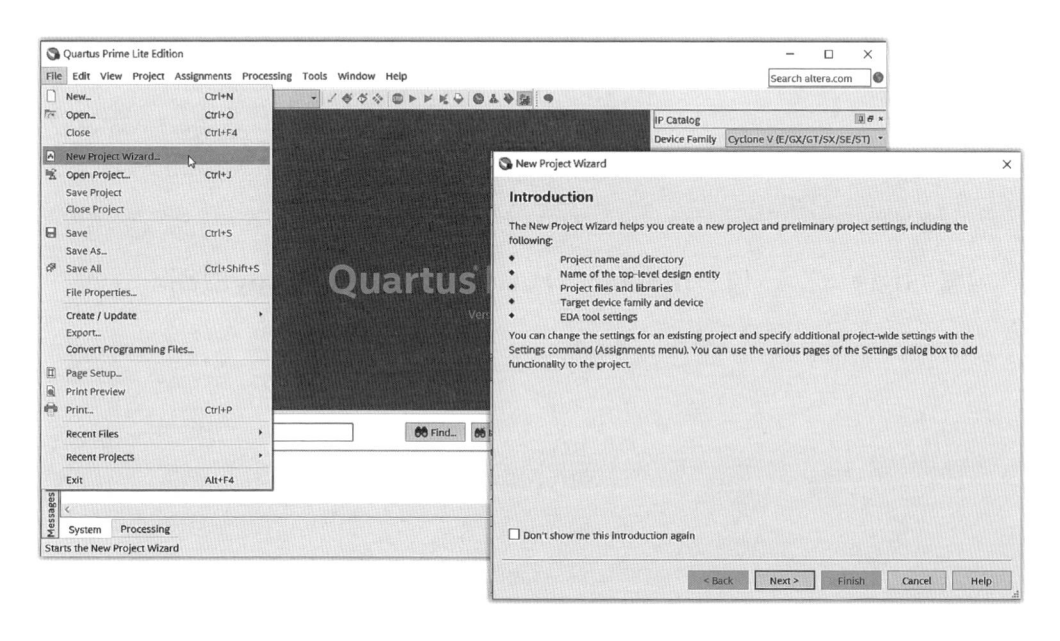

圖 7S1-1　啟動 Quartus 16.1 後,選用 File > New Project Wizard

接下來要讓 Project Wizard 知道 Directory 的位置及其名稱，Project 的名稱，還有 Top-Level Entity 的名稱。如圖 7S1-2 所示。

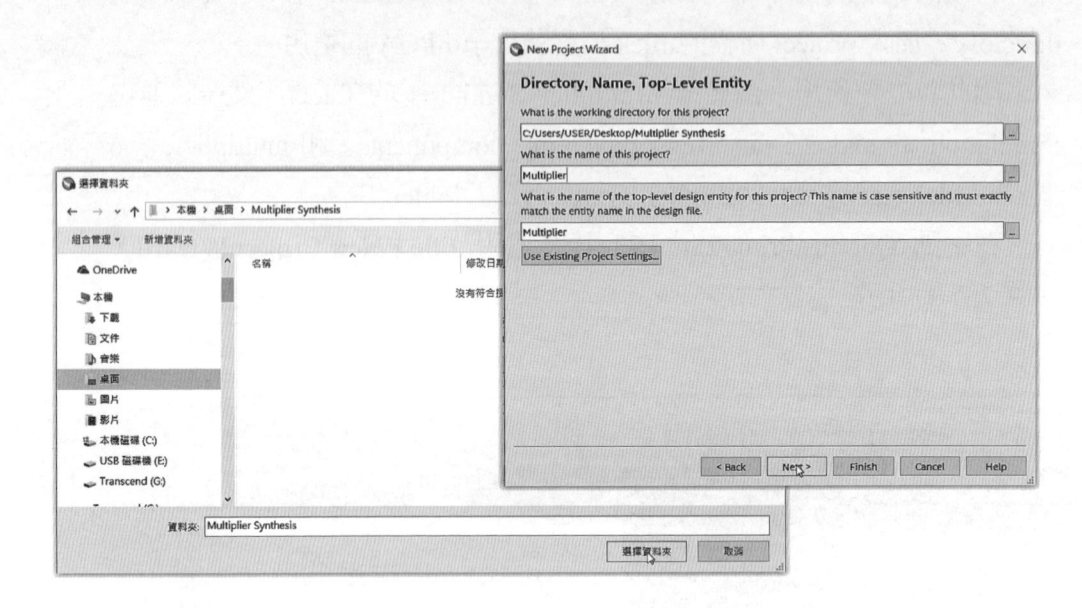

圖 7S1-2　填入 Directory，Name，和 Top-Level Entity 的名稱

第七章　串行平行乘法器　　161

再把 Multiplier Synthesis folder 中的 and_2、fau、reg、pipe、my_components、和 multiplier 等 6 個 vhd files 加入到 Project Wizard 的 Add Files 中。如圖 7S1-3 所示。

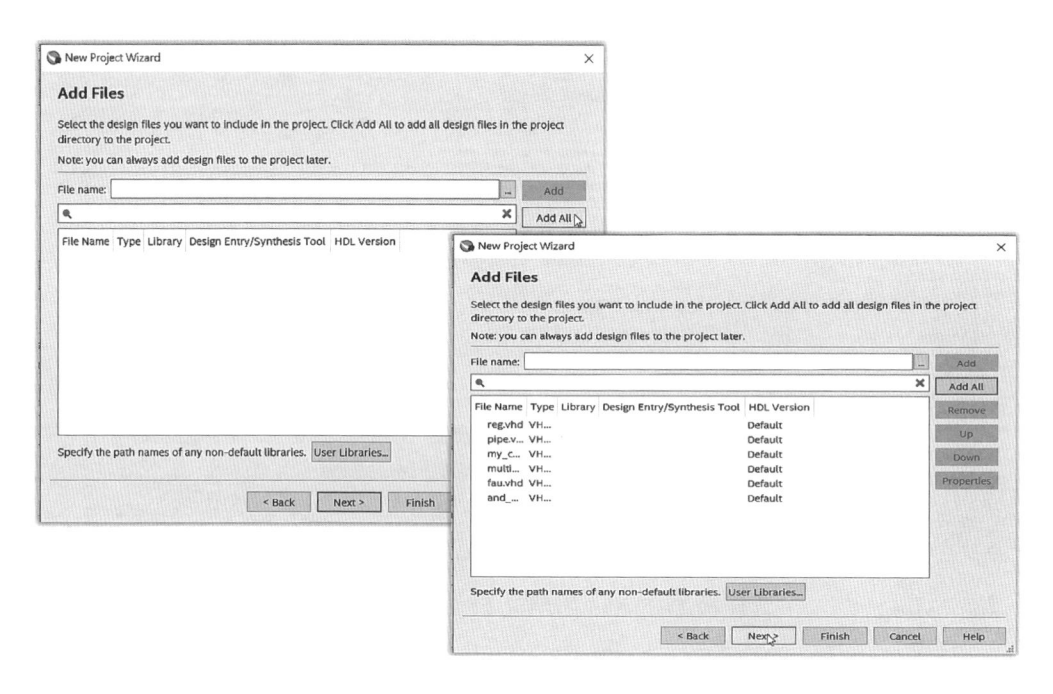

圖 7S1-3　my_components、和 multiplier 等 6 個 vhd files 加入到 Add Files 中

圖 7S1-4 為 FPGA-IC 的選用，實驗板 DE2-115 所用之 IC 為 EP4CE115F29C7。

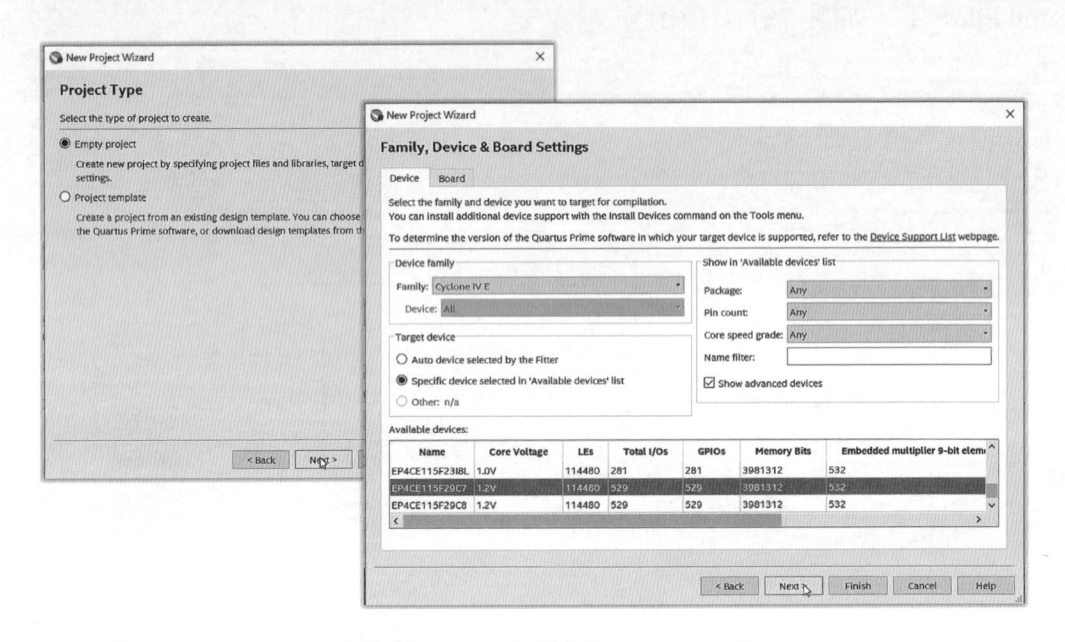

圖 7S1-4　實驗板 DE2-115 所用之 FPGA-IC 為 EP4CE115F29C7

歸納以上各種選項，總結如下圖 7S1-5 所示。完成合成軟體的第 1 部份。

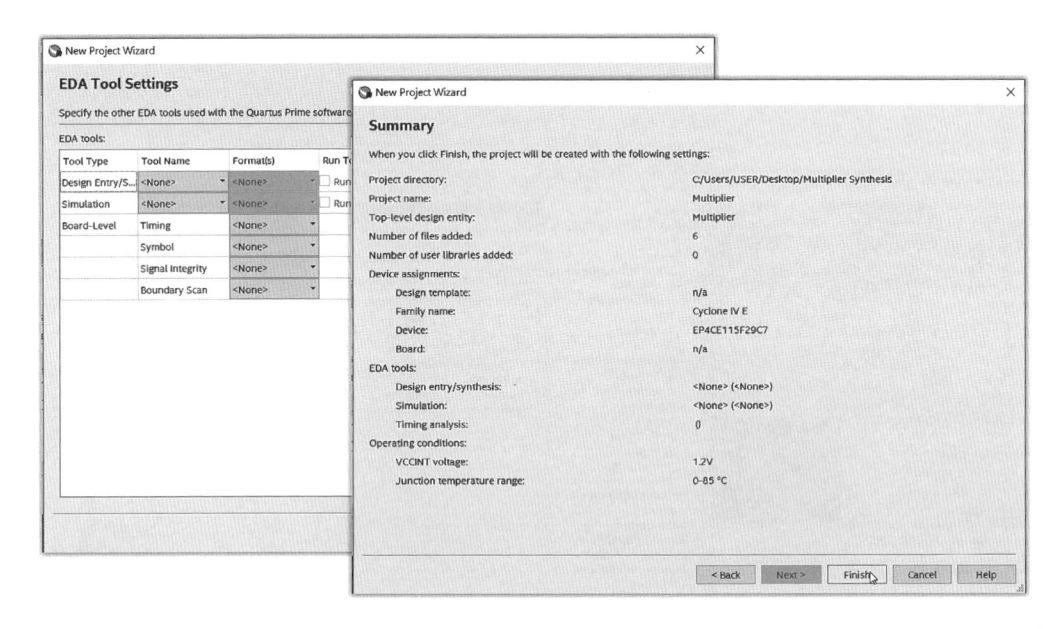

◦圖 7S1-5 ◦ 完成合成軟體的第 1 部份的總結

第 1 部份的總結是否有錯？要從 Project 視窗的 **Assignment > Settings** 入手。首先要檢查在 Settings-Multiplier > Files 中是否有 and_2、fau、reg、pipe、my_components、和 multiplier 等 6 個 vhd files？如圖 7S1-6 所示。

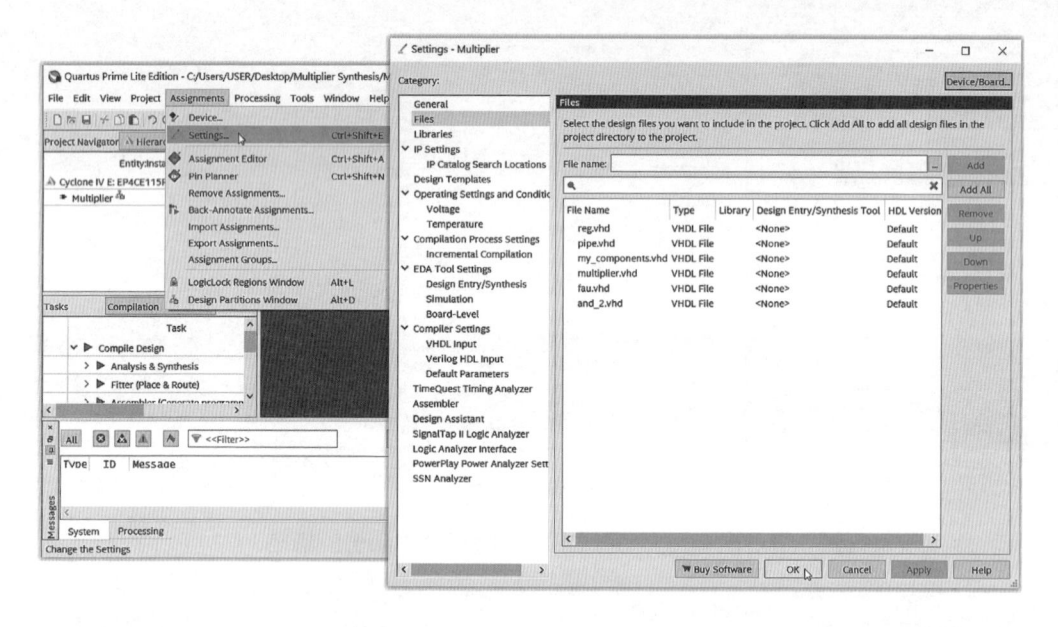

◦圖 7S1-6 ── 檢查 Assignments Settings 的 Multiplier Files 是否齊全

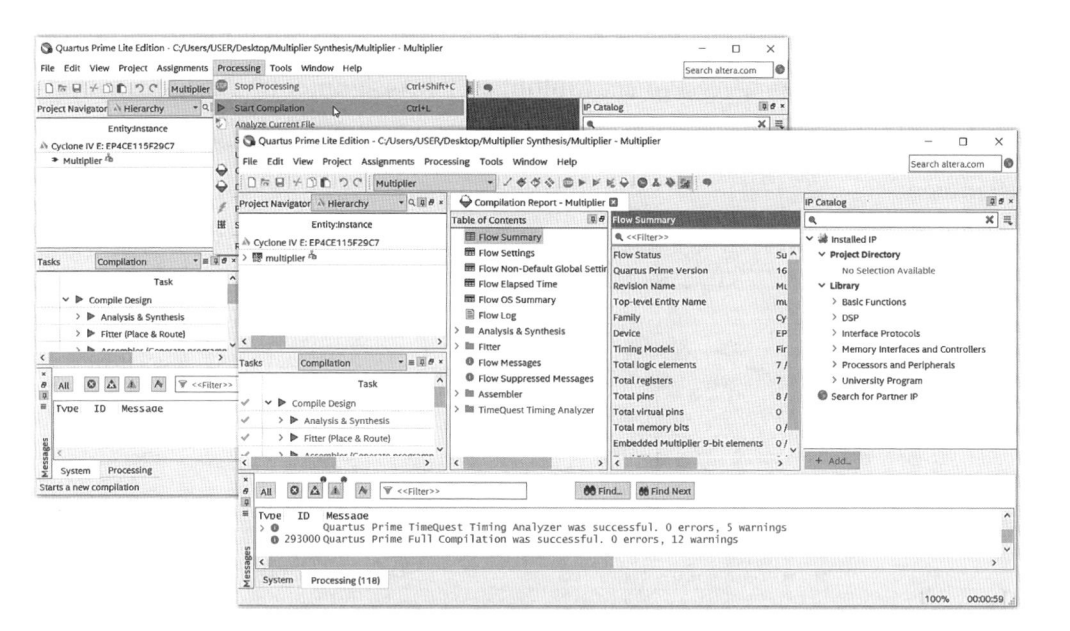

◦ 圖 7S1-7 ◦　Compiler 來檢查所有相關的 Multiplier Files 都沒有錯

通過第 1 階段 compilation 之後，接下來為第 2 階段的 FPGA 接腳 (PIN) 的設定。這部份要從 Project Multiplier 的 **Assignment > Assignment Editor** 入手。從圖 7S2-1 的 Assignment Editor 視窗，首先單擊 **New Assignment**。

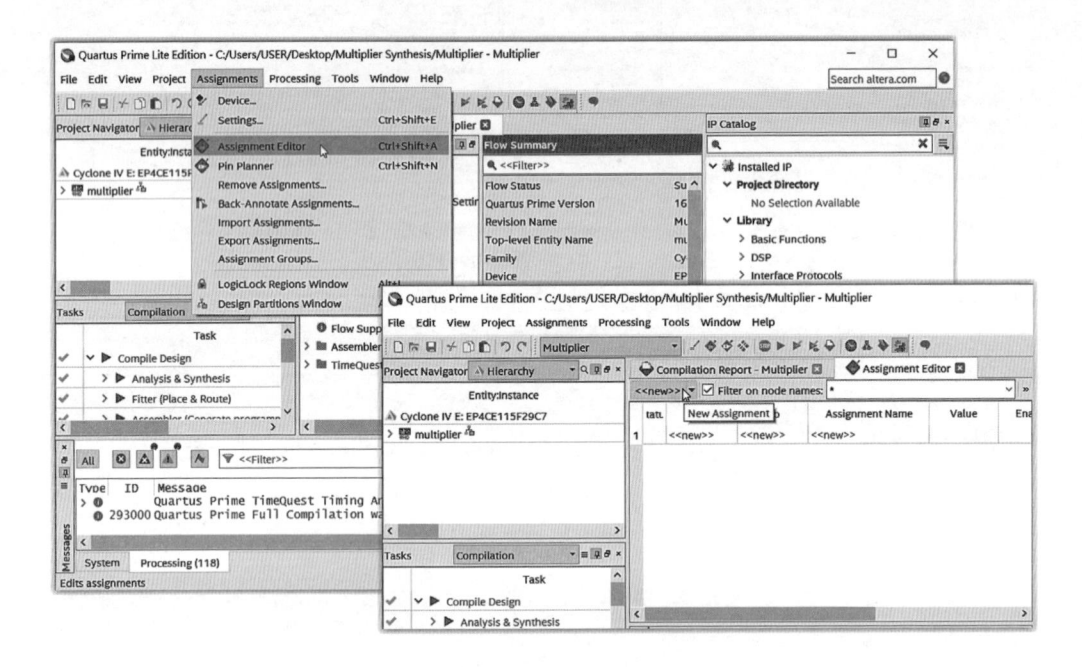

圖 7S2-1 FPGA 接腳的設定首先單擊 Assignment Editor 的 New Assignment

然後再單擊 **Assignment Editor** 的 Node Finder，如圖 7S2-2 所示。再單擊 **Node Finder** 的 List，以顯示 Project Multiplier 中的所有 Nodes。在 Matching Nodes 中選取跟 Ports 有關的所有 Signals 並移入到 Nodes Found 中，再單擊 **OK**。

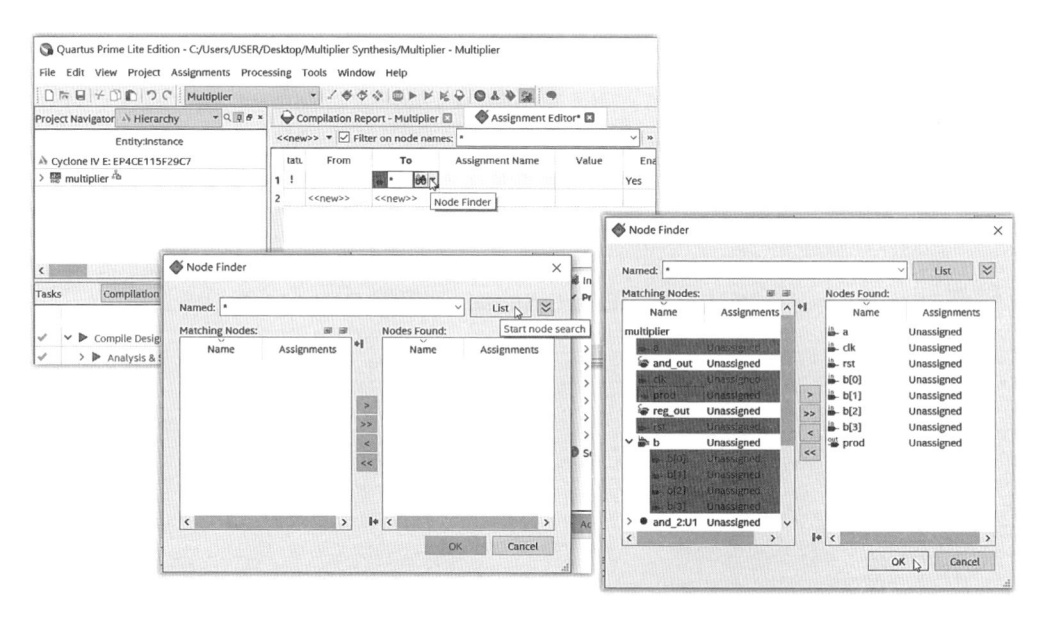

圖 7S2-2　從 Node Finder 中的 List 選取 Project 有關的所有 I/O Signals

為了易於連接 Analog Discovery 的 Digital Pattern Generator 和 Logic Analyzer，讓 Project 所有的 I/O Signals 連接到 DE2-115 的 JP5 上，如圖 7S2-3 所示。

DE-115 上 JP5 的使用，為的是易於與 Analog Discovery 的 Pattern Generator 和 Logic Analyzer 的測試接線相連接。

圖 7S2-3　Project 所有的 I/O Signals 連接到 DE2-115 的 JP5 上

Project 新加入的 pin assignments 也需要用 Compiler 來檢驗是否有錯，選用 **Project Multiplier** 的 **Processing > Start Compilation**，如圖 7S2-4 所示，0 errors。

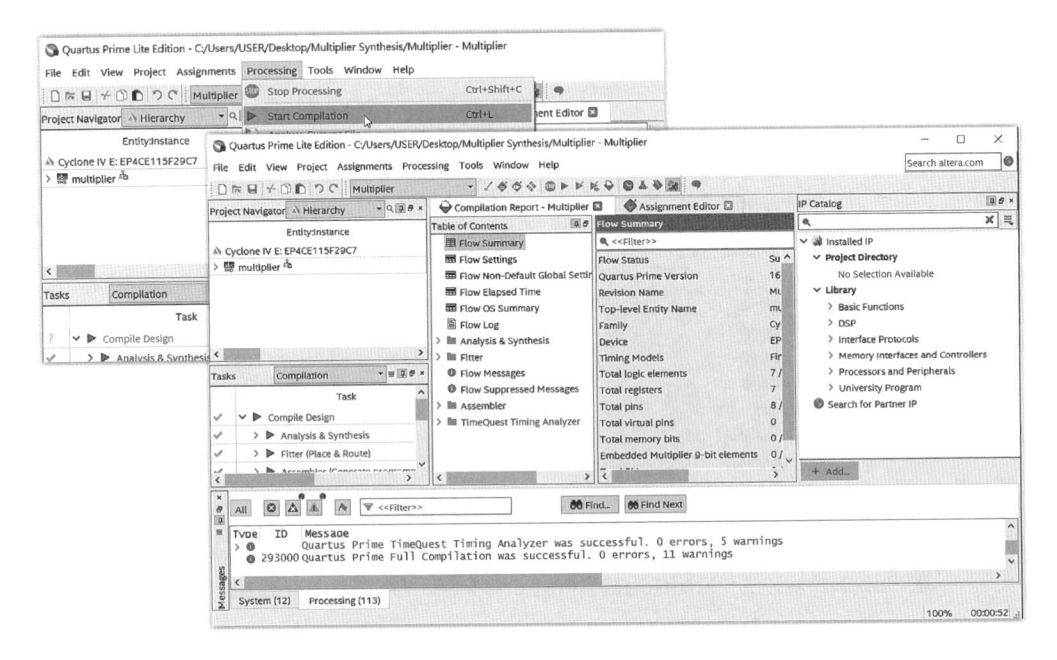

圖 7S2-4　新加入的 pin assignments 也需要用 Compiler 來檢驗是否有錯

0 Errors Compilation 代表第 2 階段的成功。這時候的 Multiplier Synthesis folder 和其中的 output_files 當如圖 7S2-5 所示。

名稱	修改日期	類型	大小
Multiplier.asm	2018/2/11 上午 07:48	RPT 檔案	5 KB
Multiplier.done	2018/2/11 上午 07:48	DONE 檔案	1 KB
Multiplier.fit	2018/2/11 上午 07:48	RPT 檔案	231 KB
Multiplier.fit	2018/2/11 上午 07:48	SMSG 檔案	1 KB
Multiplier.fit	2018/2/11 上午 07:48	SUMMARY 檔案	1 KB
Multiplier.flow	2018/2/11 上午 07:48	RPT 檔案	8 KB
Multiplier.jdi	2018/2/11 上午 07:48	JDI 檔案	1 KB
Multiplier.map	2018/2/11 上午 07:47	RPT 檔案	28 KB
Multiplier.map	2018/2/11 上午 07:47	SUMMARY 檔案	1 KB
Multiplier.pin	2018/2/11 上午 07:48	PIN 檔案	91 KB
Multiplier.sld	2018/2/11 上午 07:48	SLD 檔案	1 KB
Multiplier.sof	2018/2/11 上午 07:48	SOF 檔案	3,459 KB
Multiplier.sta	2018/2/11 上午 07:48	RPT 檔案	61 KB
Multiplier.sta	2018/2/11 上午 07:48	SUMMARY 檔案	1 KB

圖 7S2-5　Multiplier Synthesis folder 與其 output_files

第七章　串行平行乘法器　171

其中 Multiplier.sof 可以放到 **Tools > Programmer** 中，在 USB-Blaster [USB-1] 下 run 所設計的 Multiplier program。如圖 7S2-6 所示。

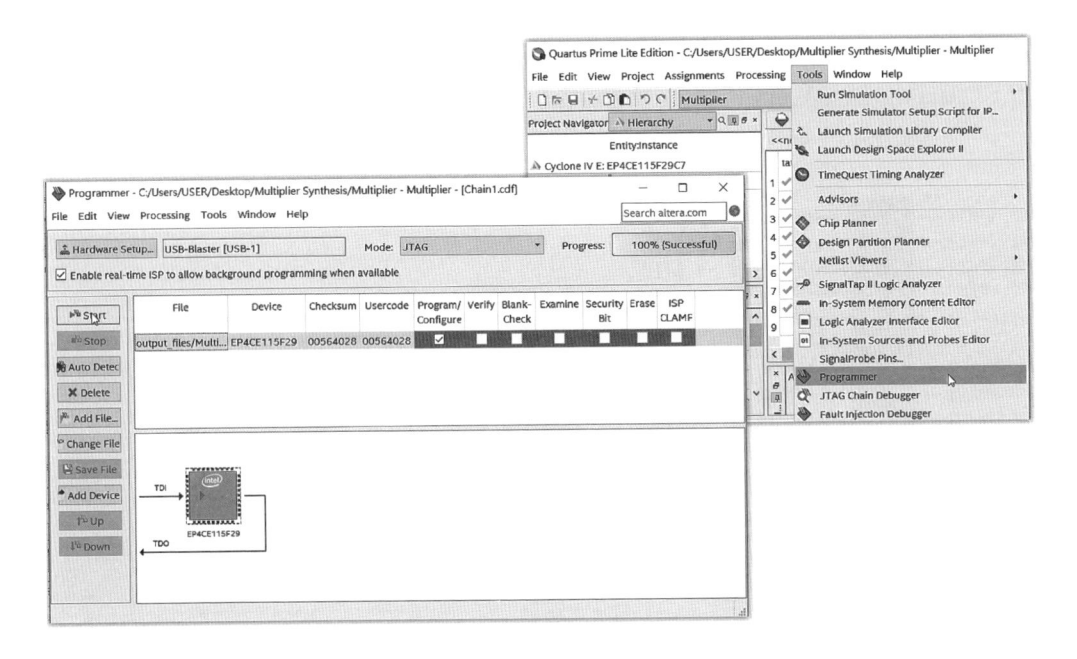

◎圖 7S2-6 ◎～～ Multiplier.sof 可以放到 Tools > Programmer 中去 run

7-4　串行平行乘法器合成電路的硬體測試部份

硬體測試部份使用 Analog Discovery 的 Digital Pattern Generator 和 Logic Analyzer 來測試 DE2-115 上 FPGA 所合成的串行平行乘法器。它的 JP5 與 Analog Discovery 間的完整訊號連線如圖 7T-1 所示。

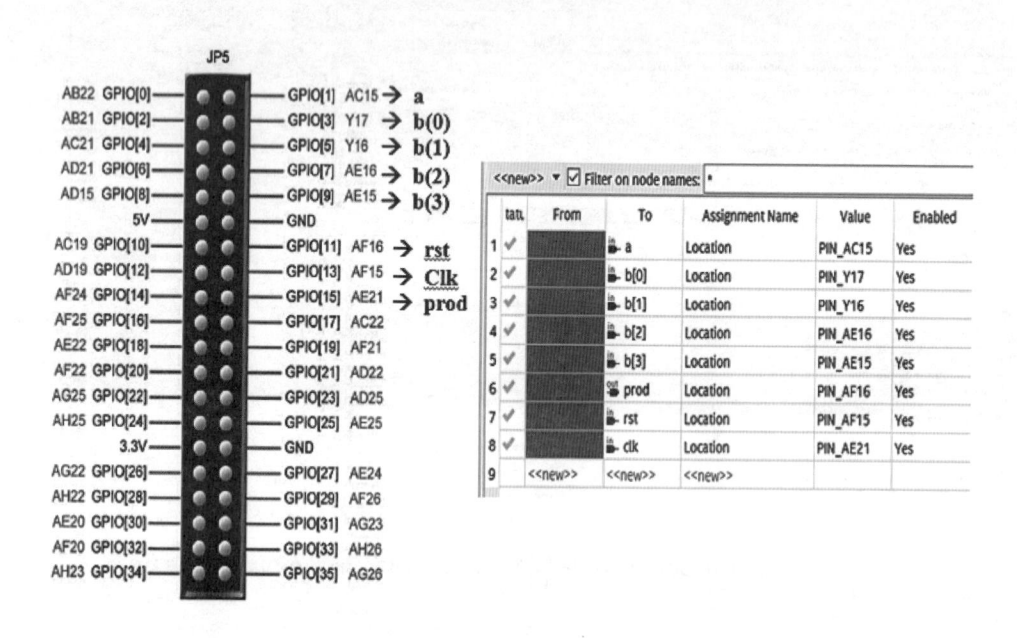

圖 7T-1　JP5 與 Analog Discovery 間的測試連線

請依照圖 7-9 模擬測試結果在 Digital Pattern Generator 設定的輸入訊號的波形。再 copy 輸入訊號的波形到 Logic Analyzer 中，再加上輸出訊號 pord。先 run Digital Pattern Generator，再 Single (run) Logic Analyzer。如圖 7T-2 所示。其輸出訊號 prod 與圖 7-9 模擬測試結果，完全相同。

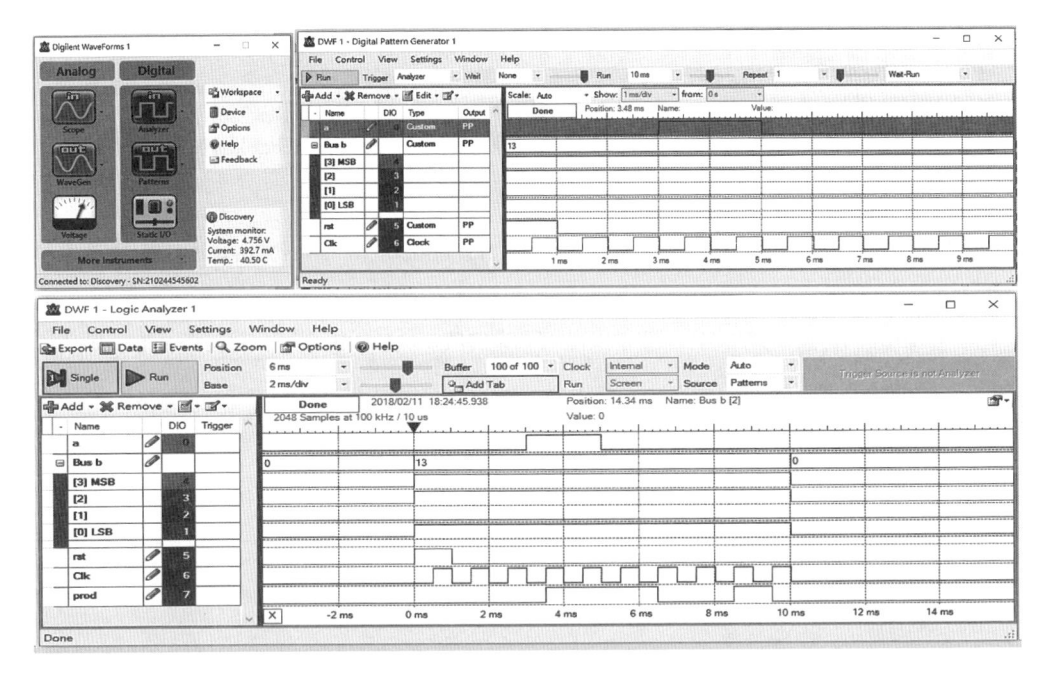

第七章　串行平行乘法器　　173

圖 7T-2　設定與 run Digital Pattern Generator 和 Logic Analyzer

7-5　課外練習

(1) 建構 Digital System 時，在何種情況下需要用到 Package，Package 包含的內容是什麼？

(2) 試擴大串行平行乘法器為 8 bit，令 a( )="11000000", b( )= "11000001" 寫出它的 Device code 及 test bench code。並用 Modelsim simulator 來測試，並用 Quartus II 及 DE2-115 的 FPGA 來做合成。AD1 Pattern Generator/Logic Analyzer 做硬體測試。

第八章　並行乘法器

◎ 8-1　並行乘法器的設計

　　圖 8-1 為一個 4-bits 並行乘法器電路，並行乘法器所有的輸入位都同時加入到系統。 因此，不需要儲存器 DFF。 注意圖 8-1 中只需用 AND gate 和 FAU (1bit full adder)。 操作數是 a( ) 和 b( ) 都是四位數，prod( ) 則為八位數。

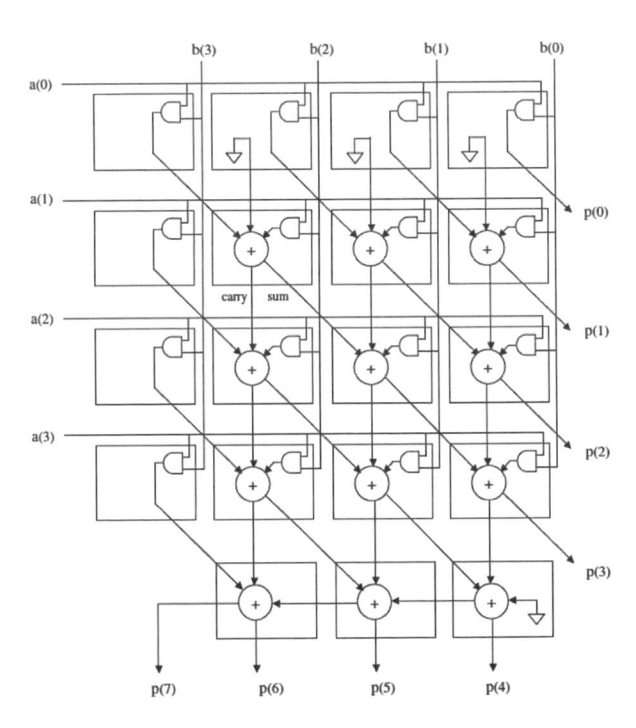

　　　　　　　　◎ 圖 8-1　　　4-bits 並行乘法器電路

為了組成電路的代碼的簡潔，可以將 a(0) 這一行寫成名為 top_row.vhd，其代碼如圖 8-2 所示。top_row.vhd 祇用到了 4 只 2 input AND gate，它的代碼如圖 8-2 所示。

```vhdl
1  ------ and_2.vhd (component): --
2  LIBRARY ieee;
3  USE ieee.std_logic_1164.all;
4  ---------------------------------
5  ENTITY and_2 IS
6      PORT ( a, b: IN STD_LOGIC;
7             y: OUT STD_LOGIC);
8  END and_2;
9  ---------------------------------
10 ARCHITECTURE and_2 OF and_2 IS
11 BEGIN
12     y <= a AND b;
13 END and_2;
14 ---------------------------------
```

圖 8-2　2 input AND gate 的代碼

Top_row.vhd 的代碼如圖 8-3 所示。

```vhdl
1  ------ top_row.vhd (component): --------------
2  LIBRARY ieee;
3  USE ieee.std_logic_1164.all;
4  USE work.my_components.all;              --
5  ---------------------------------------------
6  ENTITY top_row IS
7      PORT ( a: IN STD_LOGIC;
8             b: IN STD_LOGIC_VECTOR (3 DOWNTO 0);
9             sout, cout: OUT STD_LOGIC_VECTOR (2 DOWNTO 0);
10            p: OUT STD_LOGIC);
11 END top_row;
12 ---------------------------------------------
13 ARCHITECTURE structural OF top_row IS    -- only 1 row at the top
14 BEGIN
15     U1: COMPONENT and_2 PORT MAP (a, b(3), sout(2));
16     U2: COMPONENT and_2 PORT MAP (a, b(2), sout(1));
17     U3: COMPONENT and_2 PORT MAP (a, b(1), sout(0));
18     U4: COMPONENT and_2 PORT MAP (a, b(0), p);
19     cout(2)<='0'; cout(1)<='0'; cout(0)<='0';
20 END structural;
21 ---------------------------------------------
```

圖 8-3　Top_row.vhd 的 vhdl 代碼

a(1)、a(2)、a(3) 屬於中間項，它們除了使用 2 input gates 和 1 bit full adders，它的代碼如圖 8-4 所示。

```vhdl
------ fau.vhd (component): ------------
LIBRARY ieee;
USE ieee.std_logic_1164.all;
----------------------------------------
ENTITY fau IS
    PORT ( a, b, cin: IN STD_LOGIC;
           s, cout: OUT STD_LOGIC);
END fau;
----------------------------------------
ARCHITECTURE fau OF fau IS
BEGIN
    s <= a XOR b XOR cin;
    cout <= (a AND b) OR (a AND cin) OR (b AND cin);
END fau;
----------------------------------------
```

圖 8-4　1 bit full adders 的 vhdl 代碼

中間項 mid_row.vhd 的代碼如圖 8-5 所示。

```vhdl
------ mid_row.vhd (component): ------------
LIBRARY ieee;
USE ieee.std_logic_1164.all;
USE work.my_components.all;          --
----------------------------------------
ENTITY mid_row IS
    PORT ( a: IN STD_LOGIC;
           b: IN STD_LOGIC_VECTOR (3 DOWNTO 0);
           sin, cin: IN STD_LOGIC_VECTOR (2 DOWNTO 0);
           sout, cout: OUT STD_LOGIC_VECTOR (2 DOWNTO 0);
           p: OUT STD_LOGIC);
END mid_row;
----------------------------------------
ARCHITECTURE structural OF mid_row IS    -- there are 3 row at the middle
    SIGNAL and_out: STD_LOGIC_VECTOR (2 DOWNTO 0);
BEGIN
    U1: COMPONENT and_2 PORT MAP (a, b(3), sout(2));
    U2: COMPONENT and_2 PORT MAP (a, b(2), and_out(2));
    U3: COMPONENT and_2 PORT MAP (a, b(1), and_out(1));
    U4: COMPONENT and_2 PORT MAP (a, b(0), and_out(0));
    U5: COMPONENT fau PORT MAP (sin(2), cin(2), and_out(2), sout(1), cout(2));
    U6: COMPONENT fau PORT MAP (sin(1), cin(1), and_out(1), sout(0), cout(1));
    U7: COMPONENT fau PORT MAP (sin(0), cin(0), and_out(0), p, cout(0));
END structural;              --    a,      b,      cin,    s,     cout,
----------------------------------------
```

圖 8-5　中間項 mid_row.vhd 的代碼

最下面的一層，稱之謂 lower_row.vhd，它祇用到 1 bit full adders，代碼如圖 8-6 所示。

```
1 -------- lower_row.vhd (component): ----------
2 LIBRARY ieee;
3 USE ieee.std_logic_1164.all;
4 USE work.my_components.all;            --
5 ------------------------------------------
6 ENTITY lower_row IS
7     PORT ( sin, cin: IN STD_LOGIC_VECTOR (2 DOWNTO 0);
8            p: OUT STD_LOGIC_VECTOR (3 DOWNTO 0));
9 END lower_row;
10 ------------------------------------------- only 1 row in the bottom
11 ARCHITECTURE structural OF lower_row IS
12     SIGNAL local: STD_LOGIC_VECTOR (2 DOWNTO 0);
13 BEGIN
14     local(0)<='0';              --  a,      b,       cin,     s,     cout,
15     U1: COMPONENT fau PORT MAP (sin(0), cin(0), local(0), p(0), local(1));
16     U2: COMPONENT fau PORT MAP (sin(1), cin(1), local(1), p(1), local(2));
17     U3: COMPONENT fau PORT MAP (sin(2), cin(2), local(2), p(2), p(3));
18 END structural;
19 ------------------------------------------
```

圖 8-6　最下面的一層 lower_row.vhd 的代碼

以上圖 8-2 到圖 8-6 都是並行乘法器系統的零件，用**包裹** (package) 將之集中起來稱為 my_components.vhd 如圖 8-7 所示。

```
---- my_components.vhd (package): -----------
LIBRARY ieee;
USE ieee.std_logic_1164.all;
--------------------------------
PACKAGE my_components IS
    --------------------
    COMPONENT and_2 IS
        PORT ( a, b: IN STD_LOGIC; y: OUT STD_LOGIC);
    END COMPONENT;
    --------------------
    COMPONENT fau IS -- full adder unit
        PORT ( a, b, cin: IN STD_LOGIC; s, cout: OUT STD_LOGIC);
    END COMPONENT;
    --------------------
    COMPONENT top_row IS
        PORT ( a: IN STD_LOGIC;
               b: IN STD_LOGIC_VECTOR (3 DOWNTO 0);
               sout, cout: OUT STD_LOGIC_VECTOR (2 DOWNTO 0);
               p: OUT STD_LOGIC);
    END COMPONENT;
    --------------------
    COMPONENT mid_row IS
        PORT ( a: IN STD_LOGIC;
               b: IN STD_LOGIC_VECTOR (3 DOWNTO 0);
               sin, cin: IN STD_LOGIC_VECTOR (2 DOWNTO 0);
               sout, cout: OUT STD_LOGIC_VECTOR (2 DOWNTO 0);
               p: OUT STD_LOGIC);
    END COMPONENT;
--------------------
    COMPONENT lower_row IS
        PORT ( sin, cin: IN STD_LOGIC_VECTOR (2 DOWNTO 0);
               p: OUT STD_LOGIC_VECTOR (3 DOWNTO 0));
    END COMPONENT;
--------------------
END my_components;
--------------------------------------------------
```

圖 8-7　設計並行乘法器系統所需之零件 my_components.vhd 代碼

由 top、mid、和 lower 等 5 個 rows 所構成的並行乘法器系統 multiplier.vhd 如圖 8-8 所示。

```
1  ------ multiplier.vhd (project): -------------
2  LIBRARY ieee;
3  USE ieee.std_logic_1164.all;
4  USE work.my_components.all;
5  ----------------------------------------
6  ENTITY multiplier IS
7      PORT ( a, b: IN STD_LOGIC_VECTOR (3 DOWNTO 0);
8             prod: OUT STD_LOGIC_VECTOR (7 DOWNTO 0));
9  END multiplier;
10 ----------------------------------------
11 ARCHITECTURE structural OF multiplier IS
12     TYPE matrix IS ARRAY (0 TO 3) OF
13          STD_LOGIC_VECTOR (2 DOWNTO 0);
14     SIGNAL s, c: matrix;
15 BEGIN
16     U1: COMPONENT top_row   PORT MAP (a(0), b, s(0), c(0), prod(0));
17     U2: COMPONENT mid_row   PORT MAP (a(1), b, s(0), c(0), s(1), c(1), prod(1));
18     U3: COMPONENT mid_row   PORT MAP (a(2), b, s(1), c(1), s(2), c(2), prod(2));
19     U4: COMPONENT mid_row   PORT MAP (a(3), b, s(2), c(2), s(3), c(3), prod(3));
20     U5: COMPONENT lower_row PORT MAP (s(3), c(3), prod(7 DOWNTO 4));
21 END structural;
22 ----------------------------------------
```

圖 8-8　　並行乘法器系統 multiplier.vhd 的代碼

8-2　並行乘法器系統的模擬測試

模擬測試檔 test_bench.vhd 如圖 8-9 所示。測試結果的邏輯時序如圖 8-9 所示。

```
---- test_bench_for_multiplier.vhd --------
LIBRARY ieee;
USE ieee.std_logic_1164.all;
USE work.my_components.all;

Entity test_bench is
End test_bench;

ARCHITECTURE stimulus OF test_bench IS
    component multiplier
        PORT ( a, b: IN STD_LOGIC_VECTOR (3 DOWNTO 0);
                 prod: OUT STD_LOGIC_VECTOR (7 DOWNTO 0));
    END component;

    Signal  a: STD_LOGIC_VECTOR(3 DOWNTO 0):= "0000";
    Signal  b: STD_LOGIC_VECTOR(3 DOWNTO 0):= "0000";
    Signal  prod: STD_LOGIC_VECTOR(7 DOWNTO 0);

begin
    DUT: multiplier port map( a, b, prod );

    -- generate signal a: --------
    process
    begin
        wait for 400 ns; a <= "0011"; -- 3
        wait for 400 ns; a <= "0110"; -- 6
        wait for 400 ns; a <= "1001"; -- 9
        wait for 400 ns; a <= "1100"; -- 12
        wait for 400 ns; a <= "1111"; -- 15
        wait for 400 ns; a <= "0000"; -- 0
        wait for 400 ns; a <= "1111"; -- 15
        wait;
    end process;
    -- generate signal b: --------
    process
    begin
        wait for 400 ns; b <= "0000"; -- 0
        wait for 400 ns; b <= "0101"; -- 5
        wait for 400 ns; b <= "0101"; -- 5
        wait for 400 ns; b <= "1010"; -- 10
        wait for 400 ns; b <= "1010"; -- 10
        wait for 400 ns; b <= "1111"; -- 15
        wait for 400 ns; b <= "1111"; -- 15
        wait;
    end process;
end stimulus;
```

◎ 圖 8-9　並行乘法器的模擬測試檔 test_bench.vhd

數並行乘法器的模擬測試的結果如圖 8-10 所示。輸入訊號 a、b 和輸出訊號都以 unsigned 數字讀出。

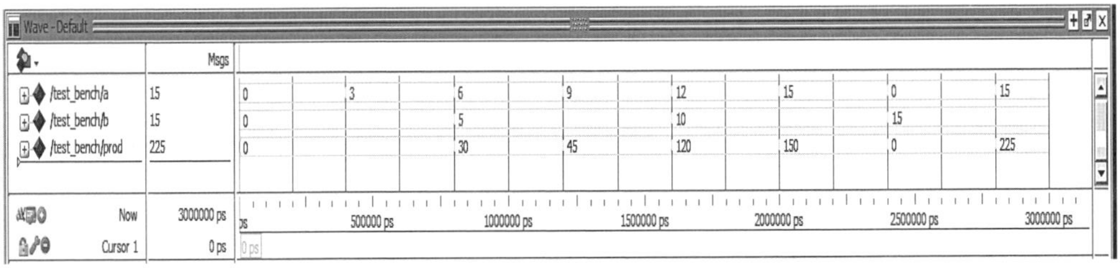

◎ 圖 8-10　模擬測試結果的 I/O 波形

8-3 平行乘法器電路合成的軟體部份

合成的軟體的第 1 部份從進入 new project wizard 開始,它要求填寫 directory,加入 project 所需的 files 和實驗板子 FPGA 的選用。

因此首先要新增一個名為 Multiplier2 Synthesis 的 folder,然後將通過 Simulation 的 and_2、fau、lower_row、mid_row、top_low、my_components、和 multiplier 等 7 個 vhd files,copy 到 Multiplier2 Synthesis 的 folder 中。

在啟動 Quartus Prime Lite 16.1 後,選用 **File > New Project Wizard**,如圖 8S1-1 所示。

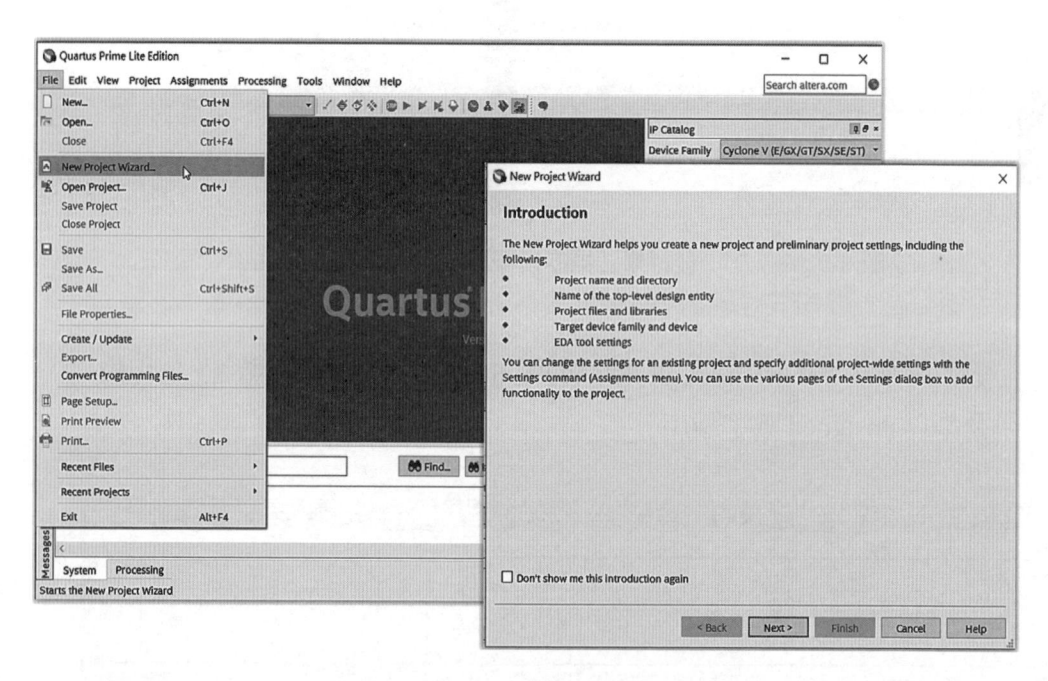

圖 8S1-1　啟動 Quartus 16.1 後,選用 File > New Project Wizard

第八章　並行乘法器　183

接下來要讓 Project Wizard 知道 Directory 的位置及其名稱，Project 的名稱，還有 Top-Level Entity 的名稱。如圖 8S1-2 所示。

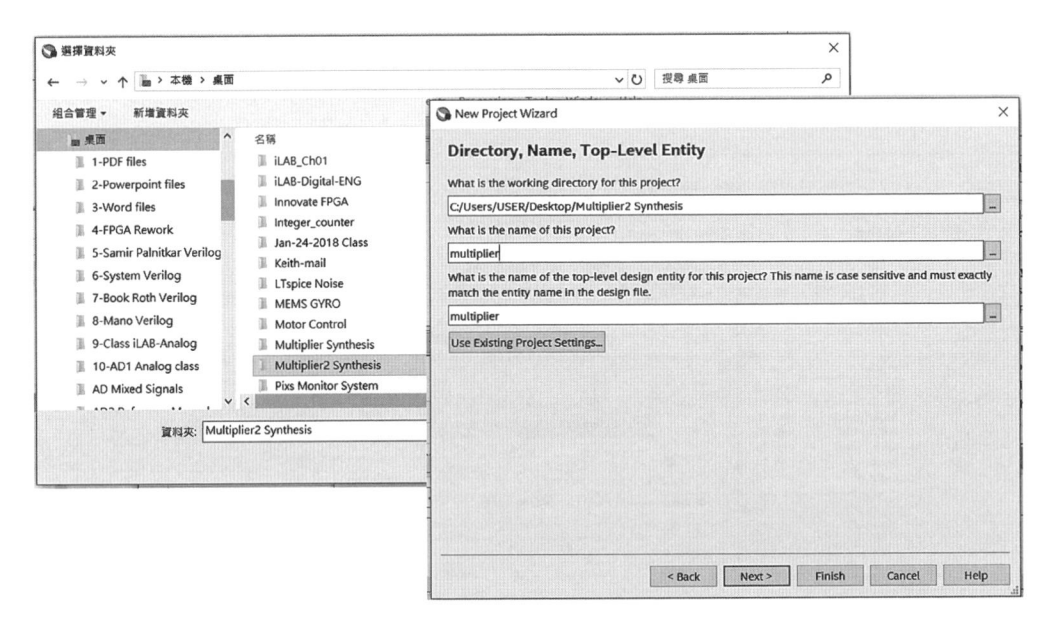

◦ 圖 8S1-2 ◦　　填入 Directory，Name，和 Top-Level Entity 的名稱

再把 Multiplier Synthesis folder 中的 and_2、fau、、lower_row、mid_row、top_low、my_components、和 multiplier 等 7 個 vhd files 加入到 Project Wizard 的 Add Files 中。FPGA-IC 的選用，實驗板 DE2-115 所用之 IC 為 EP4CE115F29C7。如圖 8S1-3 所示。

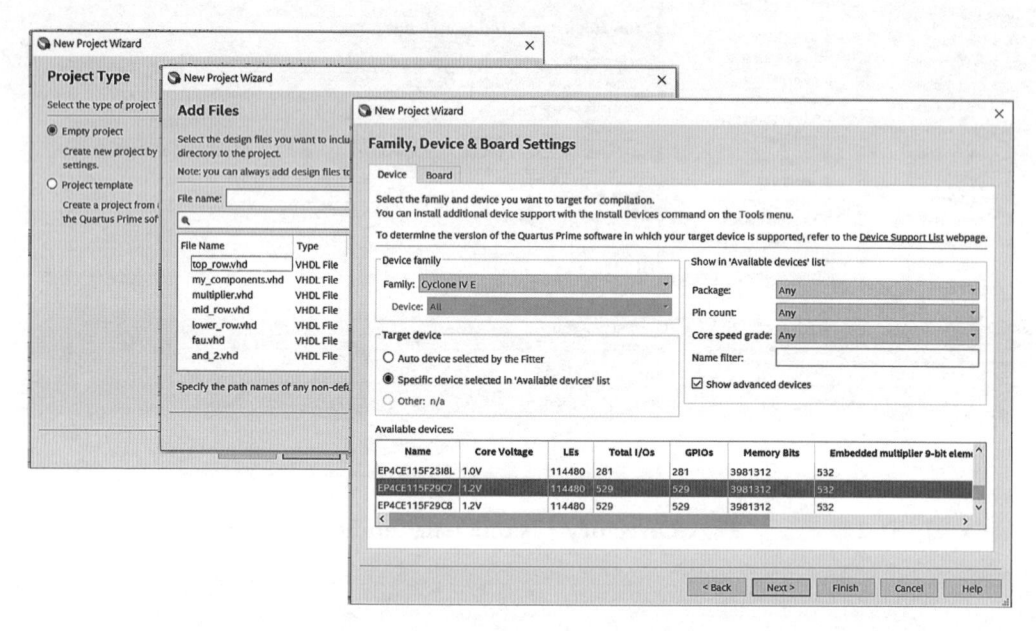

圖 8S1-3　my_components、和 multiplier 等 7 個 vhd files 加入到 Add Files 中

歸納以上各種選項，總結如下圖 8S1-4 所示。完成合成軟體的第 1 部份。

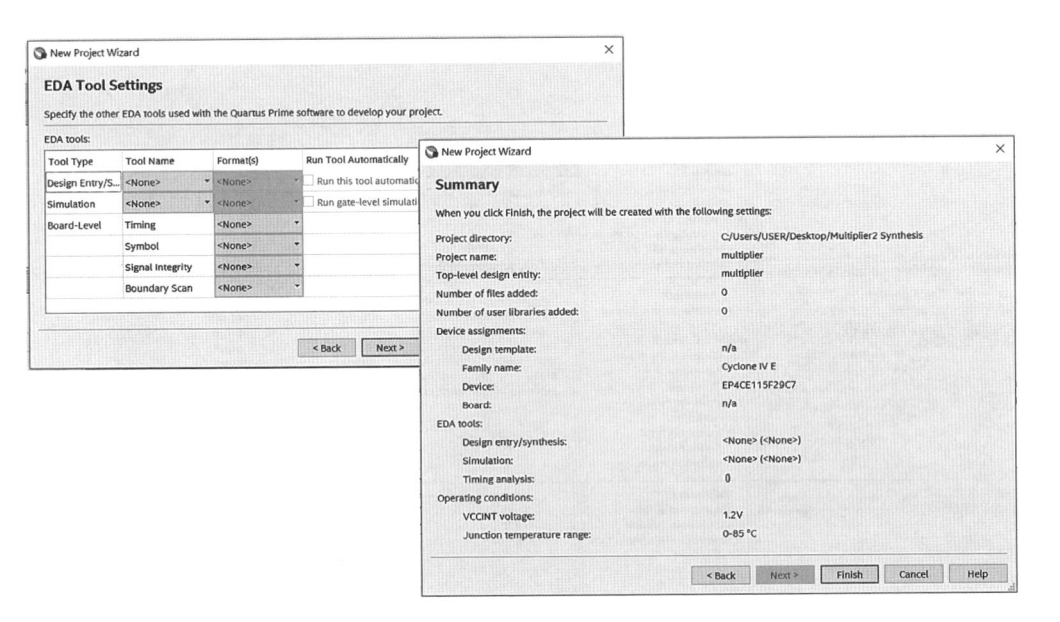

◎ 圖 8S1-4 ◎　　完成合成軟體的第 1 部份的總結

第 1 部份的總結是否有錯？要從 Project 視窗的 **Assigment > Settings** 入手。首先要檢查在 **Settings-Multiplier > File** 中是否有 and_2、fau、lower_row、mid_row、top_low、my_components、和 multiplier 等 7 個 vhd files？如圖 8S1-5 所示。

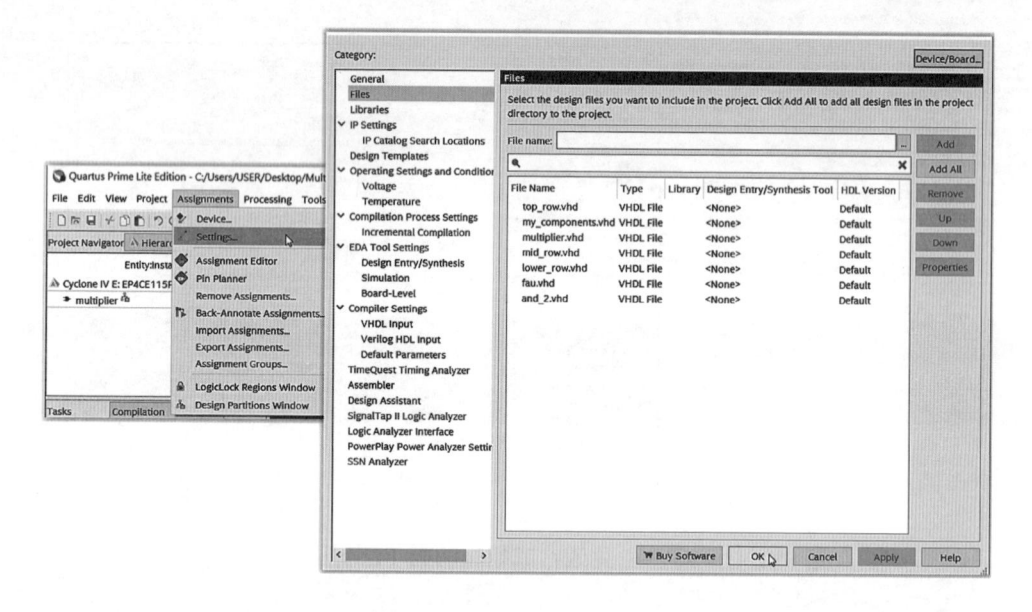

圖 8S1-5　檢查 Assignments Settings 的 Multiplier Files 是否齊全

再用 Compiler 來檢查所有相關的 Multiplier Files 都沒有錯。如圖 8S1-6 所示。

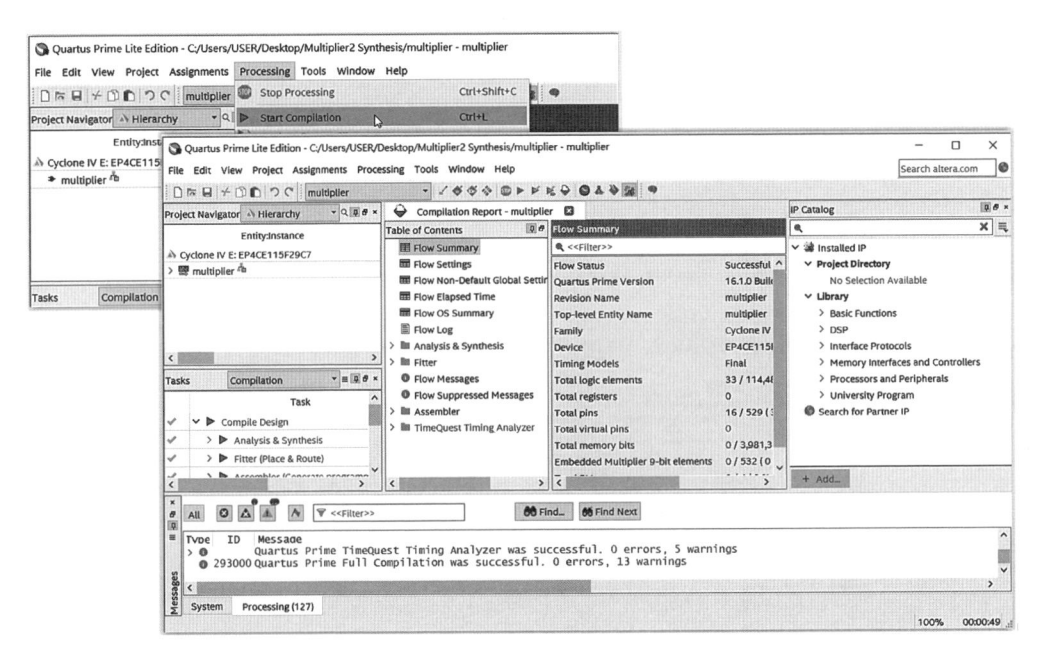

圖 8S1-6　Compiler 來檢查所有相關的 Multiplier Files 都沒有錯

通過第 1 階段 compilation 之後，接下來為第 2 階段的 FPGA 接腳 (PIN) 的設定。這部份要從 Project Multiplier 的 **Assignment > Assignment Editor** 入手。從圖 8S2-1 的 Assignment Editor 視窗，首先單擊 **New Assignment**。

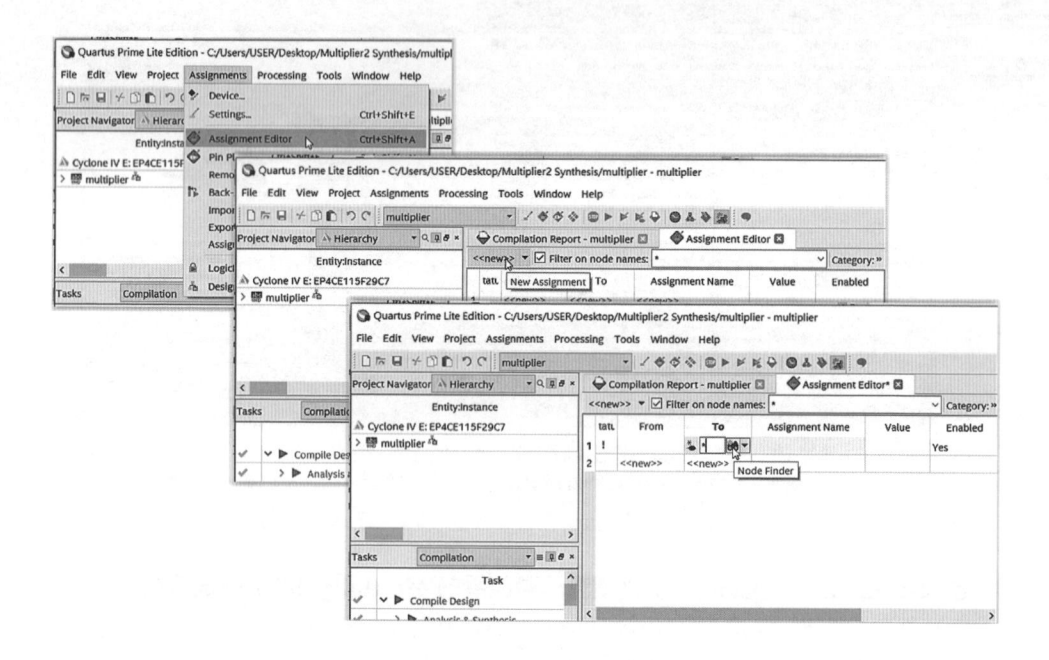

圖 8S2-1 FPGA 接腳的設定首先單擊 Assignment Editor 的 New Assignment

第八章 並行乘法器 189

然後再雙擊 **Assignment Editor** 的 Node Finder，如圖 8S2-2 所示。再單擊 **Node Finder** 的 List，以顯示 Project Multiplier 中的所有 Nodes。在 Matching Nodes 中選取跟 Ports 有關的所有 Signals 並移入到 Nodes Found 中，再單擊 **OK**。

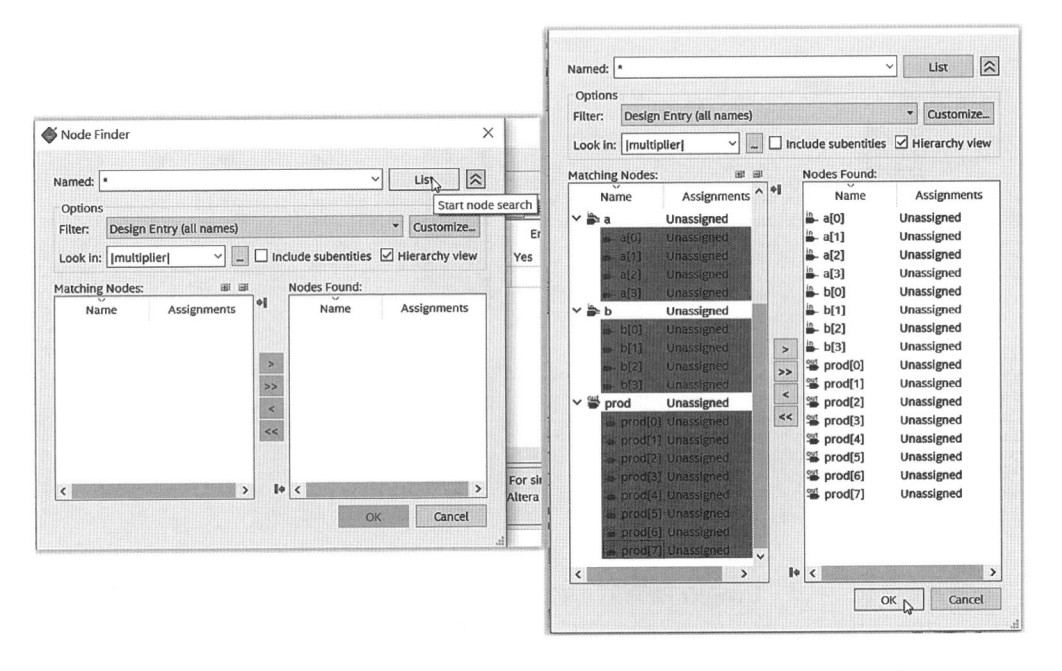

圖 8S2-2　從 Node Finder 中的 List 選取 Project 有關的所有 I/O Signals

為了易於連接 Analog Discovery 的 Digital Pattern Generator 和 Logic Analyzer，讓 Project 所有的 I/O Signals 連接到 DE2-115 的 JP5 上，如圖 8S2-3 所示。

DE-115 上 JP5 的使用，為的是易於與 Analog Discovery 的 Pattern Generator 和 Logic Analyzer 的測試接線相連接。

圖 8S2-3　Project 所有的 I/O Signals 連接到 DE2-115 的 JP5 上

第八章　並行乘法器

　　Project 新加入的 pin assignments 也需要用 Compiler 來檢驗是否有錯，選用 **Project Multiplier** 的 **Processing > Start Compilation**，如圖 8S2-4 所示，0 errors。

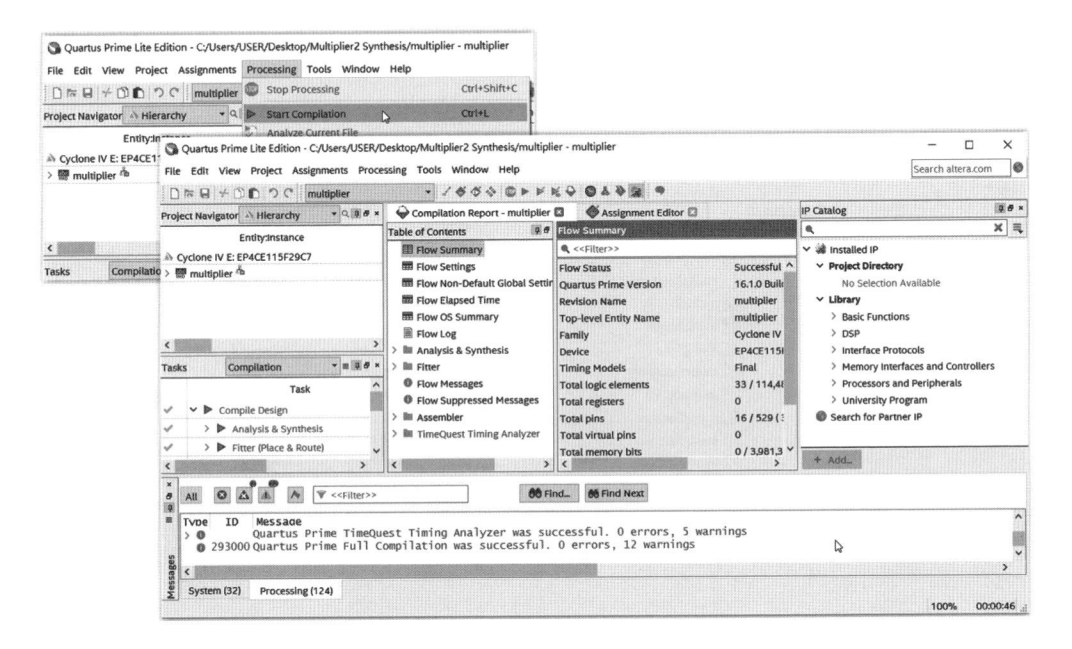

◎ 圖 8S2-4 ◎　　新加入的 pin assignments 也需要用 Compiler 來檢驗是否有錯

0 Errors Compilation 代表第 2 階段的成功。這時候的 Multiplier Synthesis folder 和其中的 output_files 當如圖 8S2-5 所示。

db	2018/2/12 下午 12:07　檔案資料夾
incremental_db	2018/2/12 上午 10:20　檔案資料夾
Multiplier2 pixs	2018/2/12 下午 12:12　檔案資料夾
output_files	2018/2/12 下午 12:07　檔案資料夾
and_2	2018/1/26 下午 05:33　硬碟映像檔　1 KB
fau	2018/1/26 下午 05:29　硬碟映像檔　1 KB
lower_row	2018/1/28 下午 09:20　硬碟映像檔　1 KB
mid_row	2018/2/4 上午 10:56　硬碟映像檔　2 KB
multiplier	2018/2/12 上午 09:20　QPF 檔案　2 KB
multiplier.qsf	2018/2/1...
multiplier	2018/1/2...
my_components	2018/1/2...
top_row	2018/1/2...

output_files

名稱	修改日期	類型	大小
multiplier.asm	2018/2/12 下午 12:07	RPT 檔案	5 KB
multiplier.done	2018/2/12 下午 12:07	DONE 檔案	1 KB
multiplier.fit	2018/2/12 下午 12:07	RPT 檔案	242 KB
multiplier.fit	2018/2/12 下午 12:07	SMSG 檔案	1 KB
multiplier.fit	2018/2/12 下午 12:07	SUMMARY 檔案	1 KB
multiplier.flow	2018/2/12 下午 12:07	RPT 檔案	8 KB
multiplier.jdi	2018/2/12 下午 12:07	JDI 檔案	1 KB
multiplier.map	2018/2/12 下午 12:07	RPT 檔案	32 KB
multiplier.map	2018/2/12 下午 12:07	SUMMARY 檔案	1 KB
multiplier.pin	2018/2/12 下午 12:07	PIN 檔案	91 KB
multiplier.sld	2018/2/12 下午 12:07	SLD 檔案	1 KB
multiplier.sof	2018/2/12 下午 12:07	SOF 檔案	3,459 KB
multiplier.sta	2018/2/12 下午 12:07	RPT 檔案	58 KB
multiplier.sta	2018/2/12 下午 12:07	SUMMARY 檔案	1 KB

圖 8S2-5　Multiplier Synthesis folder 與其 output_files

其中 Multiplier.sof 可以放到 **Tools > Programmer** 中，在 USB-Blaster [USB-1] 下 run 所設計的 Multiplier program。如圖 8S2-6 所示。

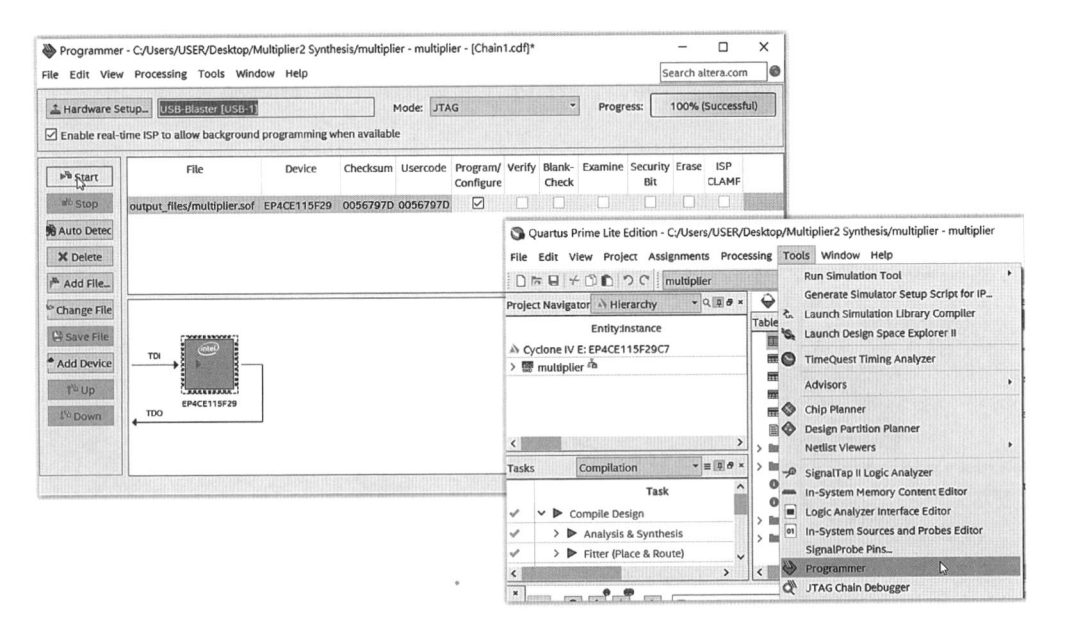

◎ 圖 8S2-6 ◎　Multiplier.sof 可以放到 Tools > Programmer 中去 run

8-4 平行乘法器合成電路的硬體測試部份

硬體測試部份使用 Analog Discovery 的 Digital Pattern Generator 和 Logic Analyzer 來測試 DE2-115 上 FPGA 所合成的串行平行乘法器。它的 JP5 與 Analog Discovery 間的完整訊號連線如圖 8T-1 所示。

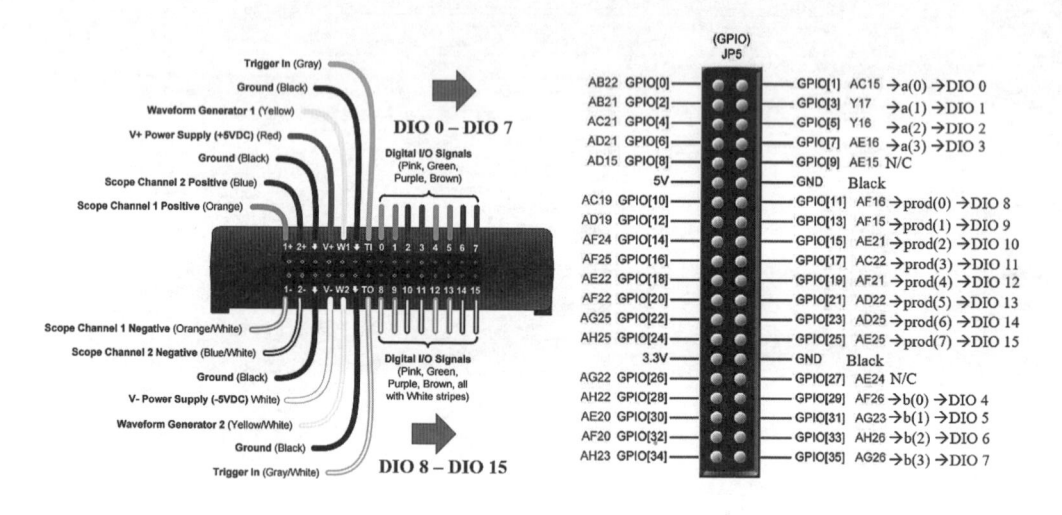

圖 8T-1　JP5 與 Analog Discovery 間的測試連線

請依照圖 8-9 模擬測試結果的 I/O 波形在 Digital Pattern Generator 設定的輸入訊號的波形。Logic Analyzer 中，再加上輸出訊號 pord。先 run Digital Pattern Generator，再 Single (run) Logic Analyzer。如圖 8T-2 所示。其輸出訊號 prod 與圖 8-9 模擬測試結果，完全相同。

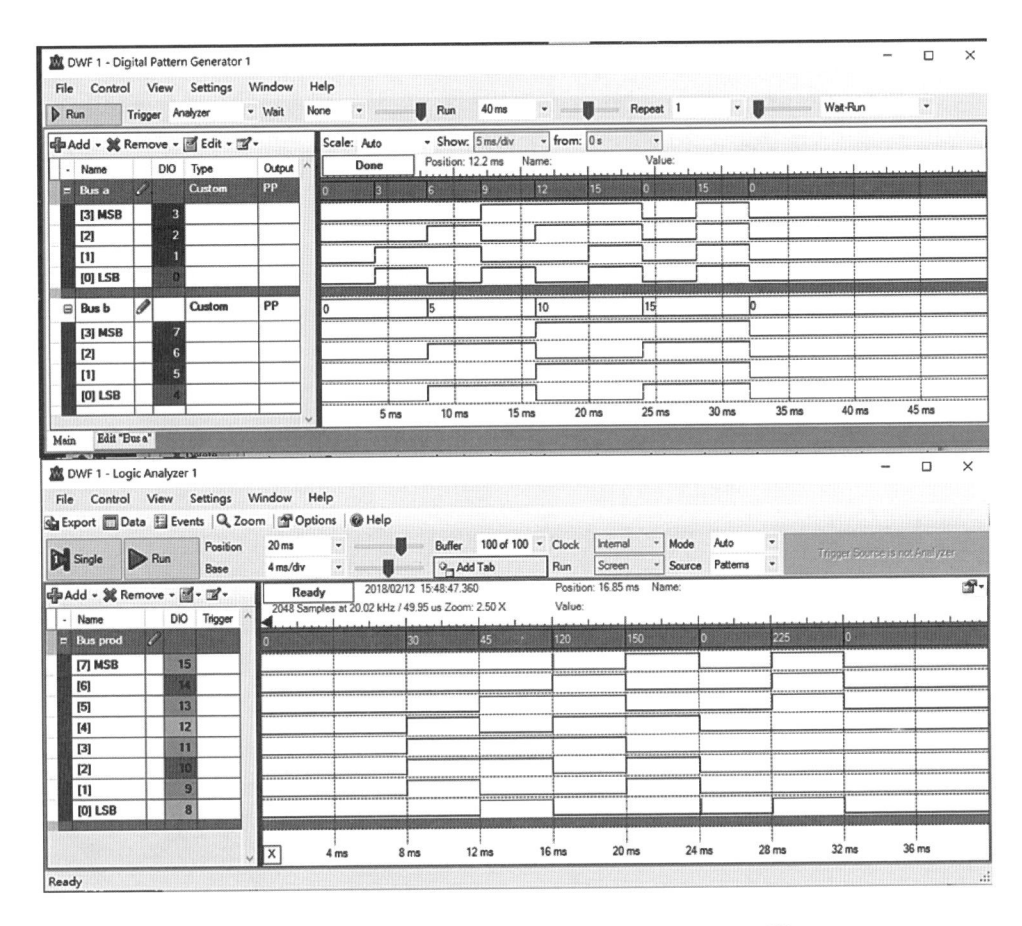

圖 8T-2　設定與 run Digital Pattern Generator 和 Logic Analyzer

8-5　課外練習

(1) 本章的例子其目的是：探索與系統相關的幾個方面來使用 VHDL 進行設計。但是，對於並行乘法器的特定情況，它可以通過 ieee.std_logic_arith 中定義的 "*"（乘法）運算符號，立即獲得乘法的結果。試用該符號改寫圖 8-8 的並行乘法器系統 multiplier.vhd 的代碼，併用 ModelSim 比較二者的結果。

(2) 試比較 "乘法器" 軟體和硬體所獲得的結果，對時間來講有何不同？在應用上，請說明軟體能或不能取代硬體的理由。

第九章 乘法-累加電路

第八章介紹了並行乘法器的一種。目的爲探索與系統有關使用 VHDL 設計方式。對於並行乘法器的特殊狀況，可以使用乘法運算符號 "＊" 簡單並快速地獲得結果。這個簡單的並行乘法器，與第八章並行乘法器效果完全相同，的它的 VHDL 代碼如下圖所示。將用在乘法-累加電路中。

```vhdl
1  -- Simple Multiplier.vhd----------------
2  LIBRARY ieee;
3  USE ieee.std_logic_1164.all;
4  USE ieee.std_logic_arith.all;
5  ----------------------------------------
6  ENTITY multiplier IS
7      PORT ( a, b: IN SIGNED(3 DOWNTO 0);
8             prod: OUT SIGNED(7 DOWNTO 0));
9  END multiplier;
10 ----------------------------------------
11 ARCHITECTURE behavior OF multiplier IS
12 BEGIN
13     prod <= a * b;
14 END behavior;
15 ----------------------------------------
```

9-1 乘法-累加電路的設計

乘法-累加電路 (Multiply-Accumulate Circuits) 是許多數位系統中常見的操作。尤其是用在高度互連的，如數位濾波器，神經網絡，數據量化器等系統。

圖 9-1 為一個典型的 MAC (乘加) 架構。它包括兩個值的相乘，然後將結果加到累加電路中，新的累加值必須被重新存儲在寄存器中以便將來累積。MAC 電路的另一個特點是必須檢查 MAC 操作數量大時可能發生的**溢出** (Overflow)。

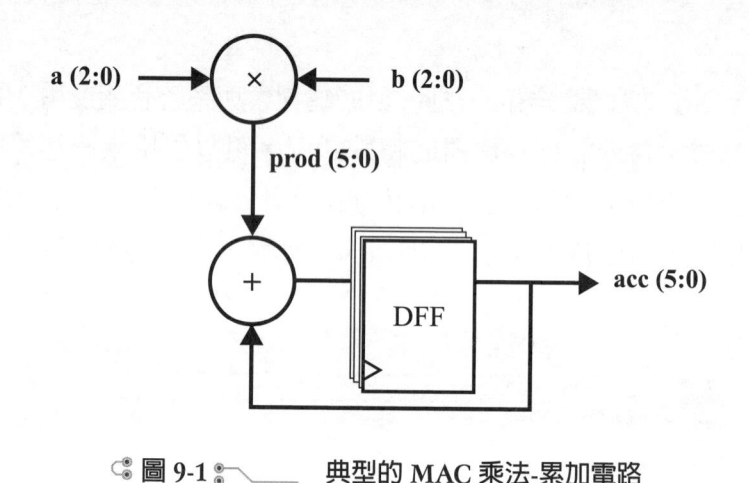

◎ 圖 9-1 ◎　　　典型的 MAC 乘法-累加電路

這個設計可以用第七章和第八章的 COMPONENTS 來完成，因為我們已經有了設計了圖 9-1 所示的每個單元零件。但由於它是一個相對簡單的電路，也可以直接設計。MAC 電路作為一個整體，可以在數字濾波器和神經網絡等應用中，當作 COMPONENT 來使用。

圖 9-2 爲 MAC 乘法-累加電路的 VHDL 代碼。它在操作 Prod 的過程中，須要處理溢出和截斷，使溢出不能發生。因此須加入 line 23 add_truncate 的**功能** (function)。

```
1  ------- Main code: ---------------------
2  LIBRARY ieee;
3  USE ieee.std_logic_1164.all;
4  USE ieee.std_logic_arith.all;
5  USE work.my_functions.all;
6  USE work.all;
7  ----------------------------------------
8  ENTITY mac IS
9      PORT ( a, b: IN SIGNED(2 DOWNTO 0);
10            clk, rst: IN STD_LOGIC;
11            acc: OUT SIGNED(5 DOWNTO 0));
12 END mac;
13 ----------------------------------------
14 ARCHITECTURE rtl OF mac IS
15     SIGNAL prod, reg: SIGNED(5 DOWNTO 0);
16 BEGIN
17     PROCESS (rst, clk)
18         VARIABLE sum: SIGNED(5 DOWNTO 0);
19     BEGIN
20         prod <= a * b;
21         IF (rst='1') THEN
22            reg <= (OTHERS=>'0');
23         ELSIF (clk'EVENT AND clk='1') THEN
24             sum := add_truncate (prod, reg, 6);
25             reg <= sum;
26         END IF;
27     acc <= reg;
28     END PROCESS;
29 END rtl;
30 ----------------------------------------
```

圖 9-2　MAC 乘法-累加電路的 VHDL 代碼

add_truncate 的功能，可以加入到功能的包裹內如圖 9-3 所示

```vhdl
-------- PACKAGE my_functions: ------------------------------
LIBRARY ieee;
USE ieee.std_logic_1164.all;
USE ieee.std_logic_arith.all;
-------------------------------------------------------------
PACKAGE my_functions IS
    FUNCTION add_truncate (SIGNAL a, b: SIGNED; size: INTEGER)
        RETURN SIGNED;
END my_functions;
-------------------------------------------------------------
PACKAGE BODY my_functions IS
    FUNCTION add_truncate (SIGNAL a, b: SIGNED; size: INTEGER)
        RETURN SIGNED IS
      VARIABLE result: SIGNED (5 DOWNTO 0);
    BEGIN
        result := a + b;
        IF (a(a'left)=b(b'left)) AND
           (result(result'LEFT)/=a(a'left)) THEN
          result := (result'LEFT => a(a'LEFT),
                     OTHERS => NOT a(a'left));
        END IF;
        RETURN result;
    END add_truncate;
END my_functions;
-------------------------------------------------------------
```

圖 9-3　功能包裹內的 add_truncate.vhd 代碼

9-2 乘法-累加電路的模擬測試

乘法-累加電路的模擬測試檔 test_bench.vhd 如圖 9-4 所示

```vhdl
------ test_bench_for_mac.vhd ---------------
LIBRARY ieee;
USE ieee.std_logic_1164.all;
USE ieee.std_logic_arith.all;
USE work.all;
---------------------------------------------
Entity test_bench is
End test_bench;
---------------------------------------------
ARCHITECTURE stimulus OF test_bench IS
component mac
    PORT ( a, b: IN SIGNED(2 DOWNTO 0);
           clk, rst: IN STD_LOGIC;
           acc: OUT SIGNED(5 DOWNTO 0));
END component;

    Signal   a: SIGNED(2 DOWNTO 0):= "000";
    Signal   b: SIGNED(2 DOWNTO 0):= "000";
    Signal   clk: std_logic := '0';
    Signal   rst: std_logic := '1';
    Signal   acc: SIGNED(5 DOWNTO 0);
begin
    DUT: mac port map( a, b, clk, rst, acc);

    -- generate signal rst: --------
    process
    begin
       wait for 100 ns; rst <= '0';
    end process;

    -- generate signal clk: ----------
    process
    begin
       wait for 200 ns; clk <= not clk;
    end process;
    -- generate signal a: ----------
    process
    begin
       wait for 400 ns; a <= "001"; -- 1
       wait for 400 ns; a <= "010"; -- 2
       wait for 400 ns; a <= "011"; -- 3
       wait for 400 ns; a <= "100"; -- -4
       wait for 400 ns; a <= "101"; -- -3
       wait for 400 ns; a <= "110"; -- -2
       wait for 400 ns; a <= "111"; -- -1
       wait;
    end process;
    -- generate signal b: ----------
    process
    begin
       wait for 400 ns; b <= "001"; -- 3
       wait for 400 ns; b <= "011"; -- 6
       wait for 400 ns; b <= "100"; -- -4
       wait for 400 ns; b <= "100"; -- -4
       wait for 400 ns; b <= "100"; -- -4
       wait for 400 ns; b <= "100"; -- -4
       wait for 400 ns; b <= "101"; -- -3
       wait;
    end process;
end stimulus;
```

圖 9-4　乘法-累加電路的模擬測試檔 test_bench.vhd 代碼

乘法-累加電路的模擬測試的結果如圖 9-5 所示。

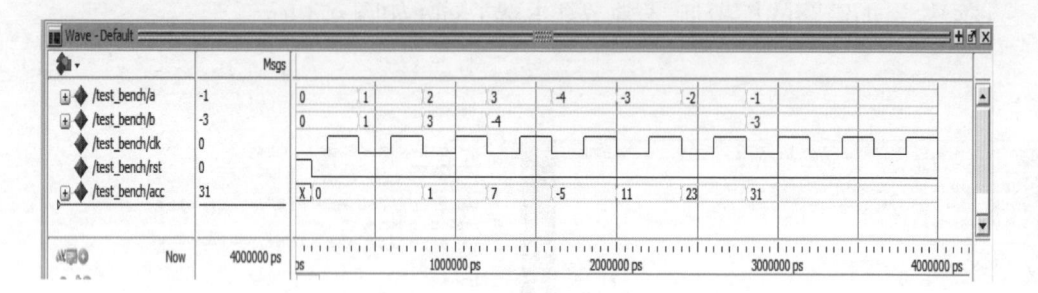

圖 9-5　累加電路的模擬測試的結果

檢示圖 9-5 的 acc(n) = a(n)*b(n) + acc(n-1) 的結果：

```
acc(0) = a(0)*b(0) + 0 = 0*0 + 0 = 0;
acc(1) = a(1)*b(1) + acc(0) = 1*1 + 0 = 1;
acc(2) = a(2)*b(2) + acc(1) = 2*3 + 1 = 7;
acc(3) = a(3)*b(3) + acc(2) = 3*(-4) + 7 = -5;
acc(4) = a(4)*b(4) + acc(3) = (-4)*(-4) + (-5) = 11;
acc(5) = a(5)*b(5) + acc(4) = (-3)*(-4) + 11 = 23;
acc(6) = a(6)*b(6) + acc(5) = (-2)*(-4) + 23 = 31;
acc(7) = a(7)*b(7) + acc(6) = (-1)*(-3) + 31 = 34;
```

acc(7) = 34（溢出），結果應該是截斷的最大正值（31）

模擬測試的結果與以上計算的結果完全相同。

第九章　乘法-累加電路　203

◎ 9-3　乘法-累加電路的合成

　　合成的軟體的第 1 部份從進入 new project wizard 開始，它要求填寫 directory，加入 project 所需的 files 和實驗板子 FPGA 的選用。

　　因此首先要新增一個名為 mac_ckt 的 folder，然後將通過 Simulation 的 mac、和 my_function 2 個 vhd files，copy 到 mac_ckt 的 folder 中。

　　在啟動 Quartus Prime Lite 16.1 後，選用 **File > New Project Wizard**，如圖 9S1-1 所示。

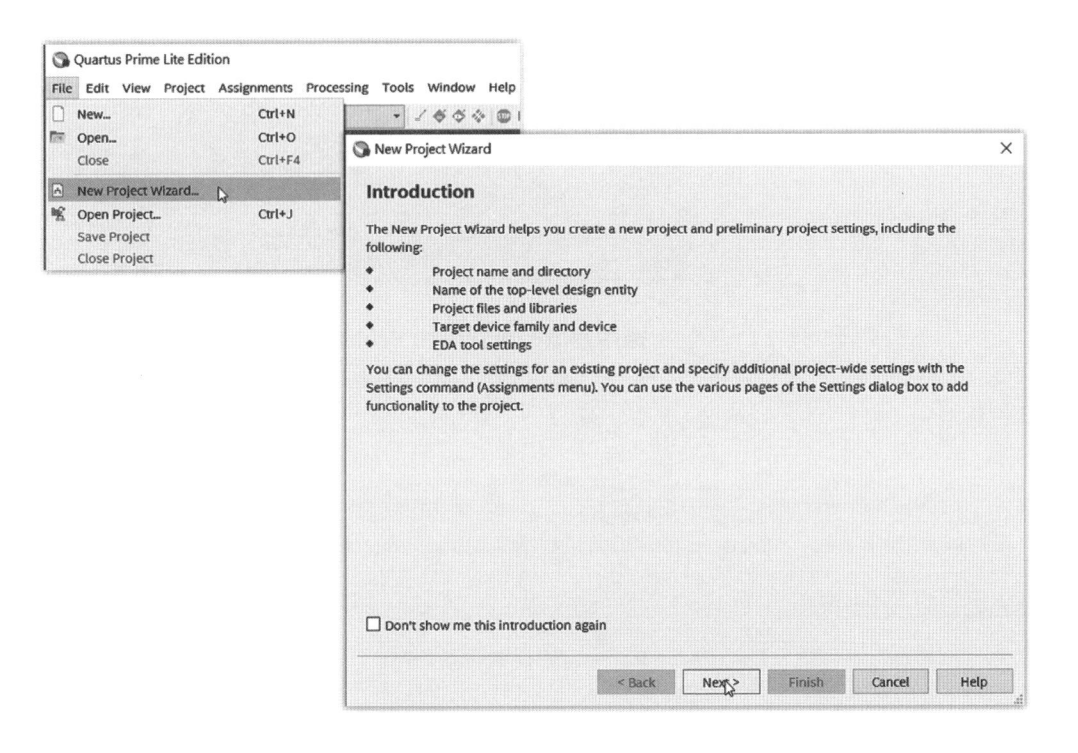

圖 9S1-1　　啟動 Quartus 16.1 後，選用 File > New Project Wizard

接下來選用合成的資料夾和 Project Type。如圖 9S1-2 所示。

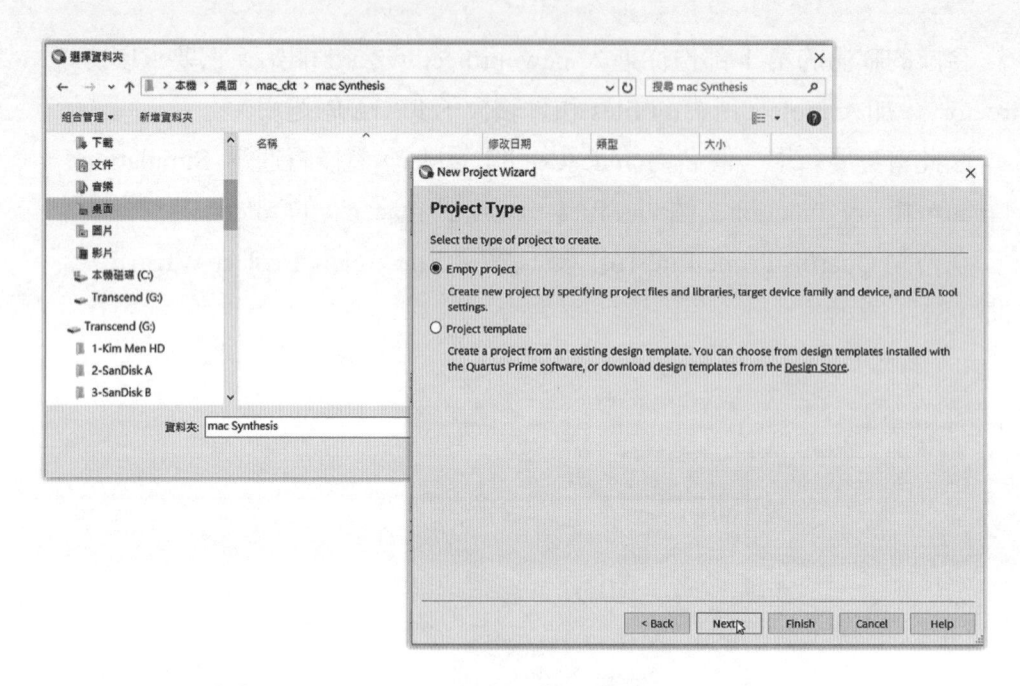

◎ 圖 9S1-2　　選用合成的資料夾和 Project Type

第九章　乘法-累加電路　205

　　然後填入 Directory，Name，和 Top-Level Entity 的名稱，並且加入 2 個相關的檔 my_functions.vhd 和 mac.vhd。如圖 9S1-3 所示。

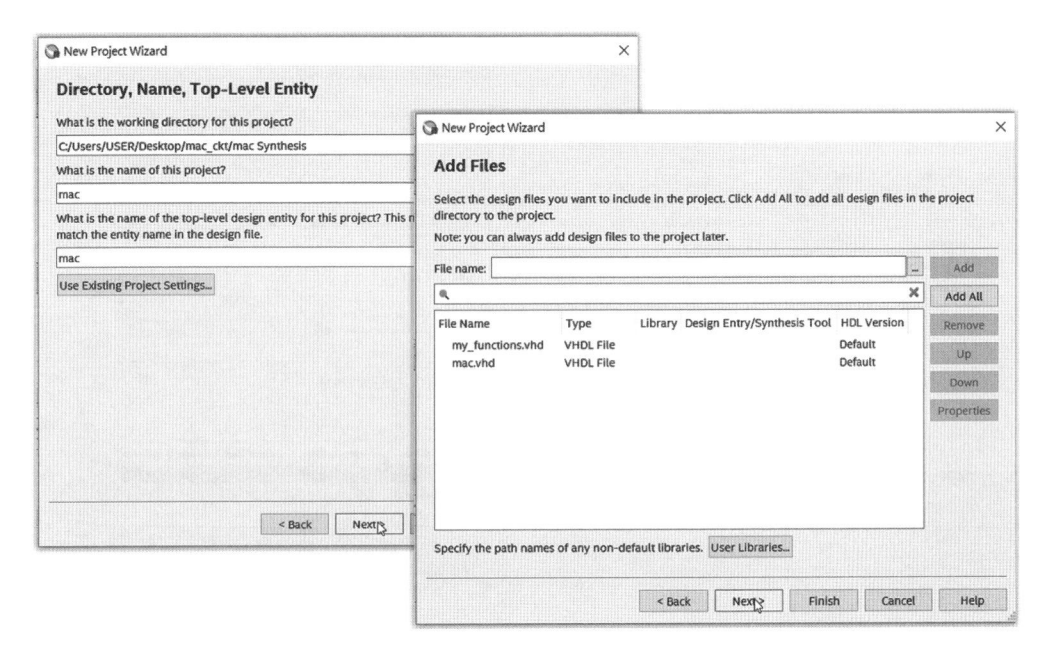

◦ 圖 9S1-3 ◦　　填入 Directory，Name，和 Top-Level Entity 的名稱和 Add Files

FPGA-IC 的選用，實驗板 DE2-115 所用的爲 Cyclone IV E 的 P4CE115F29C7。如圖 9S1-4 所示。EDA 工具的設定可跳過。

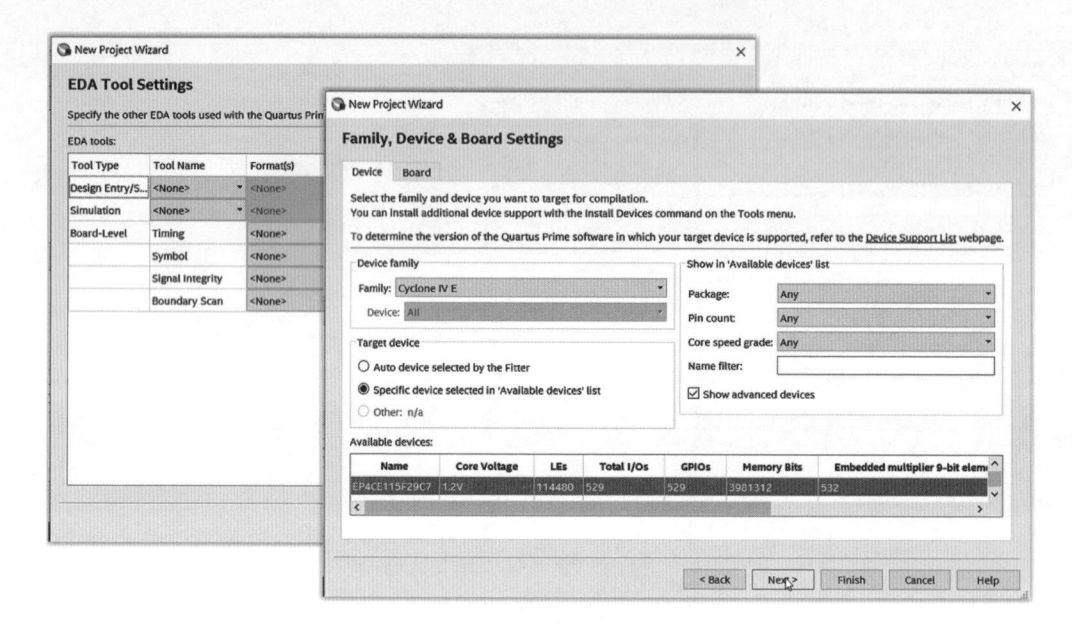

圖 9S1-4　EDA 工具的設定和 FPGA IC 的選用

歸納以上各種選項，總結如下圖 9S1-5 所示。完成合成軟體的第 1 部份。

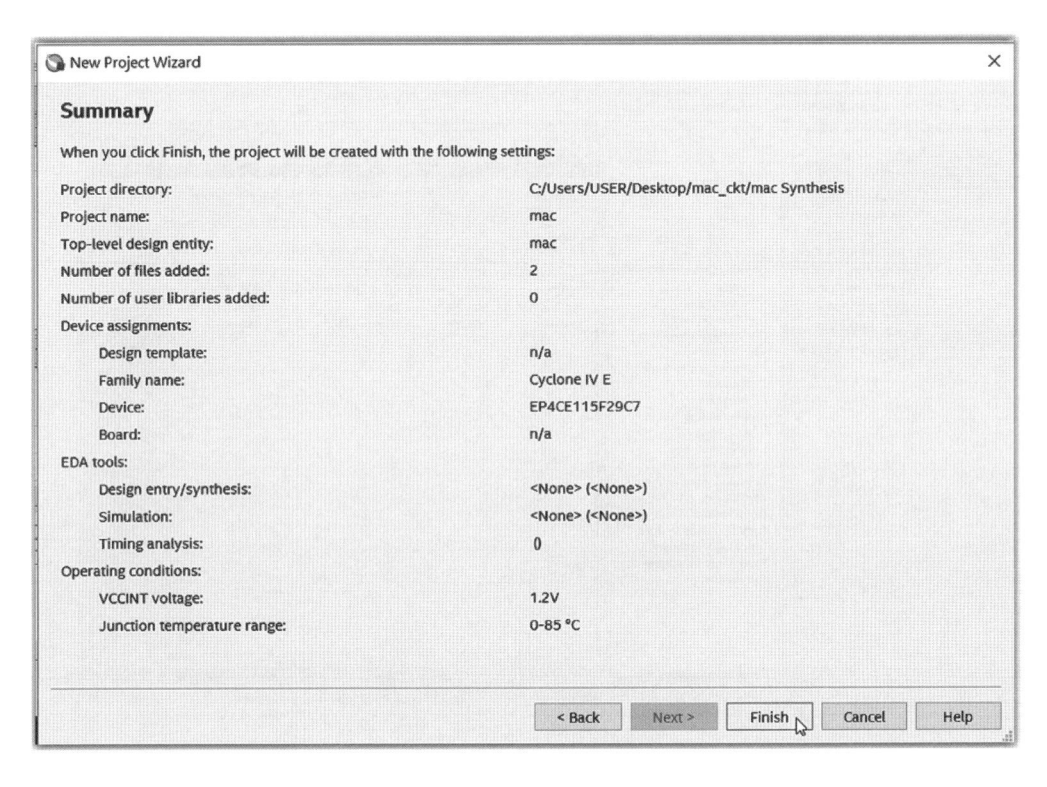

☘ 圖 9S1-5 ☘　　　完成合成軟體的第 1 部份的總結

第 1 部份的總結是否有錯？要從 Project 視窗的 **Assigment > Settings** 入手。首先要檢查在 **Settings-Multiplier > Files** 中是否有 mac.vhd 和 my_function.vhd files？如圖 9S1-6 所示。

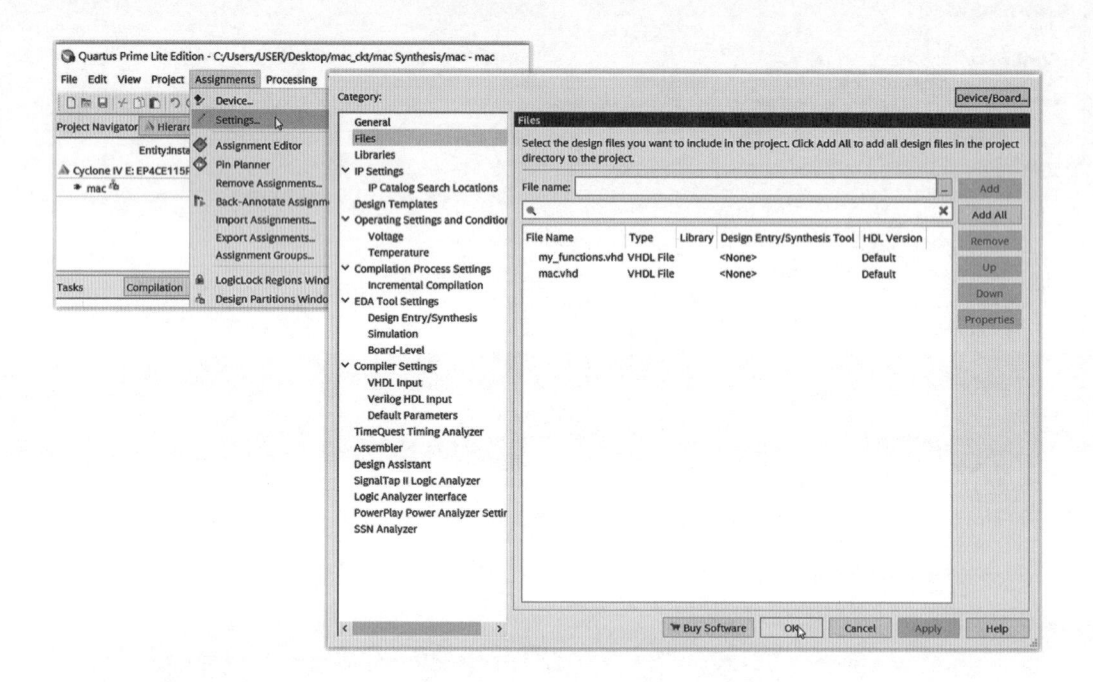

◦ 圖 9S1-6 ◦　　　　檢查 Assignments Settings 的 mac Files 是否齊全

再用 Compiler 來檢查所有相關的 Multiplier Files 都沒有錯。如圖 9S1-7 所示。

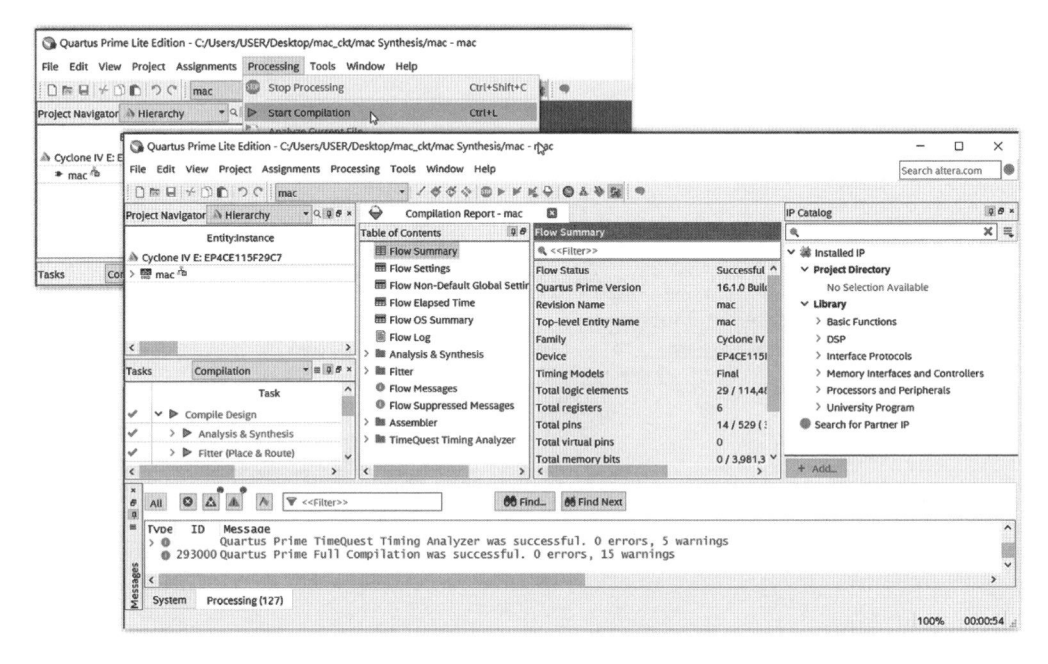

圖 9S1-7　Compiler 來檢查所有相關的 mac Files 都沒有錯

通過第 1 階段 compilation 之後，接下來為第 2 階段的 FPGA 接腳 (PIN) 的設定。這部份要從 Project mac 的 **Assignment > Assignment Editor** 入手。從圖 9S2-1 的 Assignment Editor 視窗，首先單擊 **New Assignment**。

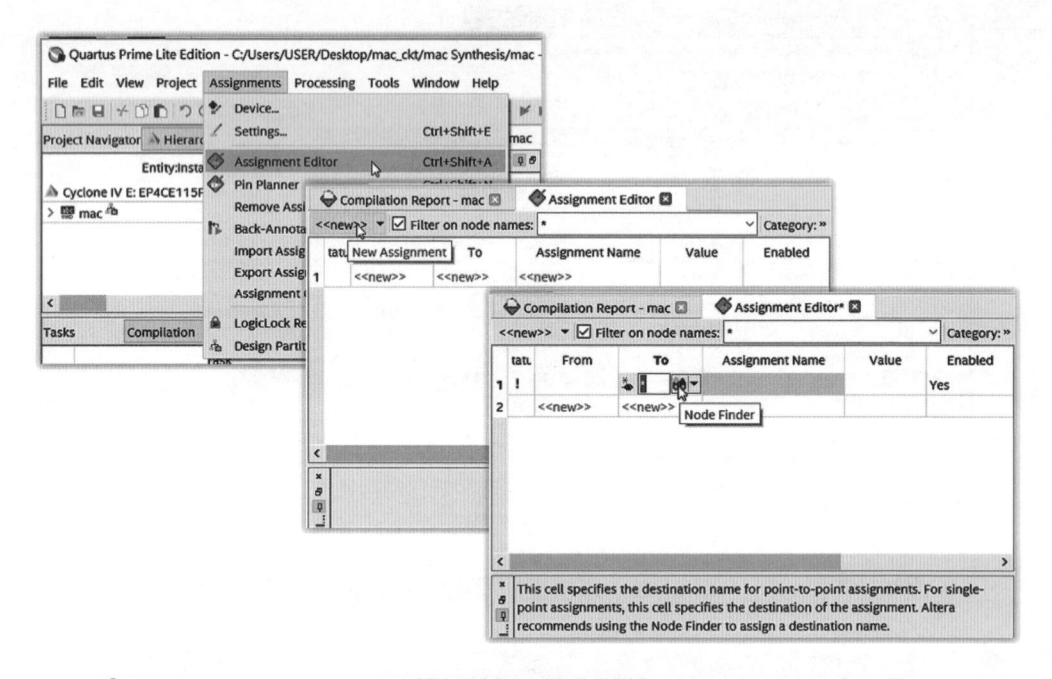

◦ 圖 9S2-1 ◦　　　FPGA 接腳的設定首先單擊 Assignment Editor 的 New Assignment

第九章　乘法-累加電路　211

然後再雙擊 **Assignment Editor** 的 Node Finder，如圖 9S2-2 所示。再單擊 Node Finder 的 List，以顯示 Project Multiplier 中的所有 Nodes。在 Matching Nodes 中選取跟 Ports 有關的所有 Signals 並移入到 Nodes Found 中，再單擊 **OK**。

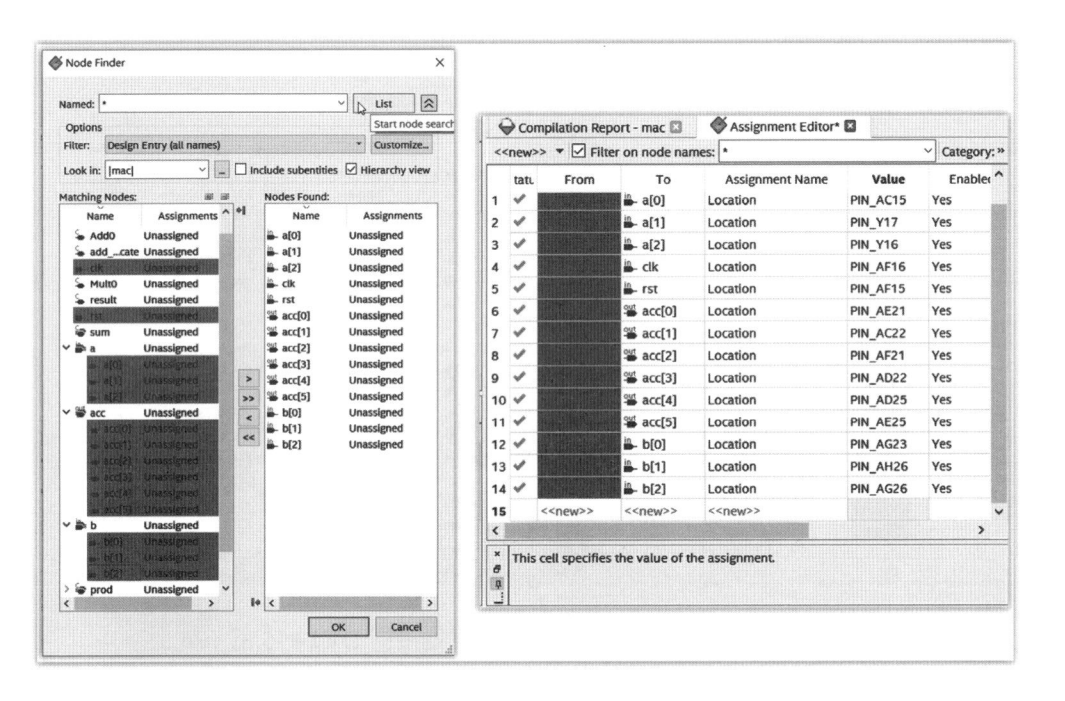

圖 9S2-2　從 Node Finder 中的 List 選取 Project 有關的所有 I/O Signals

為了易於連接 Analog Discovery 的 Digital Pattern Generator 和 Logic Analyzer，讓 Project 所有的 I/O Signals 連接到 DE2-115 的 JP5 上，如圖 8S2-3 所示。

DE-115 上 JP5 的使用，為的是易於與 Analog Discovery 的 Pattern Generator 和 Logic Analyzer 的測試接線相連接。

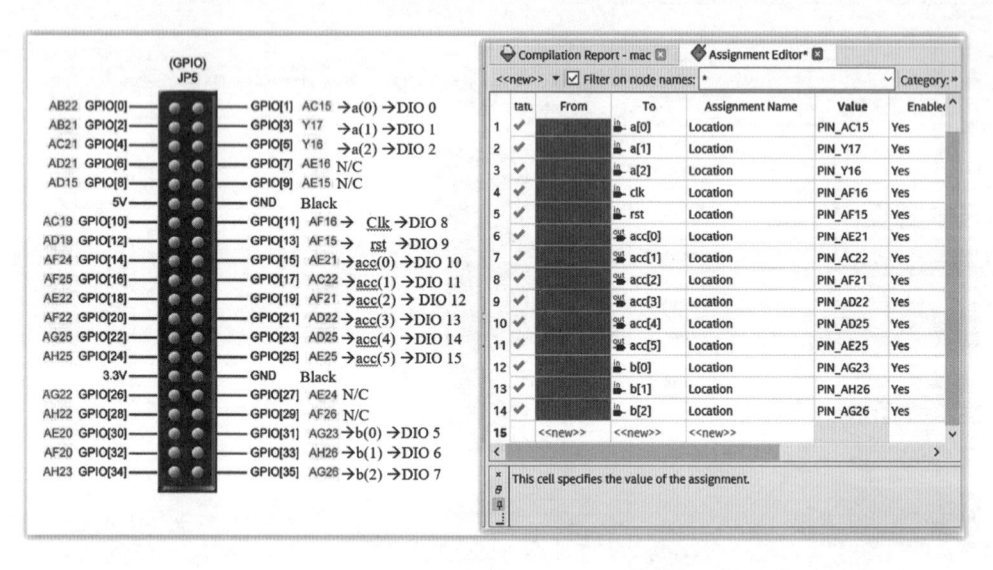

◦ 圖 9S2-3 ◦　　Project 所有的 I/O Signals 連接到 DE2-115 的 JP5 上

Project 新加入的 pin assignments 也需要用 Compiler 來檢驗是否有錯，選用 Project Multiplier 的 **Processing > Start Compilation**，如圖 9S2-4 所示，0 errors。

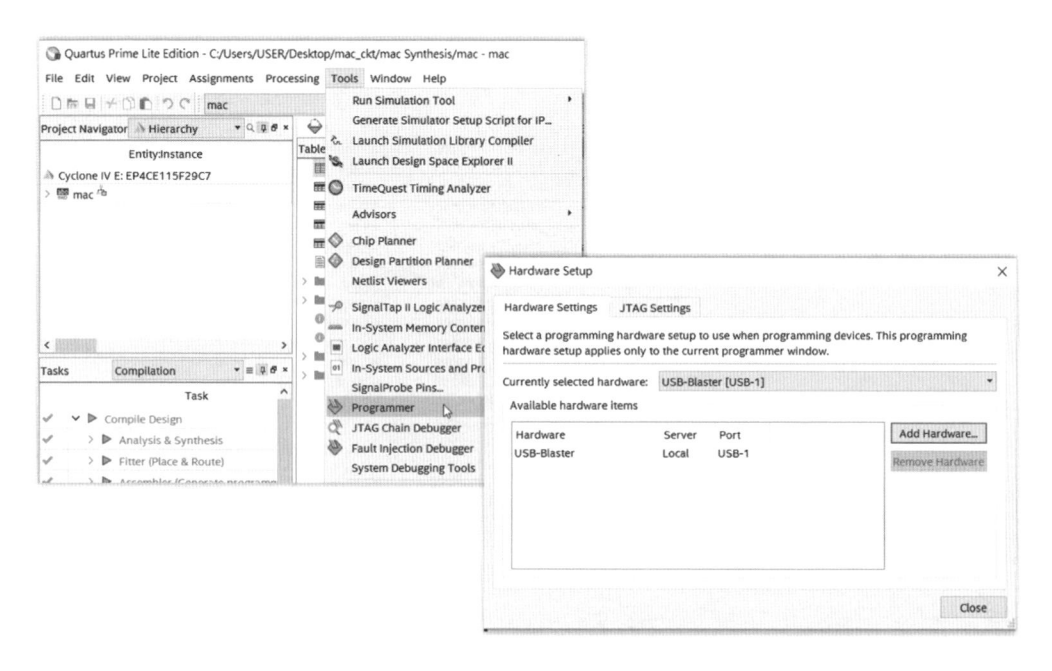

圖 9S2-4 新加入的 pin assignments 也需要用 Compiler 來檢驗是否有錯

0 Errors Compilation 代表第 2 階段的成功。這時候的 Multiplier Synthesis folder 和其中的 output_files 當如圖 9S2-5 所示。

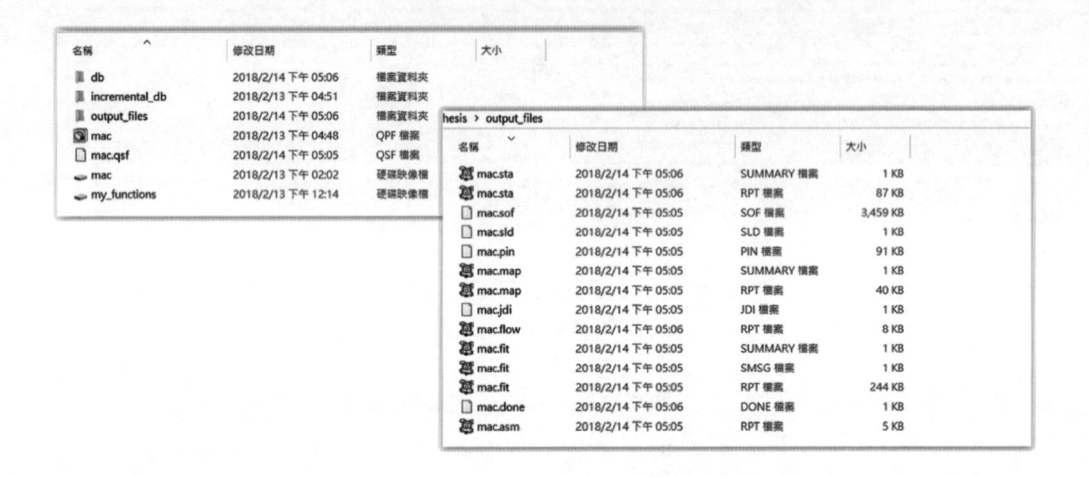

圖 9S2-5　Multiplier Synthesis folder 與其 output_files

其中 mac.sof 可以放到 **Tools > Programmer** 中，在 USB-Blaster [USB-1] 下 run 所設計的 mac program。如圖 9S2-6 所示。

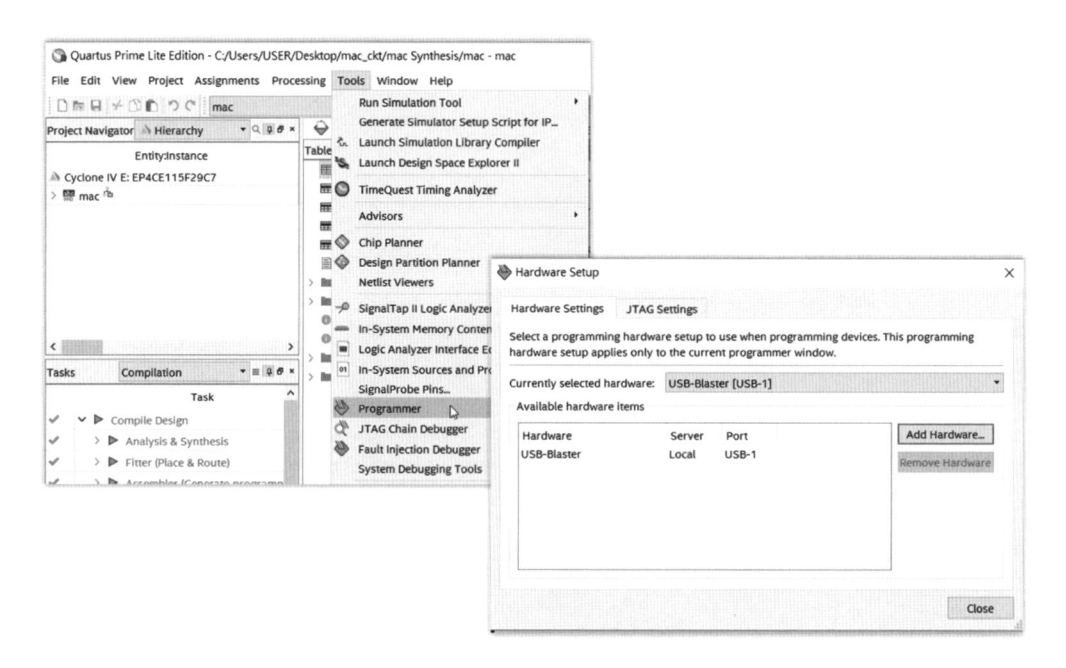

☉ 圖 9S2-6 ☉ 開啟 Analg Discovery 選用 Tools 中的 Programmer，設定 USB-Blaster

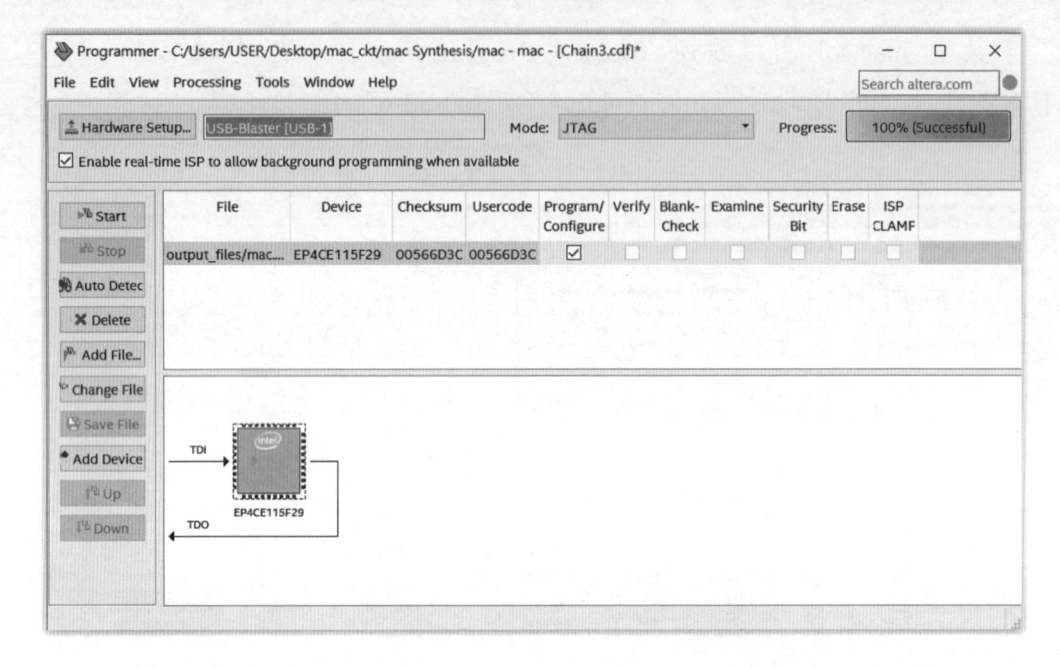

🔎 圖 9S2-7 ： mac.sof 放到 Tools > Programmer 中去 Start run

9-4 乘法-累加合成電路的硬體測試部份

mac.sof 經 Programmer run 成功之後，就可以使用 Analog Discovery 的 Digital Pattern Generator 和 Logic Analyzer 來測試 DE2-115 上 FPGA 所合成的串行平行乘法器。使令 JP5 與 Analog Discovery 間的訊號連線如圖 9T-1 所示。

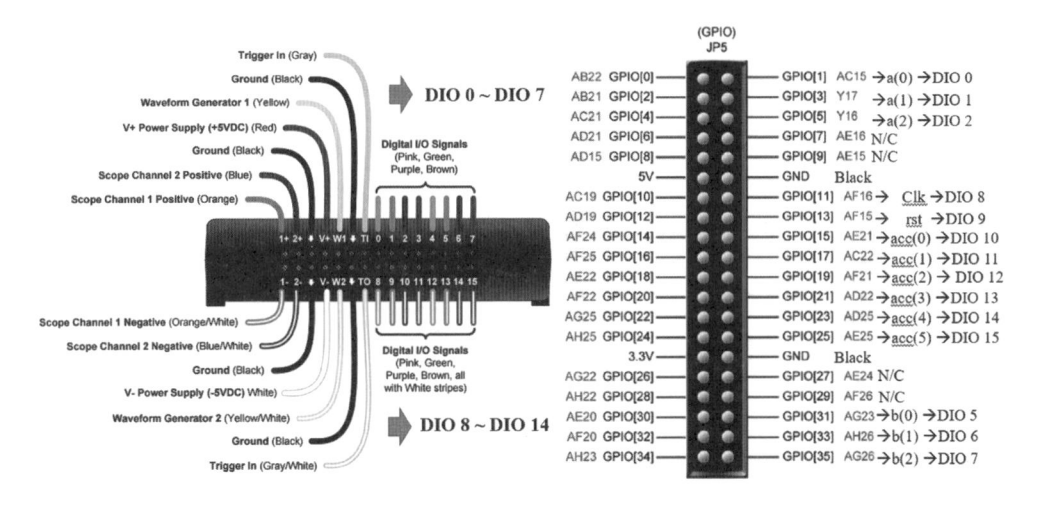

圖 9T-1　JP5 與 Analog Discovery 間的測試連線

請依照圖 9-5 模擬測試結果的 I/O 波形在 Digital Pattern Generator 設定的輸入訊號的波形。Logic Analyzer 中，再加上輸出訊號 acc。先 run Digital Pattern Generator，再 Single (run) Logic Analyzer，其輸出訊號 acc 如圖 9T-2 所示。與圖 9-5 模擬測試結果比較，完全相同。

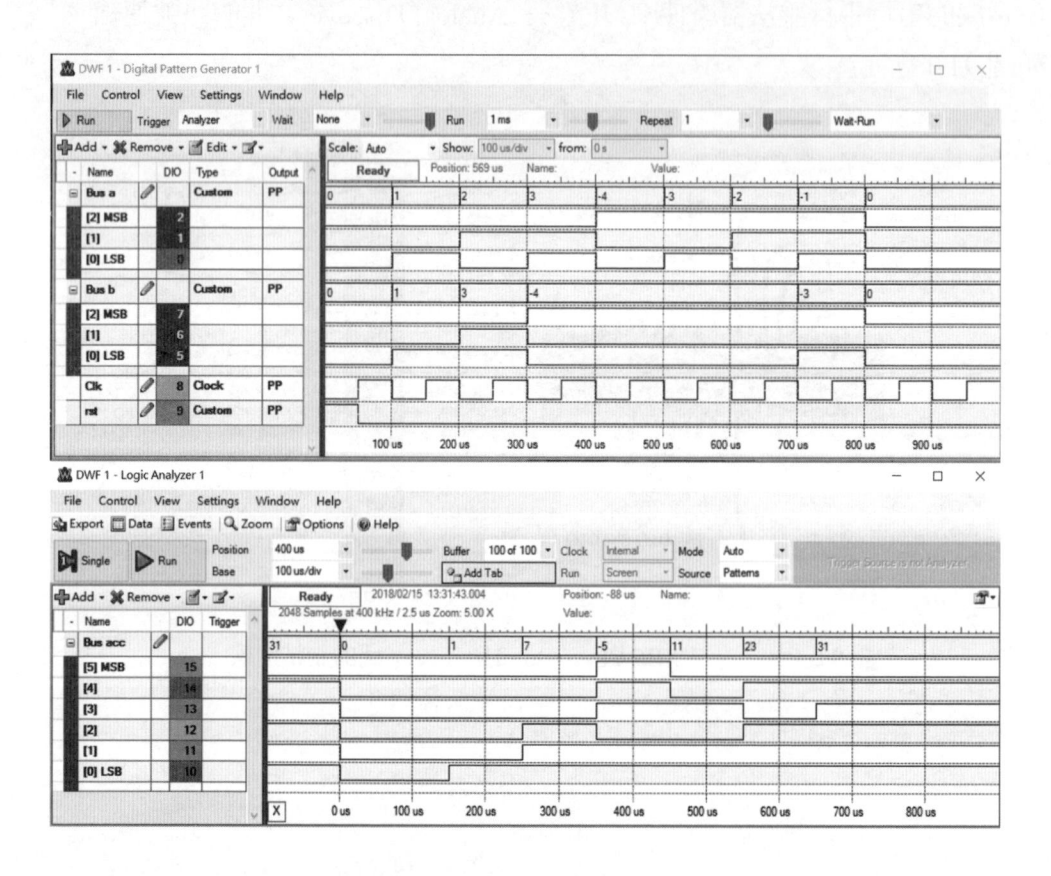

圖 9T-2　設定與 run Digital Pattern Generator 和 Logic Analyzer

9-5　課外練習

(1) 圖 P9-1 為 Fibonacci Series Generator 的邏輯電路圖。其中 Register A 和 B 及加法器均為 N = 16 bits。在初始化之後，情況是 c = 0，b = 1，並且 a = b + c = 1. 在下一個正向 clk 來到時：c = 1，b = 1，和 a = 2 的結果。 接下來，c = 1，b = 2，c = 3 依此類推。試寫出其 VHDL 代碼，並用 ModelSim 來測試之。

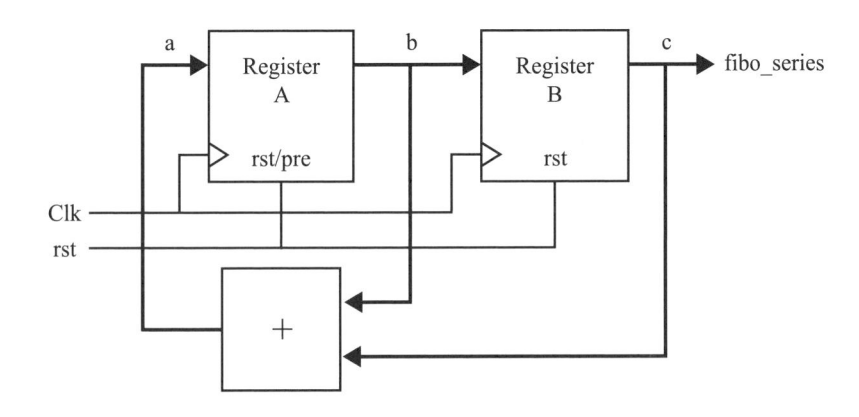

圖 P9-1：Fibonacci Series Generator 的邏輯電路圖

(2) 為了配合 Analog Discovery 的有限 Digital I/O，圖 P9-1 的 N 應減少到什麼程度？試說明並指出其缺點。

第十章　有限脈衝響應數位濾波器

10-1　有限脈衝響應數位濾波器的設計

數位濾波器為**數位信號處理** (Digital Signal Processing, DSP) 的一種。它在音頻、視頻、和通信等領域，用途很廣。圖 10-1 為數位信號處理系統與其輸入輸出訊號相位置。

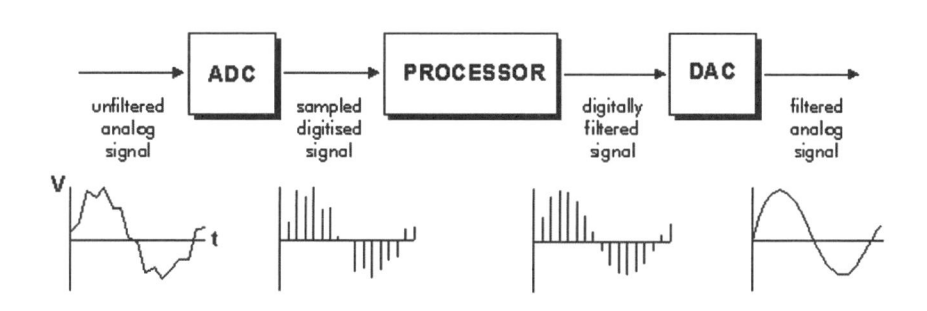

圖 10-1　數位信號處理系統與其輸入輸出訊號相關位置

有限脈衝響應 (Finite impulse response, FIR) 的應用通常是基於其 LTI (線性時不變) 特性，可用數位電路來實現。任何 LTI 系統都可以用下列的數位公式來表示：

$$y[n] = \sum_{k=0}^{M} c_k x[n-k]$$

對於具有參數 h(k) 的 N 抽頭 FIR 濾波器,其輸出被描述爲:

y(n) = h(0)x[n] + h(1)x[n-1] + h(2)x[n-2] + … h(N-1)x[n-N-1]

公式中 c_k 也就是有限脈衝響應 (FIR) 數位濾波器的參數 h( )。假定 FIR 濾波器的 4 個參數爲 h(0)~h(3),該公式可用圖 10-2 的 RTL 電路來代表。

◎ 圖 10-2　　　代表 FIR 濾波器的 RTL 電路

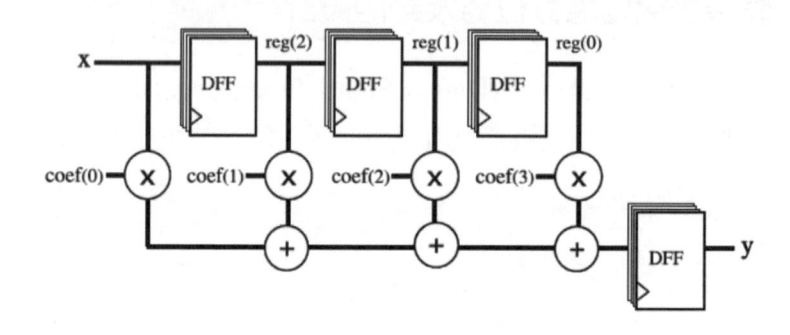

圖 10-2 中的 x 符號代表乘法器。+ 符號代表加法器。在下面的設計方案中,參數被設定爲常數,選擇的值是 coef(0) = 4,coef(1) = 3,coef(2) = 2,coef(3) = 1。選用 n = m = 4。爲了使模擬測試的結果易於觀測。在 n = m = 4 的情況下,合成電路需要 20 個 DFF (移位寄存器的每個級 4 個,加上 8 個輸出)。

圖 10-3 為圖 10-2 的 FIR 濾波器的 RTL 電路的 VHDL 代碼 fir2.vhd。

```vhdl
1  -- FIR2.VHD Pedroni Book1 p.291------------
2  LIBRARY ieee;
3  USE ieee.std_logic_1164.all;
4  USE ieee.std_logic_arith.all; -- package needed for SIGNED
5  ------------------------------------------
6  ENTITY fir2 IS
7      GENERIC (n: INTEGER := 4; m: INTEGER := 4);
8      -- n = # of coef., m = # of bits of input and coef.
9      --Besides n and m, CONSTANT (line 19) also need adjust
10     PORT ( x: IN SIGNED(m-1 DOWNTO 0);
11            clk, rst: IN STD_LOGIC;
12            y: OUT SIGNED(2*m-1 DOWNTO 0));
13 END fir2;
14 ------------------------------------------
15 ARCHITECTURE rtl OF fir2 IS
16     TYPE registers IS ARRAY (n-2 DOWNTO 0) OF
17                            SIGNED(m-1 DOWNTO 0);
18     TYPE coefficients IS ARRAY (n-1 DOWNTO 0) OF
19                            SIGNED(m-1 DOWNTO 0);
20     SIGNAL reg: registers;
21     CONSTANT coef: coefficients := ("0001", "0010", "0011",
22                            "0100");
23 BEGIN
24     PROCESS (clk, rst)
25         VARIABLE acc, prod:
26             SIGNED(2*m-1 DOWNTO 0) := (OTHERS=>'0');
27         VARIABLE sign: STD_LOGIC;
28     BEGIN
29     ----- reset: ------------------------
30     IF (rst='1') THEN
31         FOR i IN n-2 DOWNTO 0 LOOP
32             FOR j IN m-1 DOWNTO 0 LOOP
33                 reg(i)(j) <= '0';
34             END LOOP;
35         END LOOP;
36     ----- register inference + MAC: ------
37     ELSIF (clk'EVENT AND clk='1') THEN
38         acc := coef(0)*x;
39         FOR i IN 1 TO n-1 LOOP
40             sign := acc(2*m-1);
41             prod := coef(i)*reg(n-1-i);
42             acc := acc + prod;
43             ---- overflow check: ---------
44             IF (sign=prod(prod'left)) AND
45                 (acc(acc'left) /= sign)
46             THEN
47                 acc := (acc'LEFT => sign, OTHERS => NOT sign);
48             END IF;
49         END LOOP;
50         reg <= x & reg(n-2 DOWNTO 1);
51     END IF;
52     y <= acc;
53     END PROCESS;
54 END rtl;
55 ------------------------------------------
```

圖 10-3　圖 10-2 的 FIR 濾波器的 RTL 電路之 VHDL 代碼 fir2.vhd

請注意，濾波器代碼的末端為 Overflow check (溢出檢示) 部分，包含一個 MAC (乘法-累加) 流水線。這個電路與 MAC 電路密切相關。由於電路也可能發生溢出，所以在設計中必須包含一個添加/截斷過程。

10-2　FIR 數位濾波器的模擬測試

模擬測試檔 test_bench.vhd 如圖 10-4 所示。測試結果的邏輯時序如圖 10-5 所示。

```
---- test_bench_for_multiplier.vhd --------
LIBRARY ieee;
USE ieee.std_logic_1164.all;
USE ieee.std_logic_arith.all;
USE work.all;
-----------------------------------------
Entity test_bench is
    GENERIC (n: INTEGER := 4; m: INTEGER := 4);
    -- n = # of coef., m = # of bits of input and coef.
    --Besides n and m, CONSTANT (line 19) also need adjust
End test_bench;
-----------------------------------------
ARCHITECTURE stimulus OF test_bench IS
    component fir2
        PORT ( x: IN SIGNED(m-1 DOWNTO 0);
               clk, rst: IN STD_LOGIC;
               y: OUT SIGNED(2*m-1 DOWNTO 0));
END component;
-----------------------------------------
    Signal    x:   SIGNED(3 DOWNTO 0):= "0000";
    Signal    clk: std_logic := '0';
    Signal    rst: std_logic := '1';
    Signal    y:   SIGNED(2*m-1 DOWNTO 0);
begin
    DUT: fir2 port map( x, clk, rst, y);

    -- generate signal rst: --------
    process
    begin
        wait for 100 ns; rst <= '0';
    end process;
    -- generate signal clk: ---------
    process
    begin
        wait for 200 ns; clk <= not clk;
    end process;
    -- generate signal x: ---------
    process
    begin
        wait for 300 ns; x <= "0101"; -- 5
        wait for 400 ns; x <= "1010"; -- 10
        wait for 400 ns; x <= "1111"; -- 15
        wait for 400 ns; x <= "0100"; -- 4
        wait for 400 ns; x <= "1001"; -- 9
        wait for 400 ns; x <= "1110"; -- 14
        wait for 400 ns; x <= "0011"; -- 3
        wait; -- 2700 ns total run time
    end process;
end stimulus;
```

圖 10-4　fir2.vhd 的模擬測試檔 test_bench.vhd 代碼

FIR 濾波器的模擬測試的結果如圖 10-5 所示。輸入訊號 x 和輸出訊號 y 都以 Decimal (signed) 數字讀出。

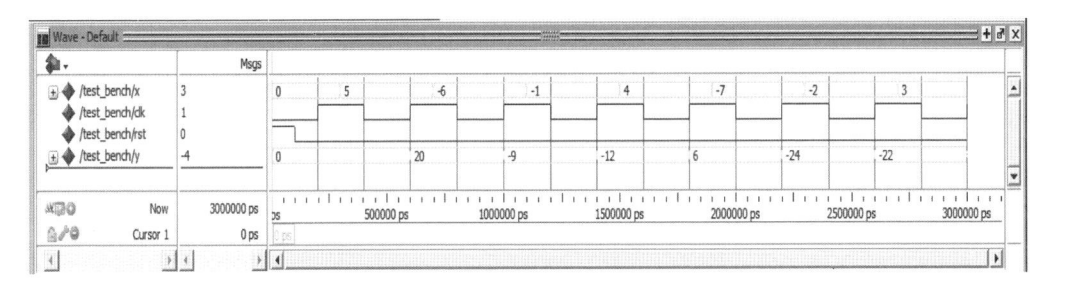

◎ 圖 10-5 ◎　　　模擬測試結果的 I/O 波形

檢示圖 10-5 的

y(n) = h(0)x[n] + h(1)x[n-1] + h(2)x[n-2] + ... h(N-1)x[n-N-1]

其中參數

h(0) = 4;　h(1) = 3;　h(2) = 2;　h(3) = 1;

其輸入訊號

x[0] = 0;　x[1] = 5;　x[2] = -6;　x[3] = -1;　x[4] = 4;　x[5] = -7;

y(0) = h(0)x[0] = 4 * 0 = 0;
y(1) = h(0)x[1] + h(1)x[0] = 4*5 + 3*0 = 20;
y(2) = h(0)x[2] + h(1)x[1] + h(2)x[0] = 4*(-6) + 3*5 + 2*0 = -24 + 15 = -9;
y(3) = h(0)x[3] + h(1)x[2] + h(2)x[1] + h(3)x[0] = 4*(-1) + 3*(-6) + 2*5 = -12;
y(4) = h(0)x[4] + h(1)x[3] + h(2)x[2] + h(3)x[1] = 4*4+3*(-1)+2*(-6) +1*5 = 6

對照圖 10-5 模擬測試結果的輸出 y，與計算出來的數值完全符合。

10-3　FIR 數位濾波器電路的合成

　　合成的軟體的第 1 部份從進入 new project wizard 開始,它要求填寫 directory,加入 project 所需的 files 和實驗板子 FPGA 的選用。

　　因此首先要新增一個名為 FIR_Filter 的 folder,然後將通過 Simulation 的 FIR2.vhd file,copy 到 FIR_Filter 的 folder 中。在啟動 Quartus Prime Lite 16.1 後,選用 **File > New Project Wizard**,首先要新增一個名為 FIR_Filter 的 folder,然後將通過 Simulation 的 FIR2.vhd 檔,copy 到 FIR_Filter 的檔案夾中。

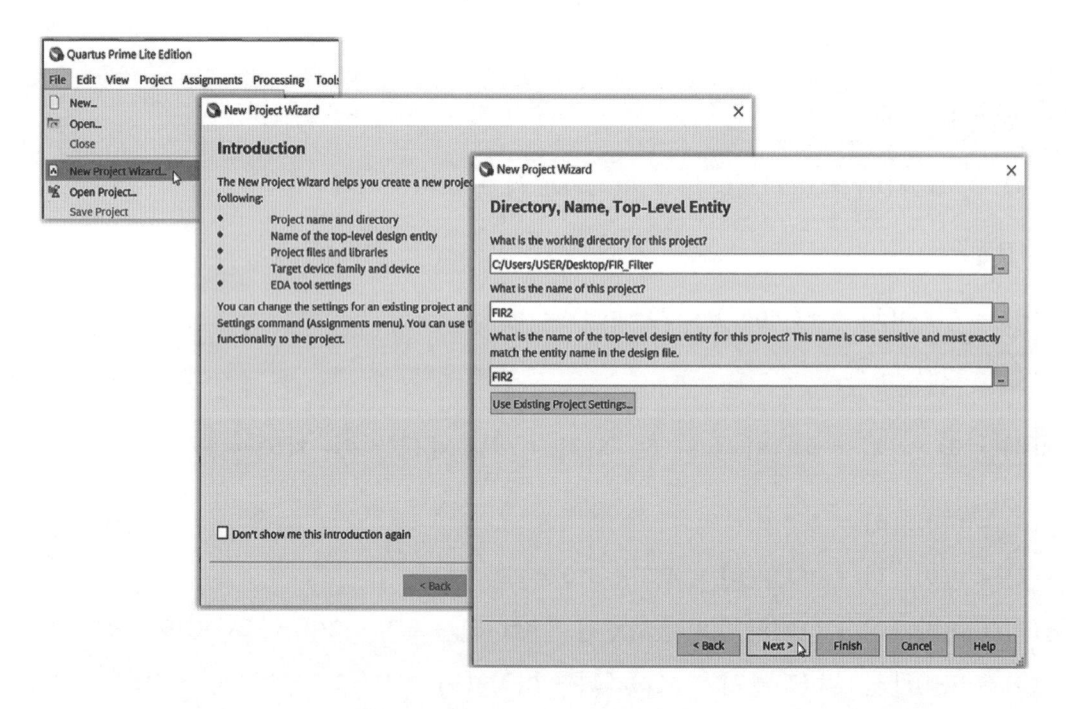

圖 10S1-1　　　　啟動 Quartus Prime,填寫 Directory 所在及 Project Name 等

第十章　有限脈衝響應數位濾波器　227

接下來選用合成的 Project 所需的 File 和使用的板子及 FPGA 的名稱。如圖 10S1-2 所示。

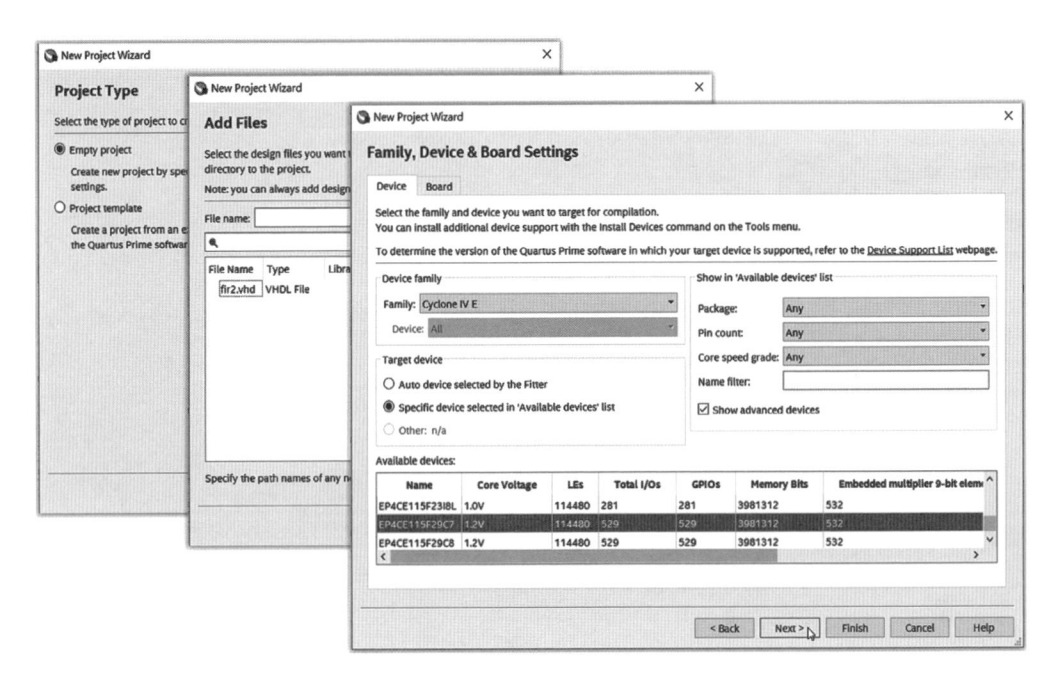

◖圖 10S1-2 ◗　　填寫合成所需的 Files，及板子和 FPGA 的名稱等資料

最後是顯示 Project 所用的工具和總結，如圖 10S1-3 所示。

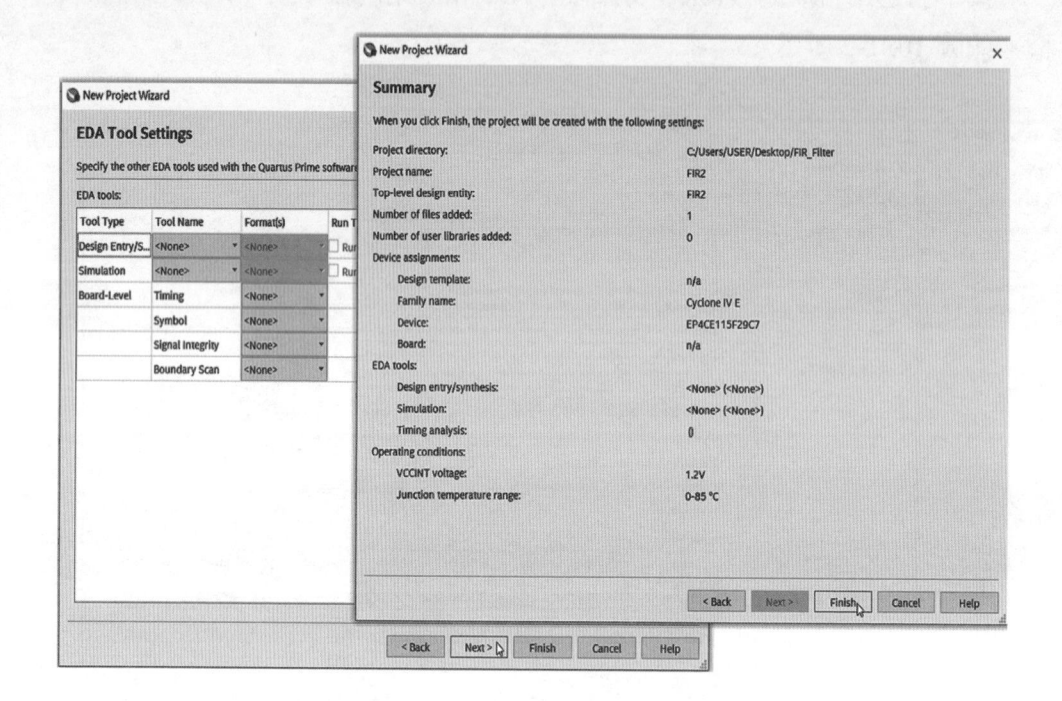

◎ 圖 10S1-3 　　Project 所用工具及 FPGA 名稱的總結顯示

在做 compile 之前還須用 **Assignments > Setting** 來查看用來合成的 FIR2.vhd 檔，和它的附隨檔都已齊備，(本合成不含附隨檔)。如圖 10S1-4 所示。

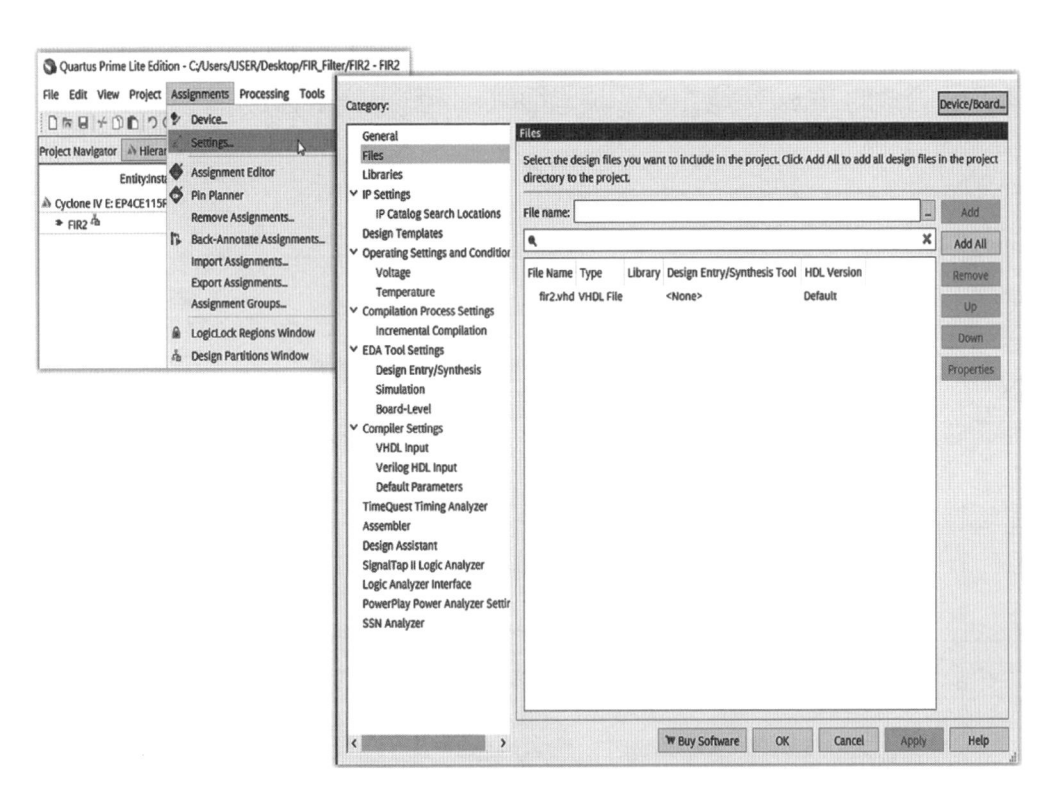

圖 10S1-4　使用 Assignments > Setting 來查看用來合成的 files 已齊備

合成的 files 已齊備，再使用 **Processing > Start Compilation** 來檢驗所有用來合成的 files 都能 100% 通過編輯。(容許 Warnings) 如圖 10S1-5 所示。

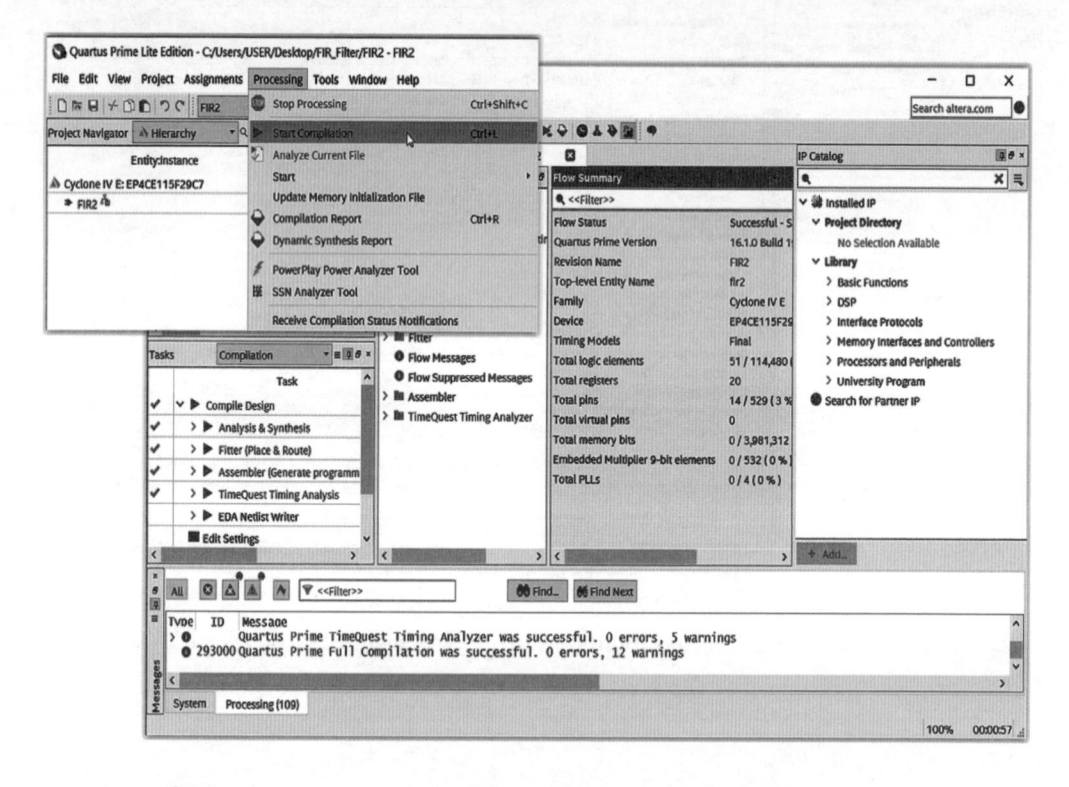

圖 10S1-5　所有參與合成的 files 100% 通過編輯的檢驗

第十章　有限脈衝響應數位濾波器　231

　　是接下來的課題。這時候須要用到 **Assignments ＞ Assignment Editor**，它須要好幾個步驟，如圖 10S2-1、首先是點擊 **Assignment Editor** 小視窗上的黑色 **<<new>>**。接下來雙擊 **To** 下面的望遠鏡圖示，在所產生的 Node Finder 視窗單擊其右上方之 List，再將左端所有的 I/O 有關 Ports 搬到右方。最後再單擊 **Node Finder** 視窗右下方之 **OK**。

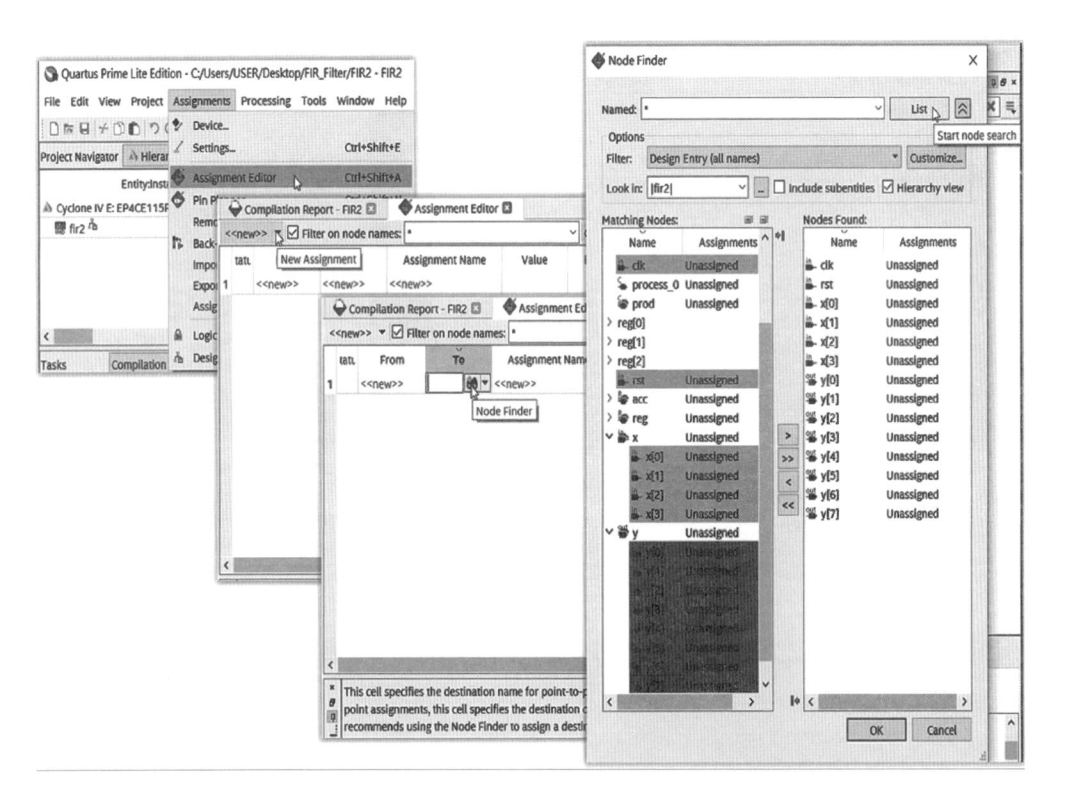

◎ 圖 10S2-1　　Project I/O Ports 對 FPGA pins 接腳選用步驟之一

FPGA pins 接腳的選用，也要注意到測試儀器與接腳連線的方便，因為測試儀器使用 Analog Discovery 的 Pattern Generator 與 Logic Analyzer，所以選用 DE2-115 的 JP5 為 Project 電路與 Analog Discovery 之間的共同連接點，它們之間的關係如圖 10S2-2 所示

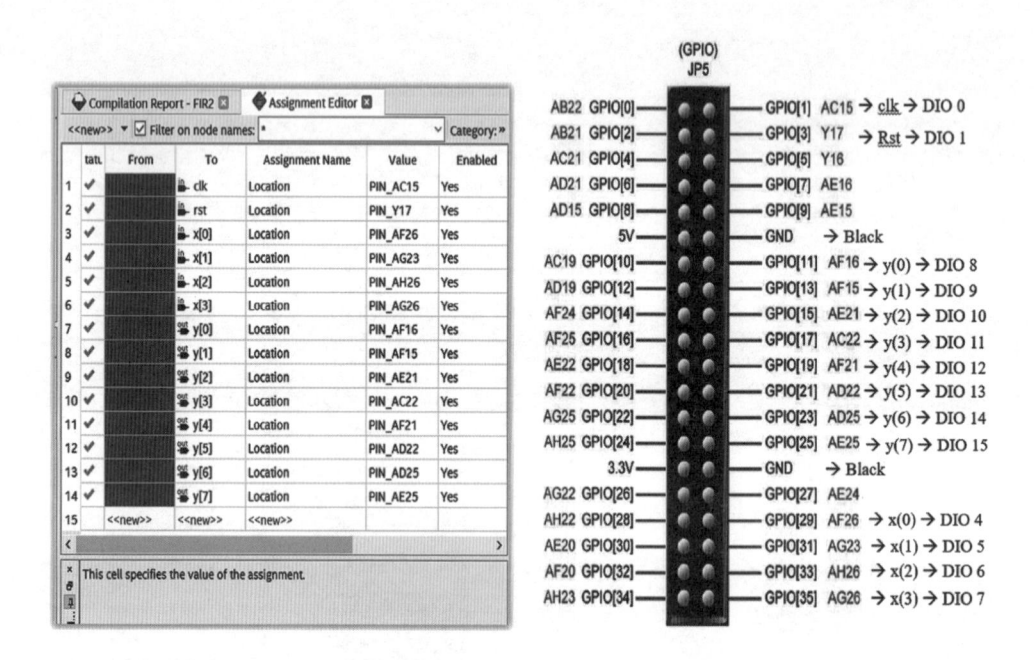

圖 10S2-2　　Project I/O Ports 對 FPGA pins 接腳選用步驟之二

第十章　有限脈衝響應數位濾波器　233

　　Assignment Editor 的 PIN_xxxx 填好之後，其中是否有錯，再須經過再一次的編輯來驗證，如圖 10S2-3 所示 100% 通過 Compilation。

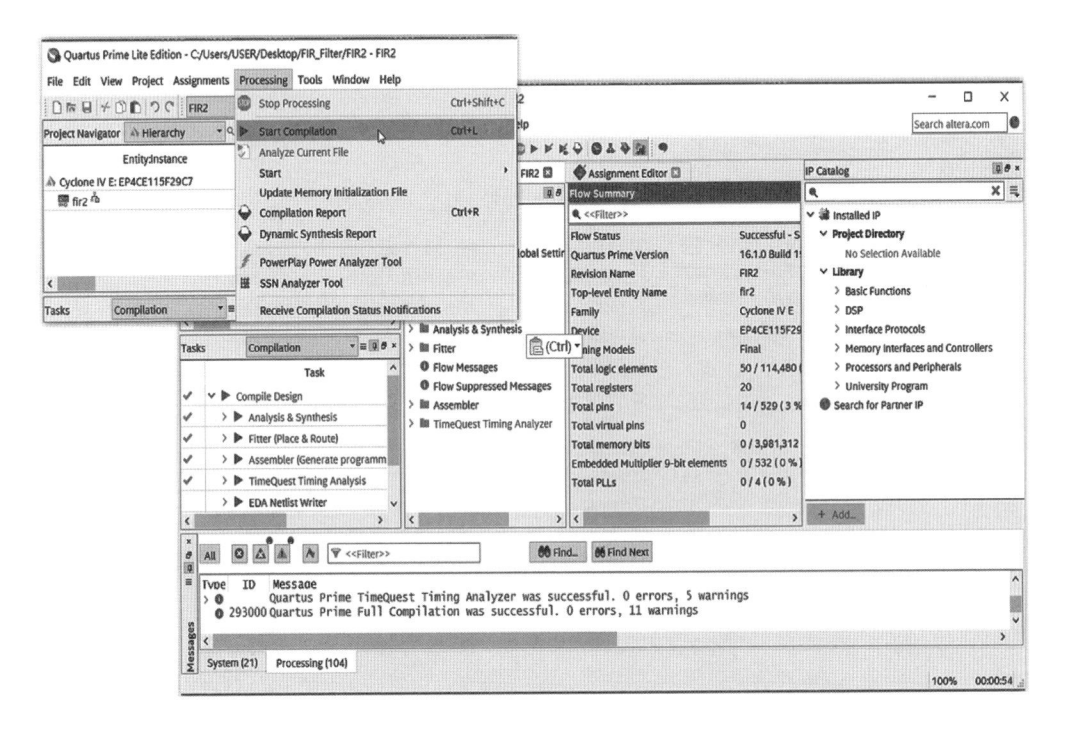

圖 10S2-3　　FPGA pins Assignment 100%通過 Compilation 驗證

0 Errors Compilation 代表第 2 階段的成功。這時候的 FIR2_Filter folder 和其中的 output_files 當如圖 10S2-4 所示。

圖 10S2-4　FIR_Filter folder 與其 output_files

其中 FIR2.sof 可以放到 **Tools > Programmer** 中，準備做 program 與 FPGA 連線的工作。如圖 10S2-5 所示。

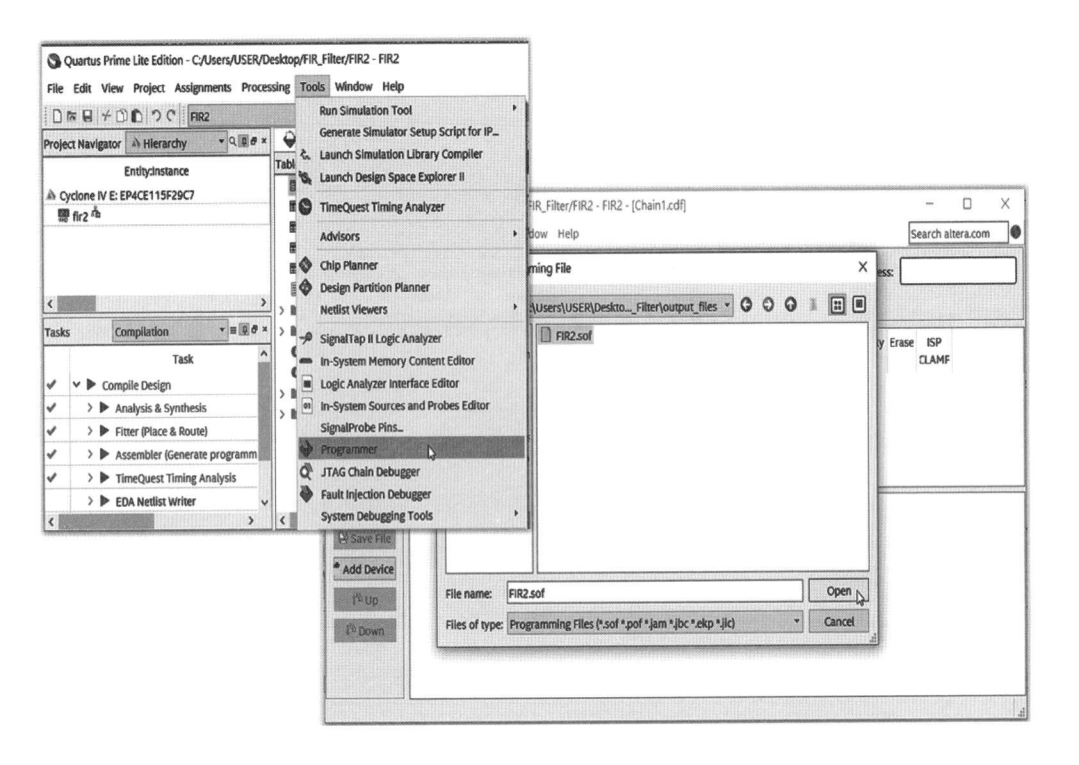

◦ 圖 10S2-5 ◦　　Programmer 選用 FIR2.sof 準備好做連線的工作

這時候當完成 DE2-115 與 PC 間的連接並開啓電源，在硬體設定 USB-Blaster [USB-0] 下，單擊 Start。如果一切正常，當如圖 10S2-6 所示，100% 成功。

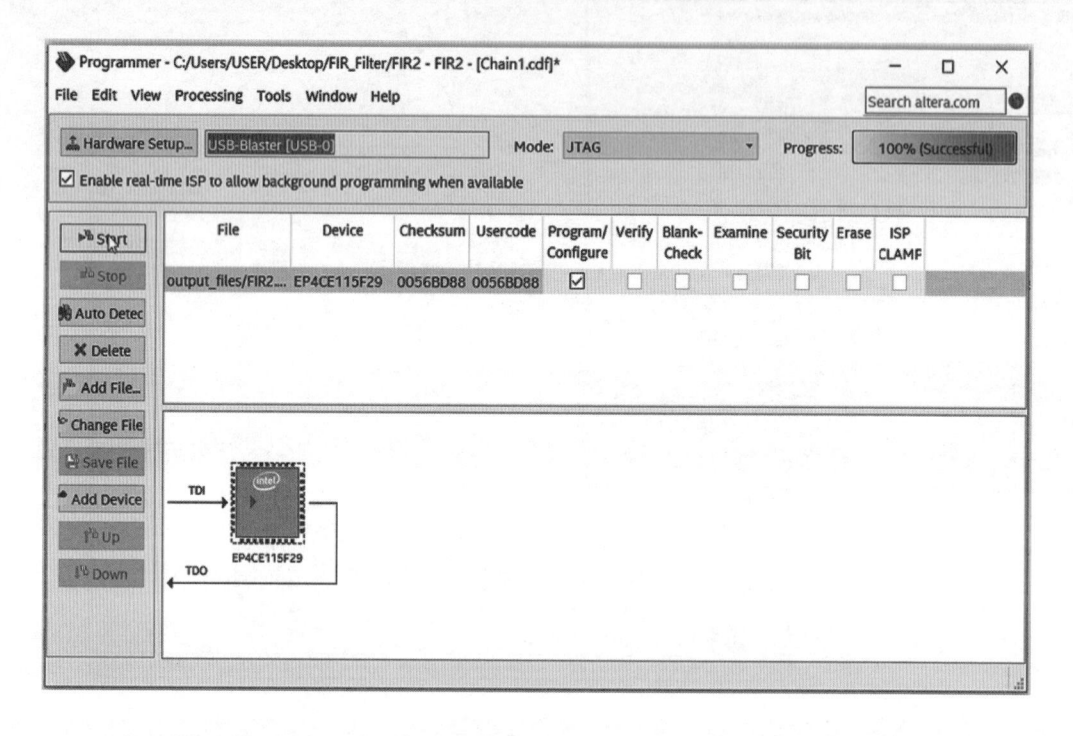

◎ 圖 10S2-6　　正常情況下 Program FIR2 與 FPGA 連線成功

使用 Analog Discovery 的 Digital Pattern Generator 和 Logic Analyzer 來測試 FIR2_Filter 的 I/O，它們之間的連線如圖 10S3-1 所示。

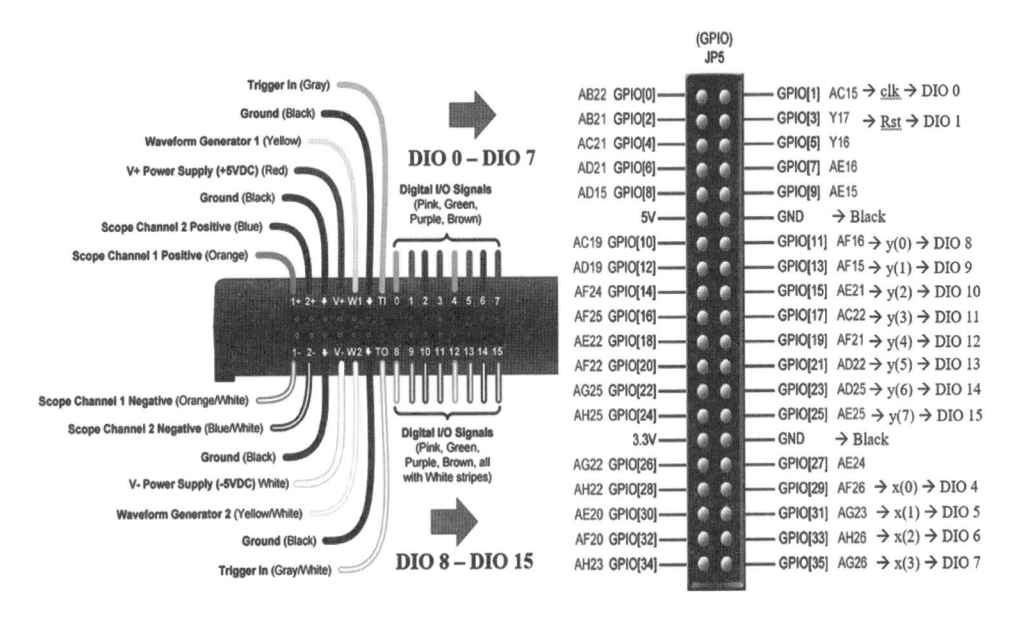

◎ 圖 10S3-1 ◎　　　Analog Discovery DIO 與 DE2-115 JP5 間的測試連線

Digital Pattern Generator 對 FIR2 filter input clk、rst 和 x Signal 的數位波形的產生，如圖 10S3-2 所示。

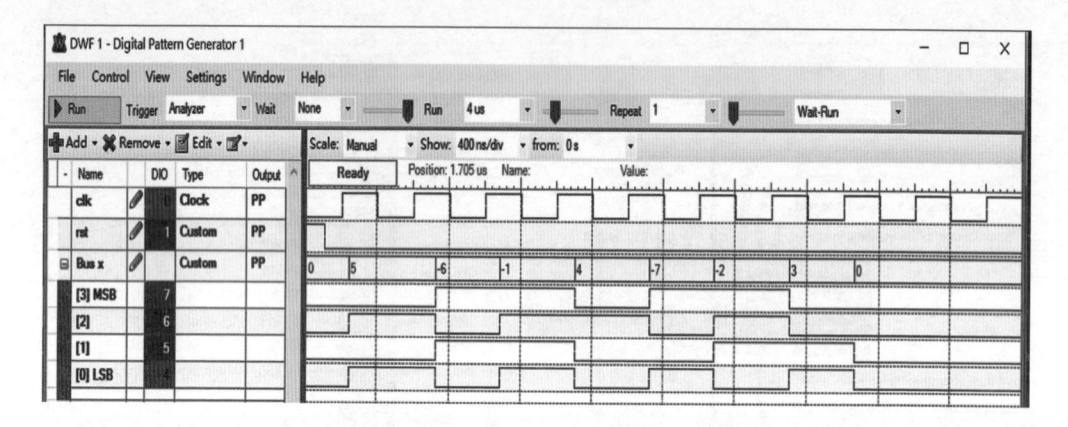

⊙ 圖 10S3-2 ⊙──── 數位訊號波形產生器提供 FIR2 數位濾波器的所有輸入波形

為了觀察輸入和輸出波形的時序關係，可以把輸入和輸出訊號一起擺到 Logic Analyzer 內。先 run Pattern Generator，然後再 (single) run Analyzer。在正常情況下，將獲得如圖 10S3-3 所示的結果。比較圖 10S3-3 所示的結果，與圖 10-5 FIR2 filter 模擬測試的結果完全相同。

⊙ 圖 10-5 ⊙──── 比較 Fir2 模擬測試的結果

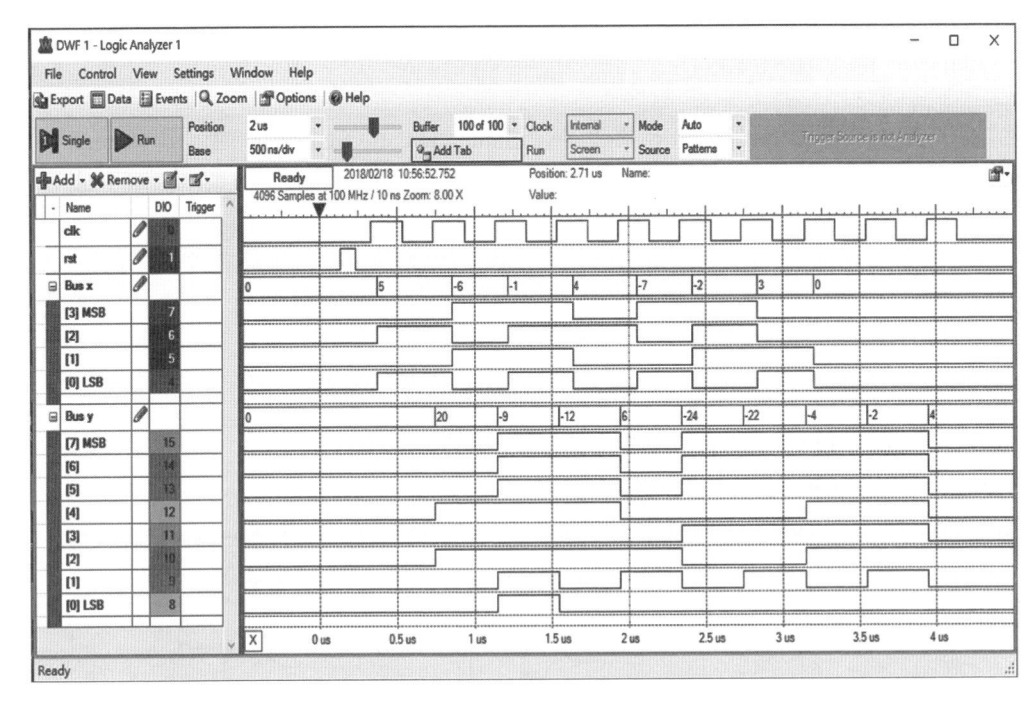

◎ 圖 10S3-3 ◎　　　(single) run Analyzer 的結果與圖 10-5 模擬測試的結果相同

　　本章所舉的例子，顯示了硬體處理 Digital Filter 樣品，實用電路要比較本例子大得多。

10-4　課外練習

(1) 圖 10-2，我們做了 FIR 濾波器的 Simulation 和 Synthesis 並且用 Pattern Generator 和 Logic Analyzer 來做了硬體的測試。這是一個完整的設計，其中濾波器的係數 (coef) 是固定的。對於一個通用濾波器 (可編程係數)，圖 P10-1 中提出了模塊化架構如下：

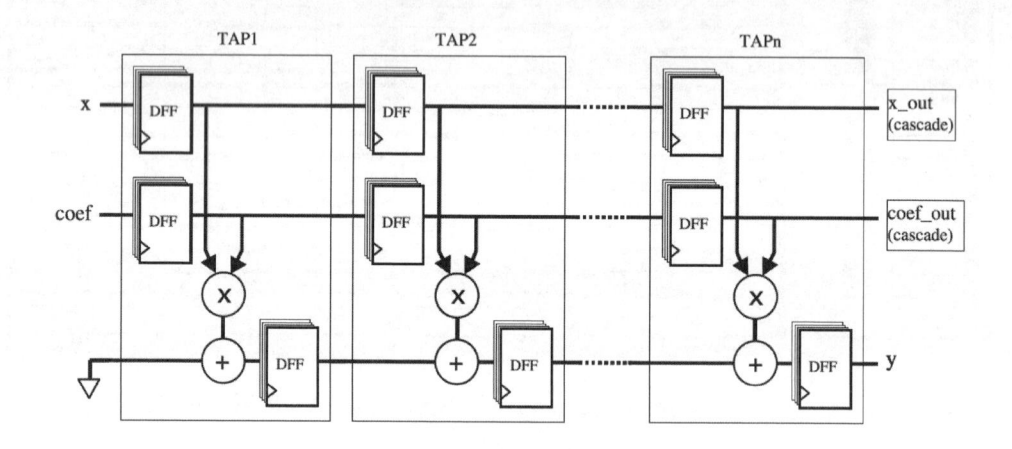

圖 P10-1　通用 FIR 濾波器的構成方塊圖

試寫出該過濾器編寫 VHDL 代碼。不要忘記在設計中包含溢出檢查。考慮從輸入 (x 和 coef) 到乘法器輸入的所有信號的位數是 m，並且從那裡開始是 2 m (即，從乘法器輸出到 y)。還要考慮抽頭 (階段) 的數量是 n。編寫盡可能通用的代碼。然後用 ModelSim 來模擬測試您的電路。

(2) 試列出測試圖 P10-1 的通用 FIR 濾波器的構成方塊圖其 Pattern Generator 和 Logic Analyzer 所需之 I/O 數目。

第十一章 Intel/Altera Qsys 系統與 NIOS-SoC 電路的設計

Altera Qsys 系統，是一個集成工具，用於設計包含處理器、存儲器、輸入/輸出接口、定時器等數位硬件系統。Qsys 工具能使設計者，用圖形來選擇電路庫中，其所需的組件。然後將所有零件連接在一起的 HDL 硬件系統。

11-1 簡介

本章將用一個簡單的例子，把硬體系統的開發流程，使用的 Qsys 工具和 Quartus®II 軟件，系統性地一步一步來說明，過程中的最後一步，為說明如何使用實際的 FPGA 板子，和它的應用程序。假定用戶使用的是 Altera DE 系列開發教育板，連接到安裝有版本 14.1 的 Quartus II 和 Nios®II 軟件的 PC 上；如果使用其他版本的軟件，圖像可能會略有不同。

系統的開發流程其內容如下：

- Nios II 系統
- Altera 的 Qsys 工具
- Nios II 的系統集成到 Quartus II 工程
- 對使用 Qsys 的 Quartus II 工程做編譯
- 使用 Altera 的監控程序，下載所設計的硬件系統，並啟動其應用程序

11-2　Altera DE2-115 教育板

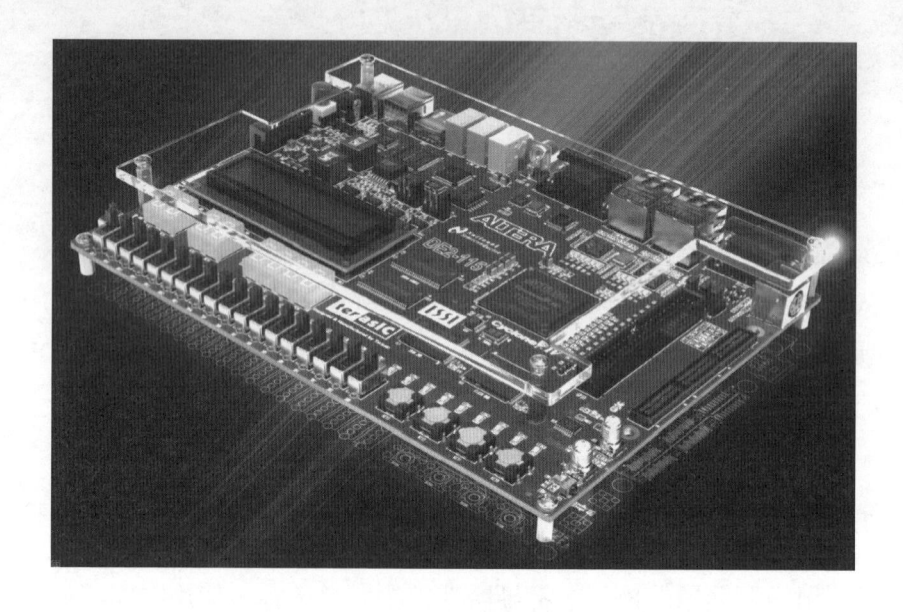

圖 11-1　開發教育板 DE2-115

11-3　數字硬件系統的例子

圖 11-2 為一個簡單的系統例子，其中它包括 Altera 的 Nios®II 嵌入式處理器，它被定義為一個硬件描述語言代碼的軟處理器模塊，是一個簡單的硬件系統。一個 Nios II 的模塊，作為較大系統的一部分，可以藉 Altera FPGA 芯片和 Quartus II 軟件來實現。

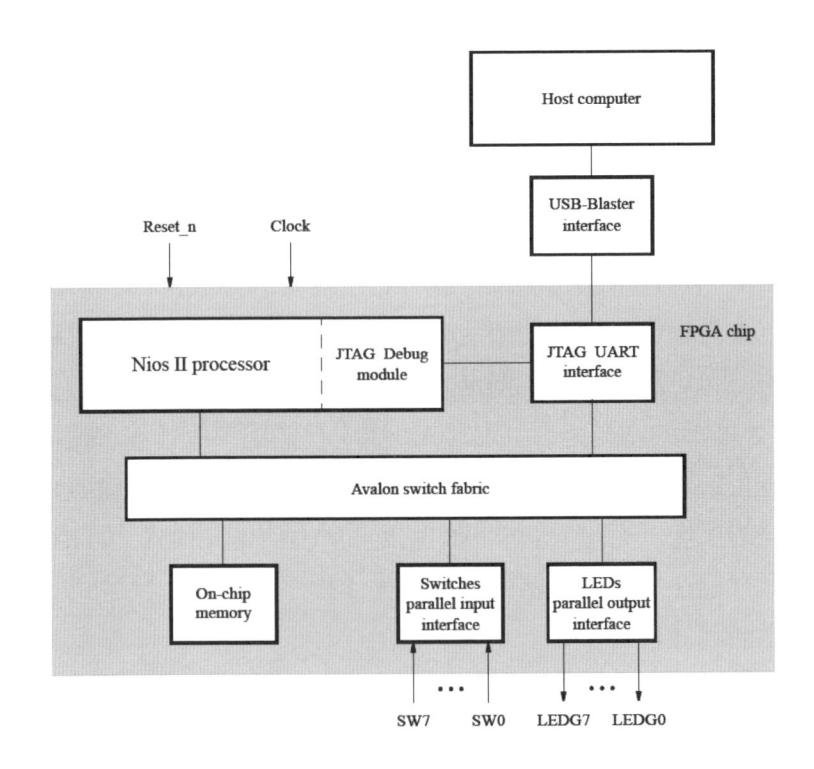

◦ 圖 11-2　　NIOS II 系統的簡單例子

如圖 11-2 中所示，Nios II 處理器是由互連裝置，經由網名為 Avalon 的交換架構，連接到存儲器和 I/O 的接口。此互連網絡是經由 Qsys 的工具所自動產生的。

本例子系統中的存儲器組件，使用的是 FPGA 芯片中的存儲器。Slide Switches 和 LED 的 I/O 是經由 Qsys 工具內的預定模塊來連接。至於 PCUSB 和 DE2-115 間的連接，由特殊的 JTAG UART 接口，經相關的軟件 USB-Blaster 來完成。另一個模塊，稱爲 JTAG Debug 模塊，它讓 PC 可以來控制 Nios II 系統，使得它可以執行包括：下載 Nios II 程序調入內存，啓動和停止這些項目的執行，設置斷點，以及檢查其內容等。

Memory 和 Nios II 的 registers

由於 Nios II 系統的所有部件，都是通過使用硬件描述語言來定義，實現在 FPGA 芯片上，要讓用戶來寫出該系統的代碼，將是一項繁重而耗時的工作。爲了免除用戶在這方面的負擔，Qsys 的工具將展示它能夠簡單地選擇所需的組件，讓每個組件能有它所需要的參數來實現系統所需的總體要求。本教程中，雖然使用 Qsys 設計的是一個非常簡單的系統，但是同樣的方法也可用於設計一個更大的系統。

圖 11-2 所示的系統是要 DE2-115 實現一個簡單的任務。使用它的 8 個 Slide Switches (SW 0~7) 來打開或關閉 8 個綠色 LEDs (LED 0~7)。爲了達到所需的操作，對應於該開關的狀態的八個 BITs 模式，經由 Nios II 處理器執行存儲在片上存儲器中的連續操作程序來進行，發送到所述輸出端口，以激活其 LED。

在下一節中，我們將使用 Qsys 工具來做如圖 11-2 所示的設計，設定 FPGA 的接腳和 DE2-115 上 的開關和 LED 之間的連接後，用編譯來檢查系統的設計。接下來在第十二章，將使用 Altera Monitor Program 來下載所設計的電路到 FPGA 上。並執行執行 Nios II 程序所需的任務。

讀者將從本教程了解到：

- 使用的 Qsys 工具來設計一個基於 Nios II 系統

- 集成設計的 Nios II 系統進入 Quartus II 工程
- 實施 DE2-115 上所設計的系統
- 運行 Nios II 處理器上的應用程序

11-4　Altera 的 Qsys 產生 HDL 的軟體工具

Qsys 的工具須跟 Quartus II CAD 軟件一起使用。用戶可以將設計做在 Nios II 處理器的系統中，通過簡單地選擇所需要的功能單元，並指定它們的參數。為了實現該系統存在於圖 11-4 中，我們必須初始化以下的功能單元：

- Nios II 處理器
- FPGA 芯片的存儲器，將指定一個 4 KB/32 BITs 的內存設置
- 兩個並行 I／O 接口
- JTAG UART 接口與 PC 主機之通訊

使用 Qsys 的工具來對 NIOS 系統的設計，應從 Quartus II 軟件入手。使用 Altera/Quantus II 軟體來合成 NIOS 系統的 S1 的步驟，請參照第一章 1-6 節 Software Synthesis ALU_simple 電路的例子。它們之間的第一階段 S1 除了 Project 的名稱不同外，其它的完全相同。如圖 11-3A 和圖 11-3B 所示。

實作使用：	Family	Device Name
DE2-115	Cyclone IV E	EP4CE115F29C7
DE1-SoC	Cyclone V	5CSEMA5F31C6
DE0-Nano	Cyclone IV E	EP4CE22F17C6
DE0-Nano-SoC	Cyclone V	5CSEMA4U23C6

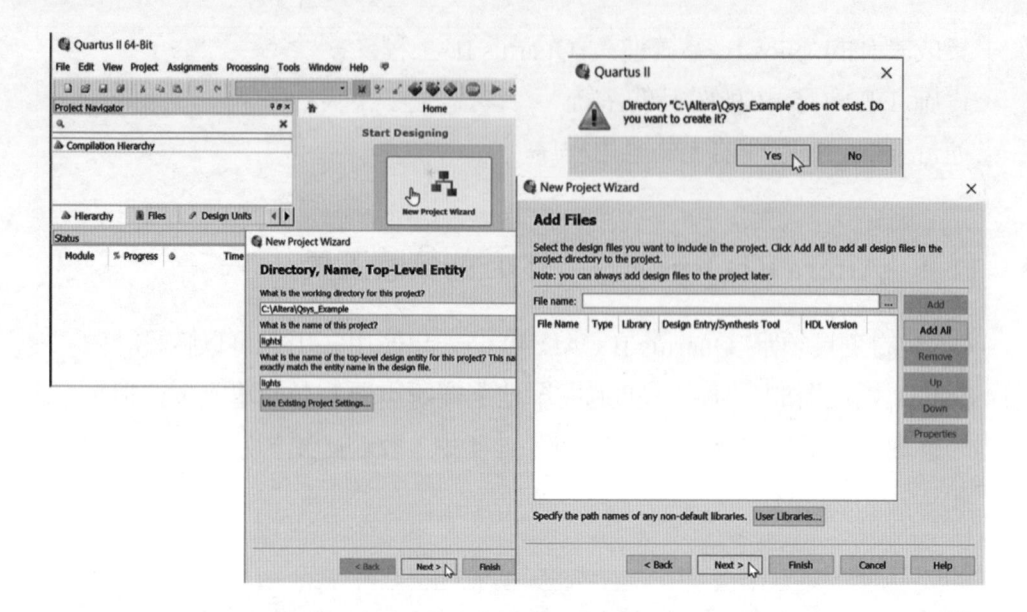

☞ 圖 11-3A ☜　　Quartus II Project 的 Directory, Name, Top-Level Entity 定名

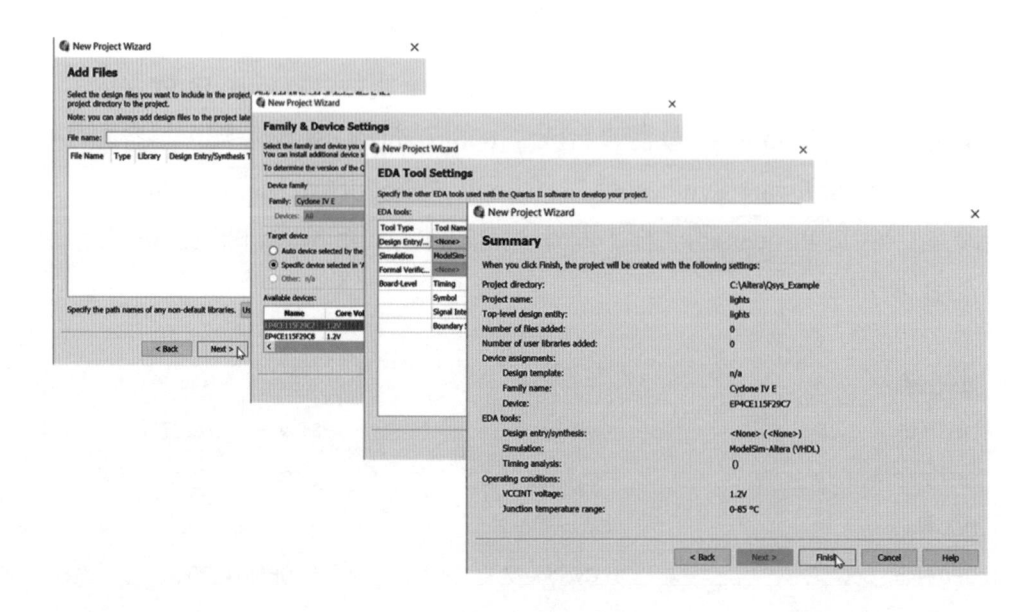

☞ 圖 11-3B ☜　　用於 Quartus II Project 的 FPGA Family 和 Device 名稱

第十一章　Intel/Altera Qsys 系統與 NIOS-SoC 電路的設計　247

接下來是在 QUARTUS II 64-bit Project 視窗選用 **Tools > Qsys**。結果將有 Qsys 視窗的出現，如圖 11-4 所示。

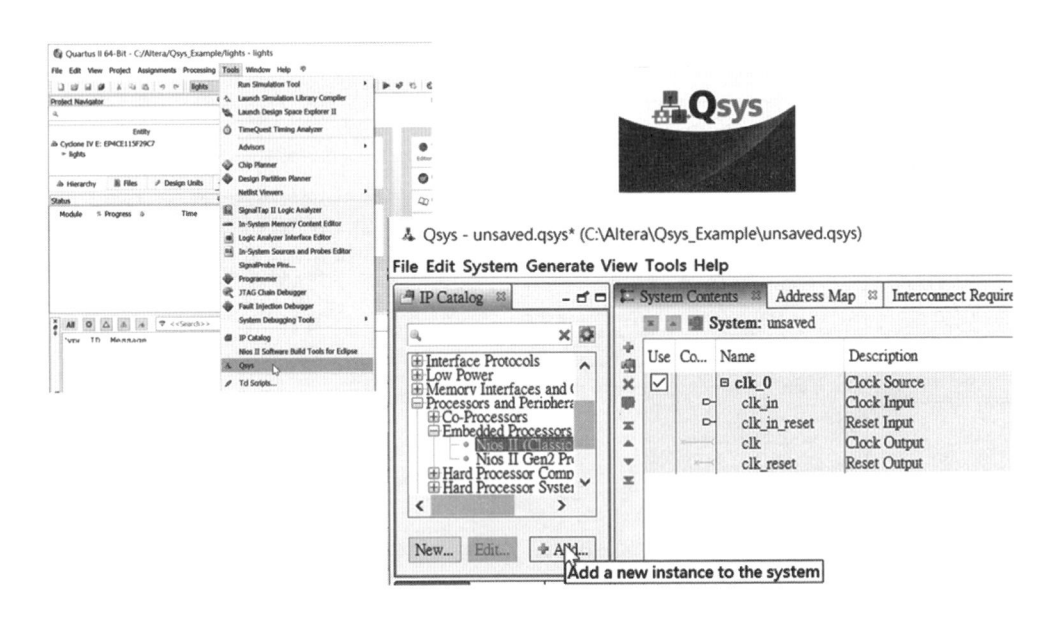

圖 11-4　Quartus Project 視窗選用 Tools > Qsys 的結果

在 Qsys 視窗的 IP Catalog，選用 **Processor and Peripherals > Embedded > Processor > Nios II(Classic) Processor**。點選 **Nios IIe**，如圖 11-5A 所示。

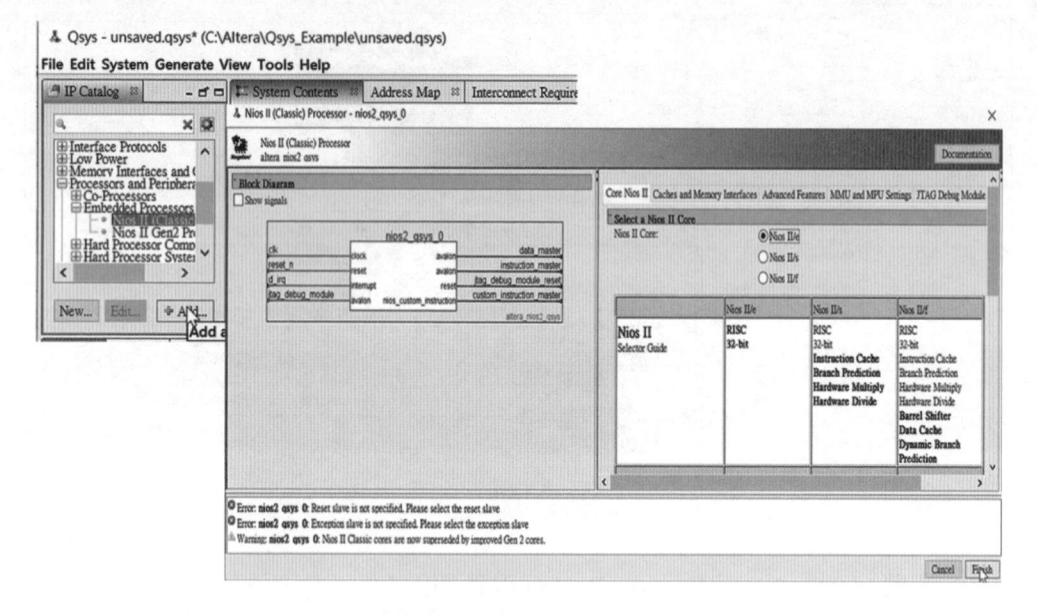

◦ 圖 11-5A　　　選用 Nios II(Classic) Processor 中的 Nios IIe

第十一章　Intel/Altera Qsys 系統與 NIOS-SoC 電路的設計　249

圖 11-5B 為 Nios IIe 加入到 C:\Altera\Qsys_Example\unsaved.qsys 中，系統所提供的名稱為 nios2_qsys_0。

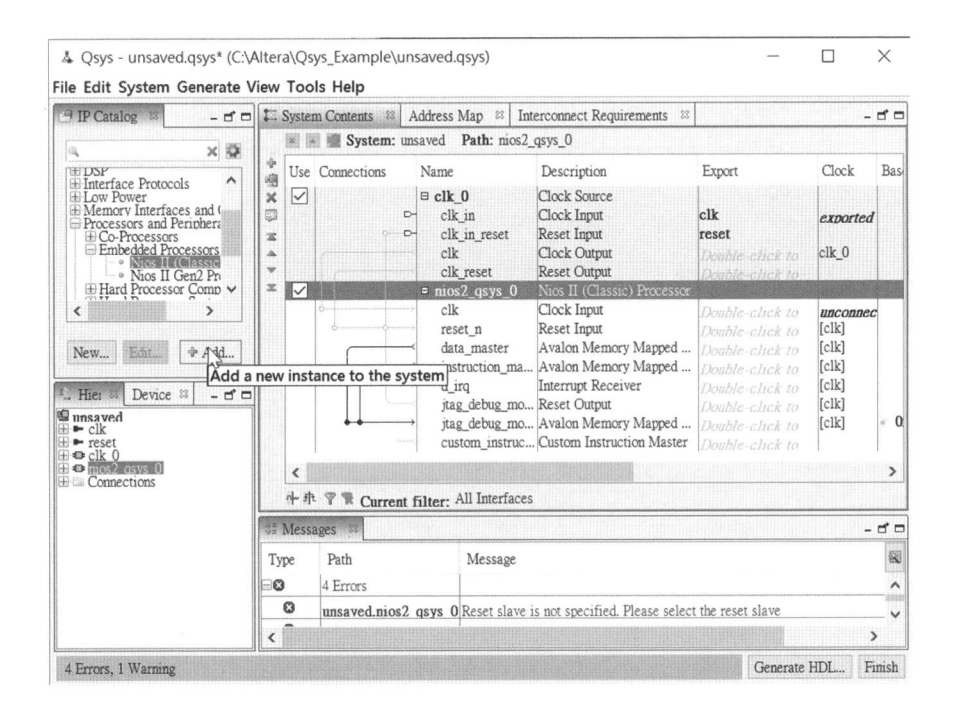

◎ 圖 11-5B　　　Nios IIe 加入到 Qsys 系統所提供的名稱為 nios2_qsys_0。

接下來從 Basic Functions 的 On Chip Memory 中,選用 **On-Chip Memory**。如圖 11-6A 所示。存儲器的大小,選用 4,096 bytes。

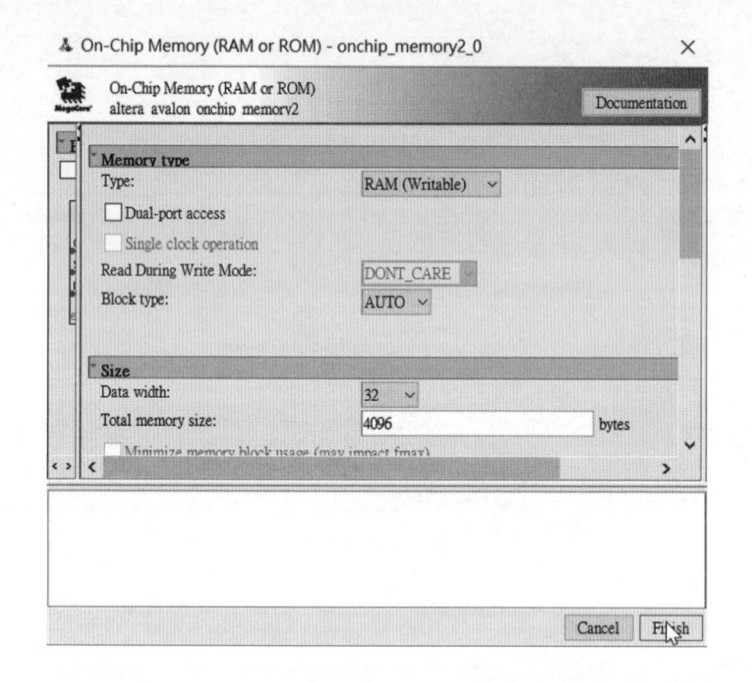

圖 11-6A　　在 **Qsys** 視窗的 **IP Catalog** 中選用 **On-Chip Memory**

第十一章　Intel/Altera Qsys 系統與 NIOS-SoC 電路的設計　251

On-Chip Memory 加入到 Qsys 系統所提供的名稱為 onchip_memory2_0，如圖 11-6B 所示。

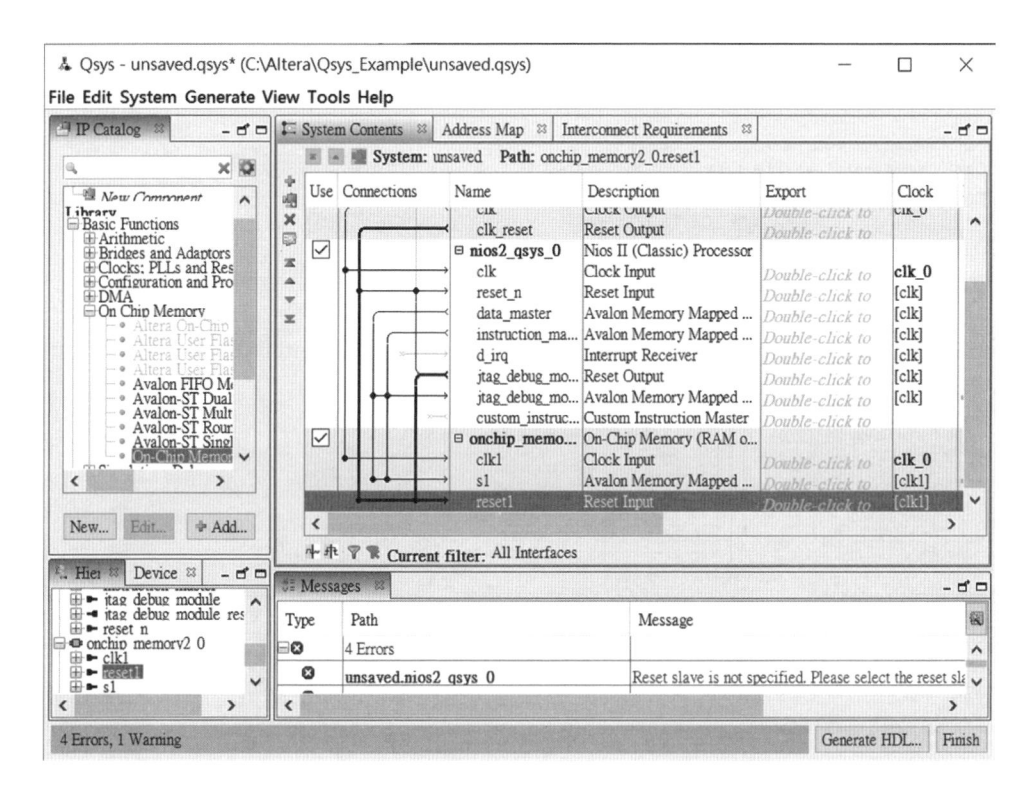

◎ 圖 11-6B　On-Chip Memory 加入到 Qsys 系統所提供的名稱為 onchip_memory2_0

再從 Processor and Peripherals 的 Peripherals 中選取 PIO(Parallel)。共選二組，一組為 8 bits Input，另一組為 8 bits Output。如圖 11-7A 所示。

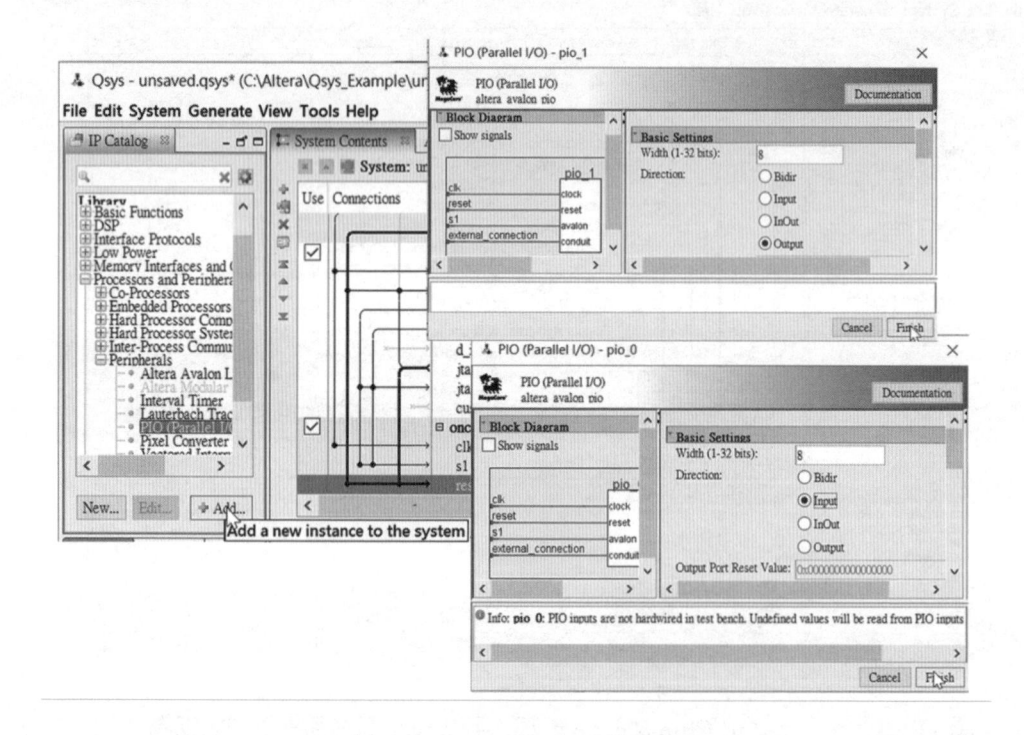

◎ 圖 11-7A　　　在 Qsys 視窗的 IP Catalog 中選用二組 PIO(Parallel)

第十一章　Intel/Altera Qsys 系統與 NIOS-SoC 電路的設計　253

二組 PIO 加入到 Qsys 系統所提供的名稱為 pio_0 和 pio_1，如圖 11-7B 所示。

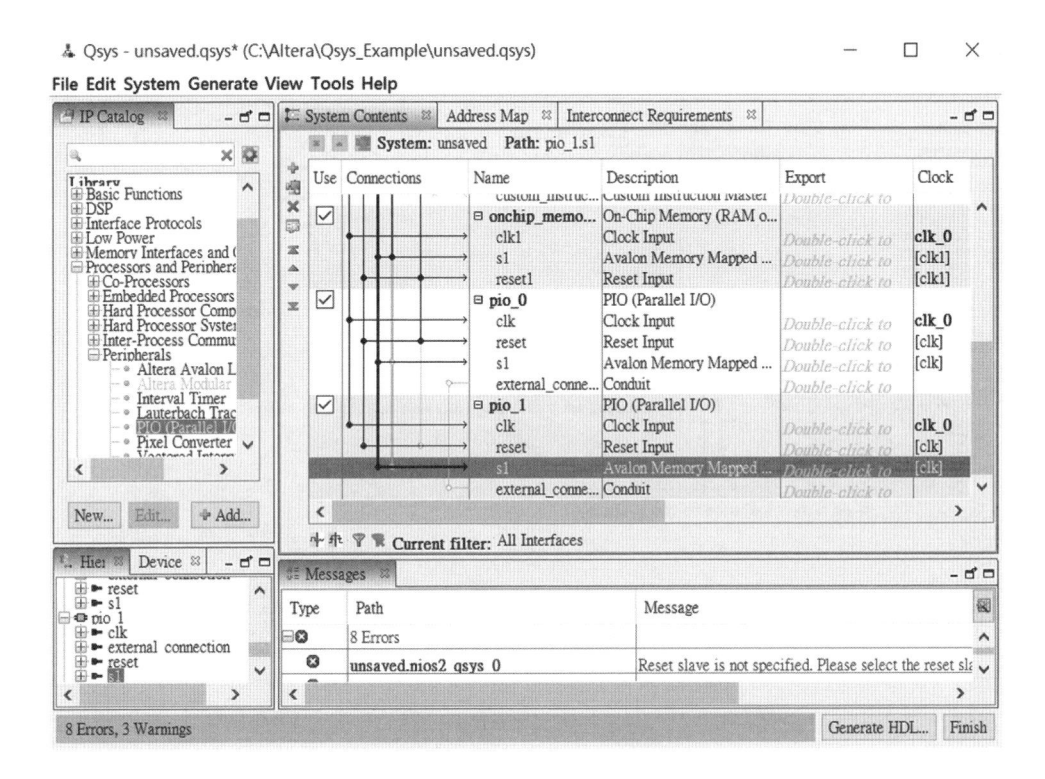

◎ 圖 11-7B ◎　　　二組 PIO(Parallel) 顯示在 System Contents 中

最後還要在 IP Catalog 中的 Interface Protocol 的 Serial 中選用 JTAG UART，如圖 11-8A 所示。

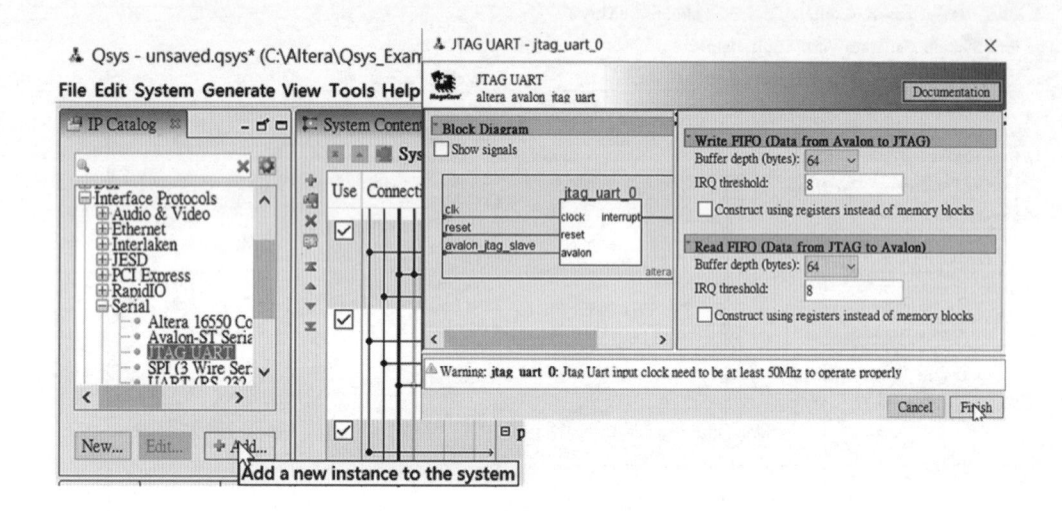

◦ 圖 11-8A　在 Qsys 視窗的 IP Catalog 中選用 JTAG UART

第十一章　Intel/Altera Qsys 系統與 NIOS-SoC 電路的設計　255

圖 11-8B 為 JTAG UART 顯示在 Qsys 系統的 System Contents 中。

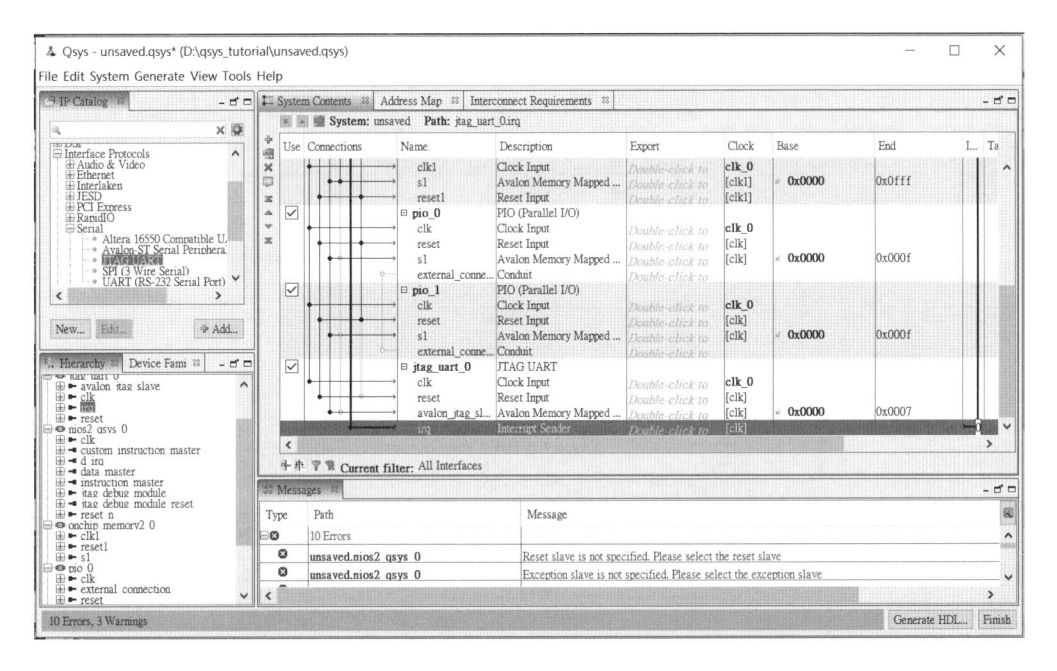

☜ 圖 11-8B ☞　　　JTAG UART 顯示在 System Contents 中

System Content 中 Name 是可以修改的，只要用滑鼠指著該 Name 再單擊滑鼠右鍵，得圖 11-9 選取其中的 Rename 來改之。

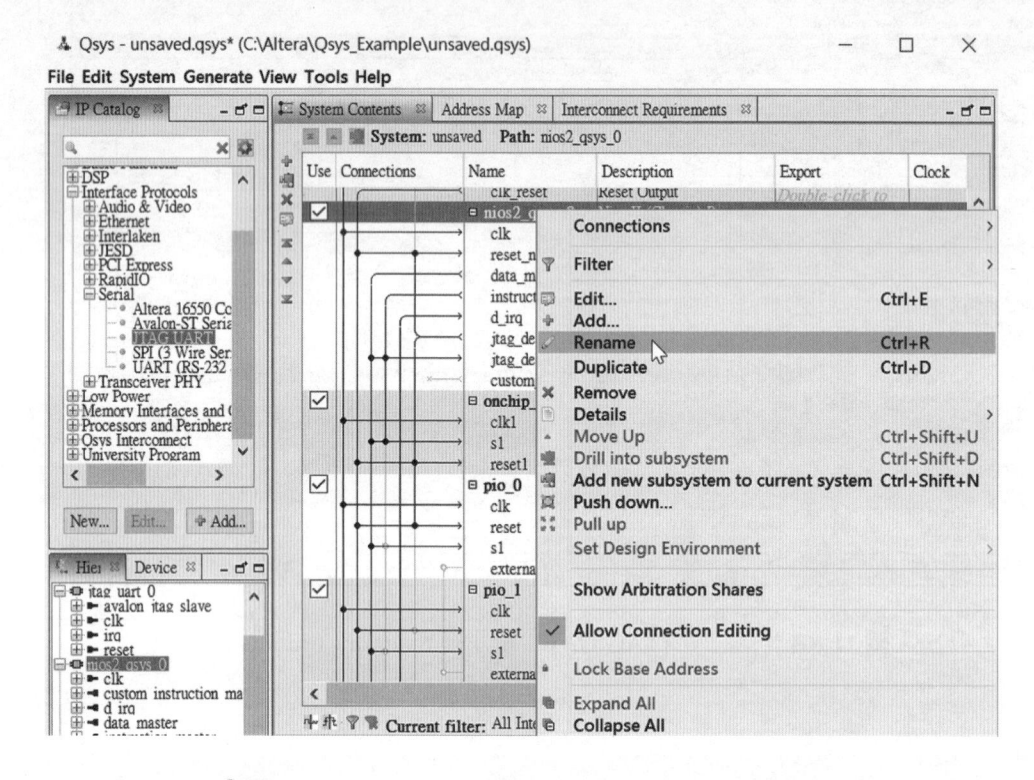

圖 11-9　System Contents 中 Name 之修改

第十一章　Intel/Altera Qsys 系統與 NIOS-SoC 電路的設計　257

如圖 11-10 將 nios_qsys_0 改為 nios2_processor，onchip_memory2_0 改為 onchip_memory，pio_0 改為 switches，pio_1 改為 LEDs。

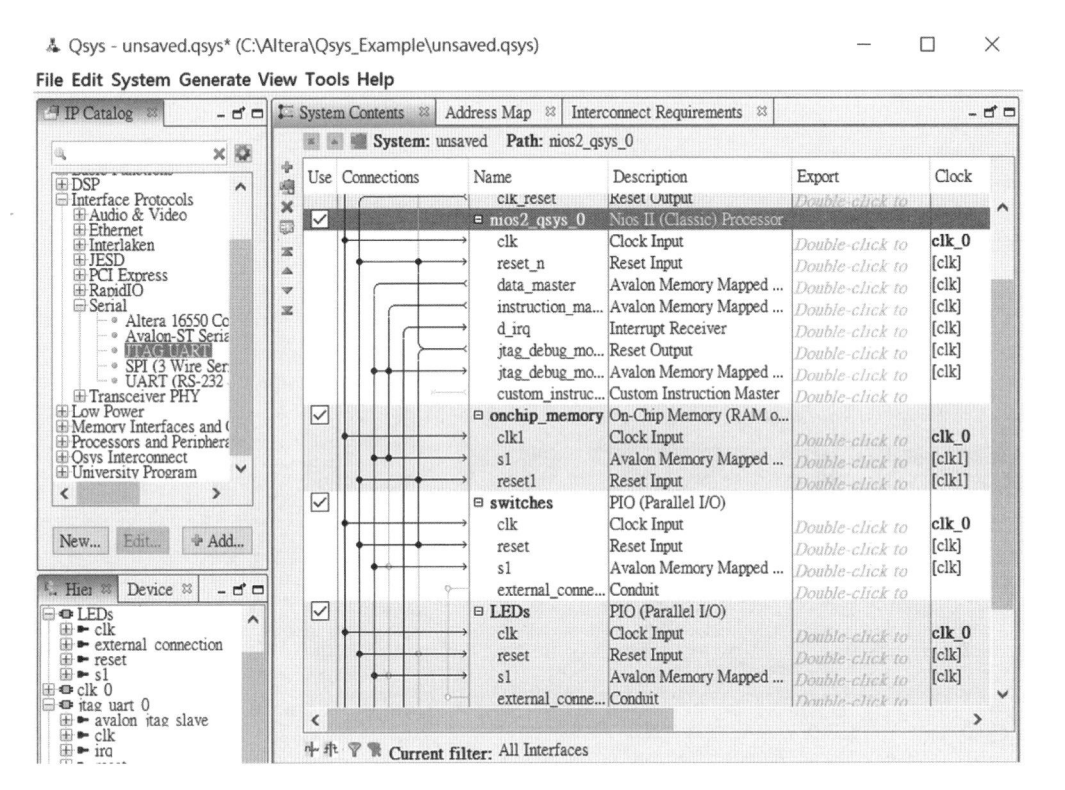

◎ 圖 11-10 ◎　　　　System Contents 中 4 個 Name 之修改

然後將滑鼠指著 nios2_processor 的 Base address，雙擊並修改為 0x0000。同時把 onchip_memory 的 Base address 改為 0x0000，並且將它**鎖住** (Lock)。如圖 11-11 所示。

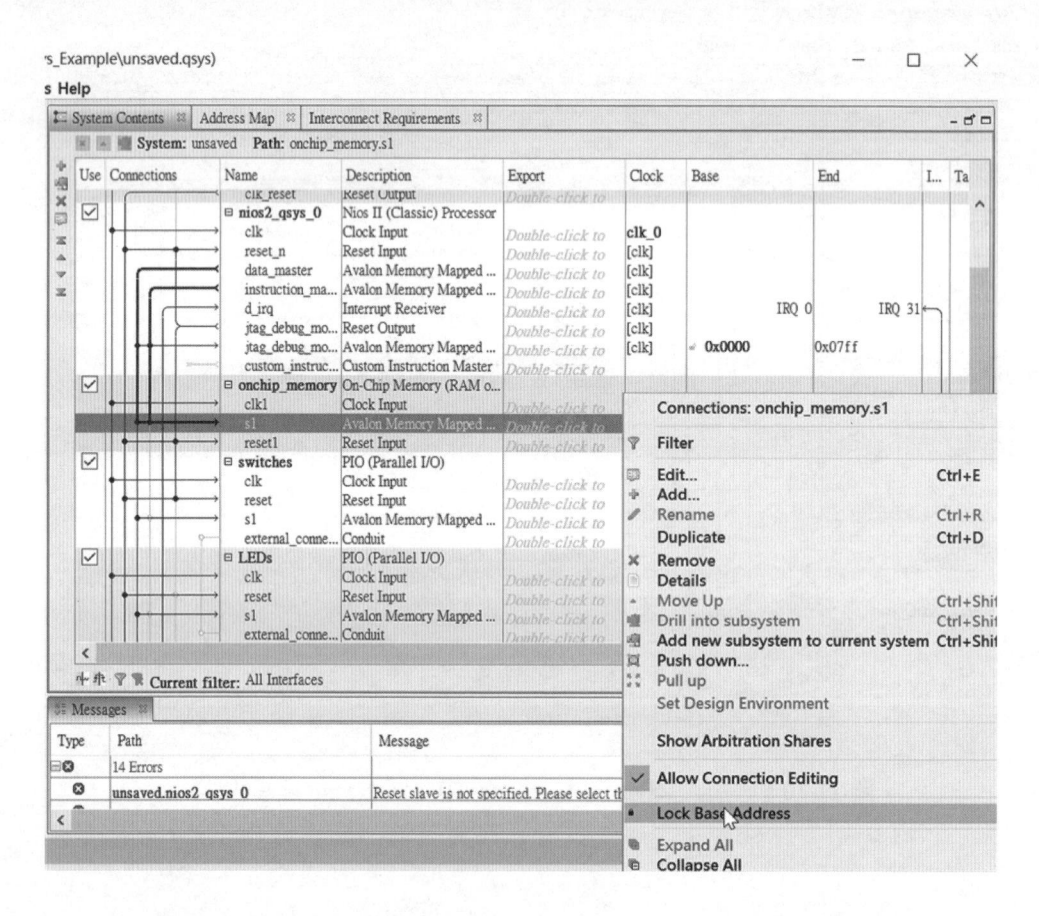

圖 11-11 System Contents 中 Base Address 的修正與鎖住

第十一章　Intel/Altera Qsys 系統與 NIOS-SoC 電路的設計

接下來讓 Qsys 用 System > Assign Base Addresses 來設定其它各組件的 Address，如圖 11-12 所示。onchip_memory 因 clock 而保持 0x0000 不變，nios2_qsys_0 的 onchip_memory 的 address 則從 0x0000 被設定為 0x1800。

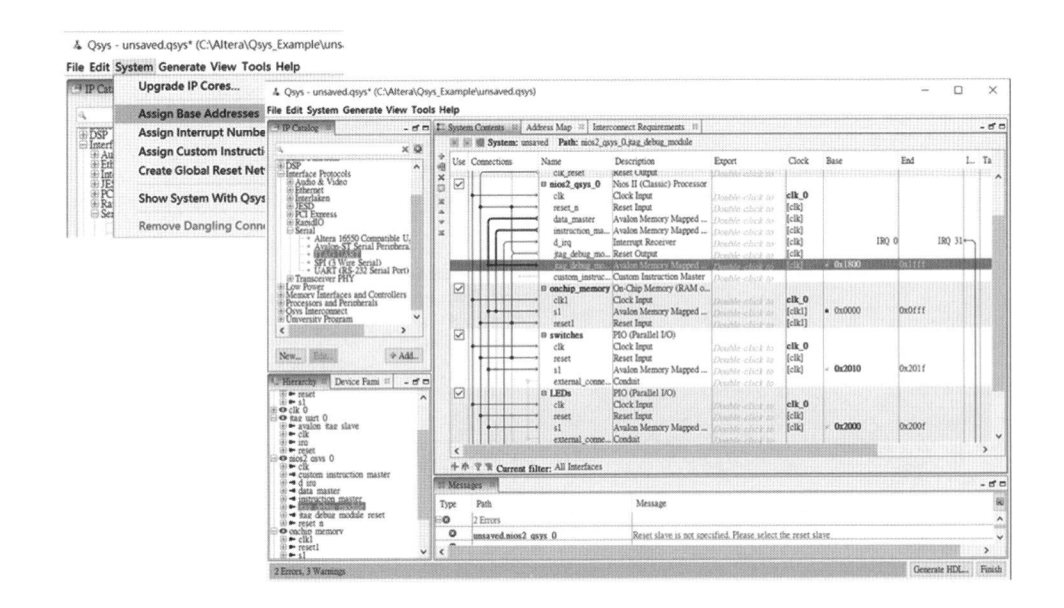

圖 11-12　Qsys 設定其它各組件的 Address

Nios II Processor 當被 reset 時,它的 reset vector 存在於 memory 的 Reset Vector 和 memory 的 Exception Vector 中,因此可用滑鼠指著 nio2_processor 並雙擊滑鼠左鍵,得圖 11-13。將 Reset Vector Memory 和 Exception Vector Memory 的選項全部為 onchip_memory.s1。如圖 11-13 所示。

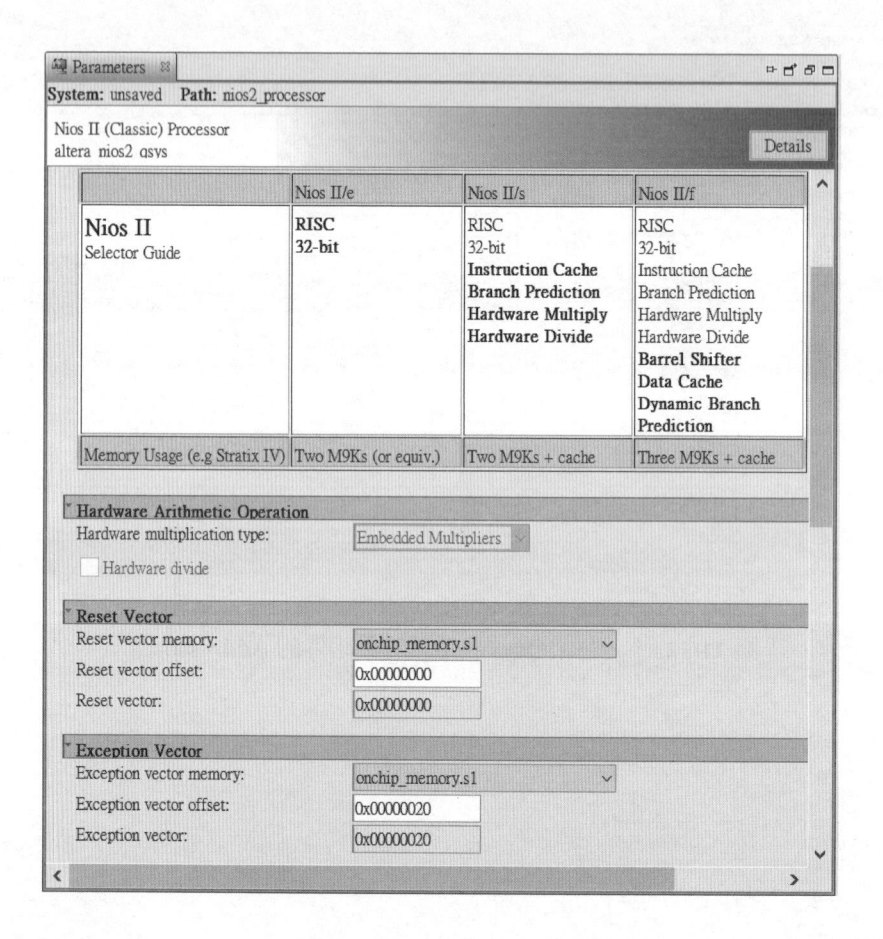

☆ 圖 11-13 ☆　　　nio2_processor 的 Reset Vector 和 Exception Vector 的設定

第十一章　Intel/Altera Qsys 系統與 NIOS-SoC 電路的設計

到目前為止，已經完成了 nios_system 電路的內部連線。現在必須對連接到外部的 Switches 和 LEDs 部加以設定。如圖 11-14 所示，請雙擊在 Switches 和 LEDs 的 external_connection 和 Export 相交叉處，結果產生了二個信號：swtches_external_connection 和 leds_external_connection。

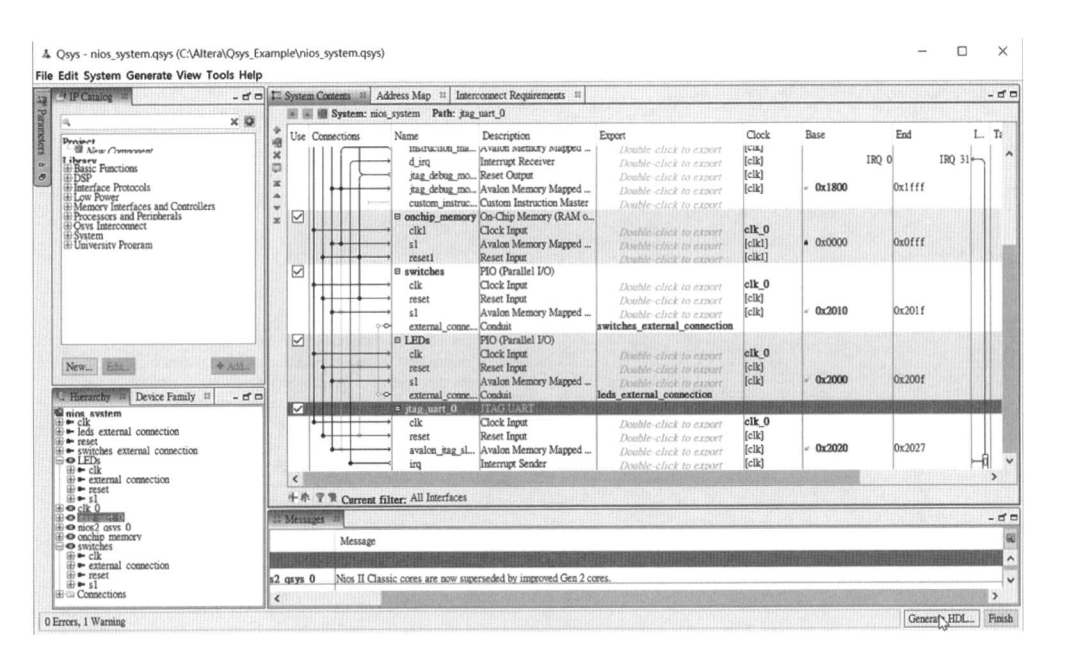

圖 11-14　nios_system 電路連接到外部的 Switches 和 LEDs 信號的設定

請在做下個操作之前，把這個 nios_system 電路，儲存起來。然後再按照圖 11-15 的 Qsys 選用 **Generate > Generate HDL**，在 Create HDL design file for synthesis 選用 VHDL。然後單擊 **Generate**，當 Generate completed 出現時，必須 Error Free。

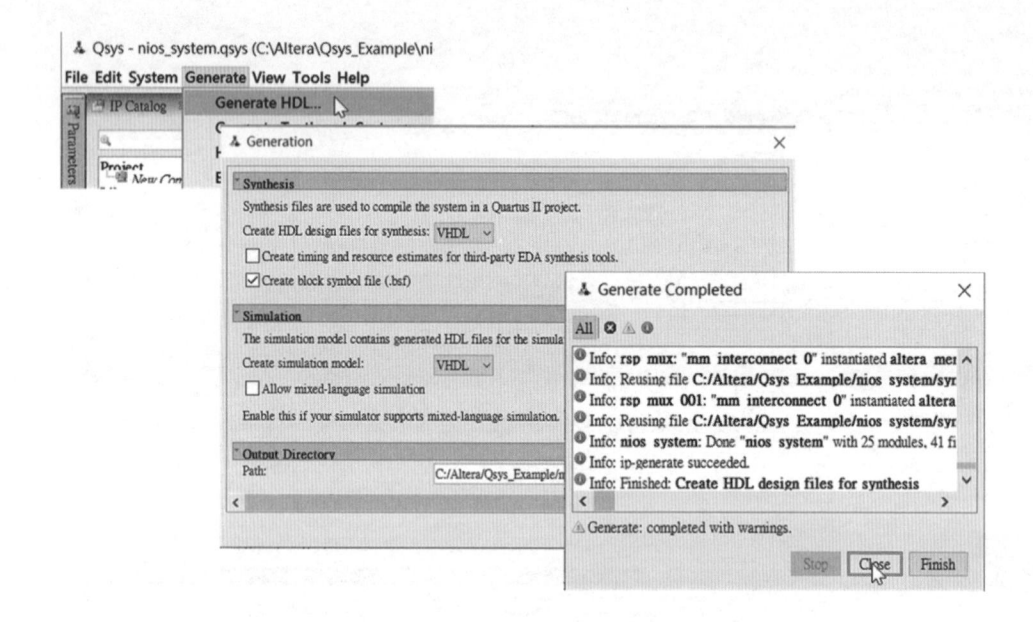

◦圖 11-15◦　　Qsys 產生 nios_system.vhd 檔

nios_system.vhd 儲存在 C:/Altera/Qsys_Example/nios_system/synthesis/ 中。共 51 KB、625 Lines。

11-5　如何將 Nios II 系統納入 Quartus II Project

為了完成硬體的設計，須做下列 4 個步驟：

- 啟用由 Qsys 所產生之模塊，加入到 Quartus II project 中。
- FPGA pin 接腳的設定。
- 編輯所設計的電路檔。
- 計劃和配置在 DE2-115 板子上的 FPGA

圖 11-16 所示為 top-level VHDL 模塊，啟用了 Nos II system，名為 lights.vhd

```vhdl
-- lights.vhd (top-level VHDL module)
-- CLOCK_50 is the system clock; KEY0 is the active-low system reset
-- I/Os: SW7-0 are parallel port inputs; LEDG7-0 are parallel port outputs
LIBRARY ieee;
USE ieee.std_logic_1164.ALL;
USE ieee.std_logic_unsigned.ALL;
-----------------------------------------------------------
ENTITY lights IS
PORT ( CLOCK_50 : IN STD_LOGIC;
            KEY : IN STD_LOGIC_VECTOR (0 DOWNTO 0);
             SW : IN STD_LOGIC_VECTOR (7 DOWNTO 0);
           LEDG : OUT STD_LOGIC_VECTOR (7 DOWNTO 0));
END lights;
-----------------------------------------------------------
ARCHITECTURE lights_rtl OF lights IS
    COMPONENT nios_system
        PORT (
            SIGNAL clk_clk: IN STD_LOGIC;
            SIGNAL reset_reset_n : IN STD_LOGIC;
            SIGNAL switches_external_connection_export : IN STD_LOGIC_VECTOR (7 DOWNTO 0);
            SIGNAL leds_external_connection_export : OUT STD_LOGIC_VECTOR (7 DOWNTO 0)
        );
    END COMPONENT;
BEGIN
NiosII : nios_system
    PORT MAP(
        clk_clk => CLOCK_50,
        reset_reset_n => KEY(0),
        switches_external_connection_export => SW(7 DOWNTO 0),
        leds_external_connection_export => LEDG(7 DOWNTO 0));
END lights_rtl;
```

圖 11-16　啟用了 Nos II system 模塊的 lights.vhd

在 Quartus II 64-Bit project window 上選 Assignments > Setting 將 lights.vhd 匯入 Project 中,如圖 11-17 所示。

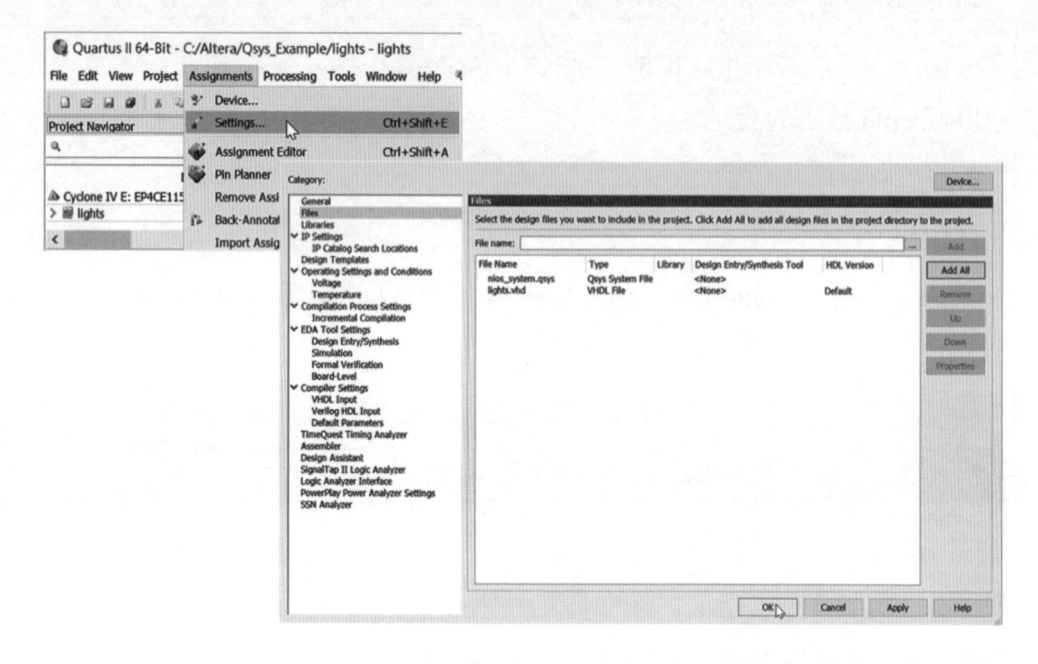

圖 11-17　Assignments > Setting 將 lights.vhd 匯入 Project 中

第十一章　Intel/Altera Qsys 系統與 NIOS-SoC 電路的設計

在尚未 Assign Pins 之前先用編輯來查錯，如圖 11-18 所示，結果 OK。

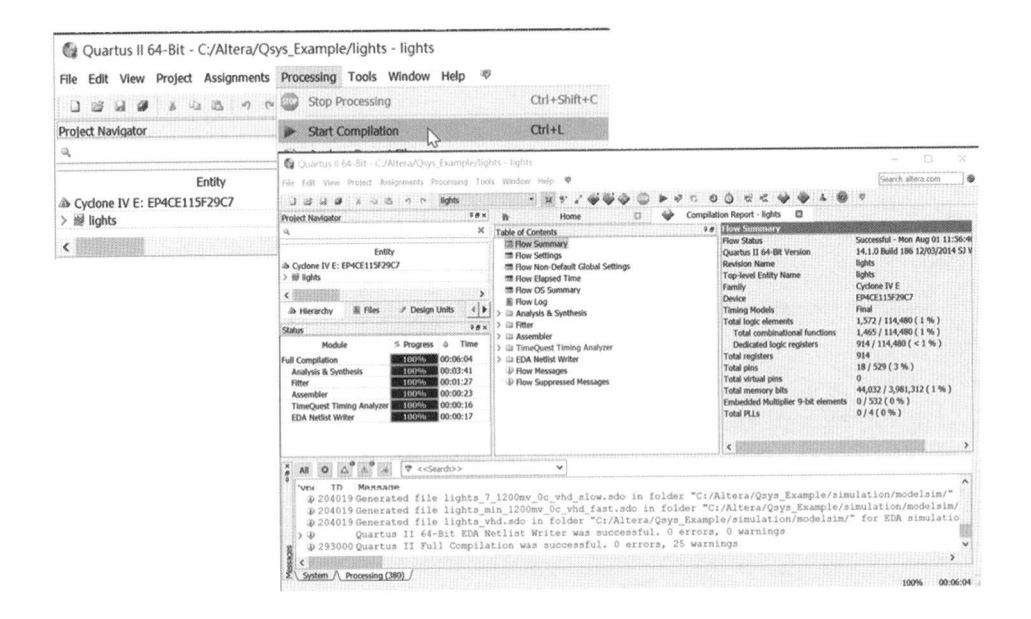

◦ 圖 11-18 ◦　　　尚未 Assign Pins 之前先用編輯來查錯

在 Quartus II 64-Bit project window 上選 **Assignments > Assignment Editor** 來完成 FPGA 的 Pin Assignments 如圖 11-19 所示。

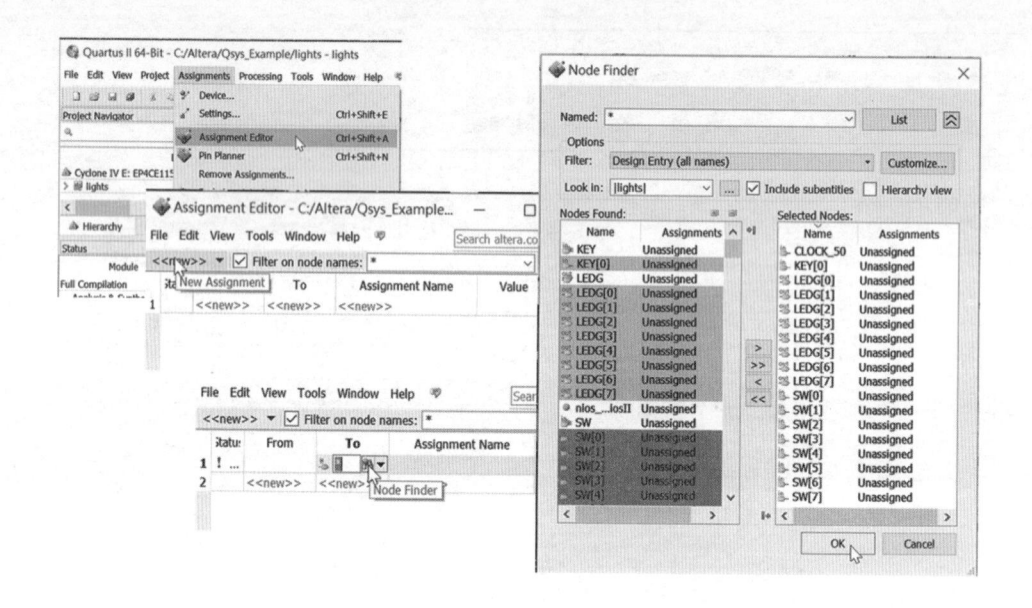

圖 11-19　用 Assignment Editor 來完成 FPGA 的 Pin Assignments

第十一章　Intel/Altera Qsys 系統與 NIOS-SoC 電路的設計　267

Pin Assignments 完成之後，要將它儲存起來，如圖 11-20 所示。

Status	From	To	Assignment Name	Value	Enabled	Entity
1 ✓		CLOCK_50	Location	PIN_Y2	Yes	
2 ✓		KEY[0]	Location	PIN_M23	Yes	
3 ✓		LEDG[0]	Location	PIN_AB28	Yes	
4 ✓		LEDG[1]	Location	PIN_AC28	Yes	
5 ✓		LEDG[2]	Location	PIN_AC27	Yes	
6 ✓		LEDG[3]	Location	PIN_AD27	Yes	
7 ✓		LEDG[4]	Location	PIN_AB27	Yes	
8 ✓		LEDG[5]	Location	PIN_AC26	Yes	
9 ✓		LEDG[6]	Location	PIN_AD26	Yes	
10 ✓		LEDG[7]	Location	PIN_AB26	Yes	
11 ✓		SW[0]	Location	PIN_E21	Yes	
12 ✓		SW[1]	Location	PIN_E22	Yes	
13 ✓		SW[2]	Location	PIN_E25	Yes	
14 ✓		SW[3]	Location	PIN_E24	Yes	
15 ✓		SW[4]	Location	PIN_H21	Yes	
16 ✓		SW[5]	Location	PIN_G20	Yes	
17 ✓		SW[6]	Location	PIN_G22	Yes	
18 ✓		SW[7]	Location	PIN_G21	Yes	
19	<<new>>	<<new>>	<<new>>			

圖 11-20　　完成 Pin Assignments

然後再用編輯來查錯，如圖 11-21 所示，結果必須是 0 errors！

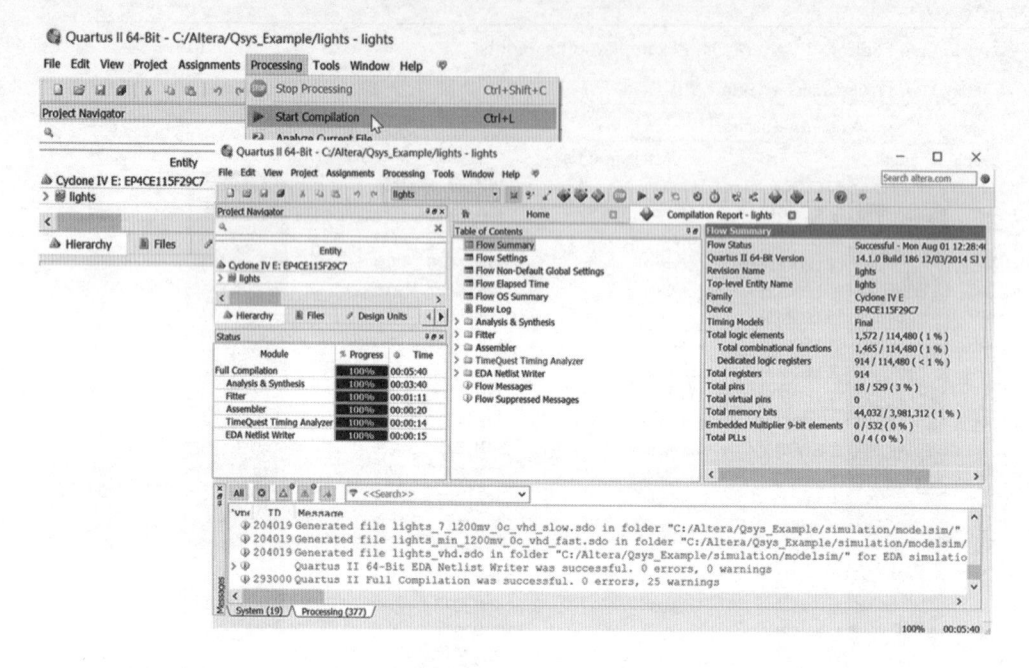

◎ 圖 11-21 ◎　　完成 Pin Assignments 再用編輯來查錯

這時候在 C:/Altera/Qsys_Example/output_files 將見到 lights.sof 檔的存在。如圖 11-22 所示。

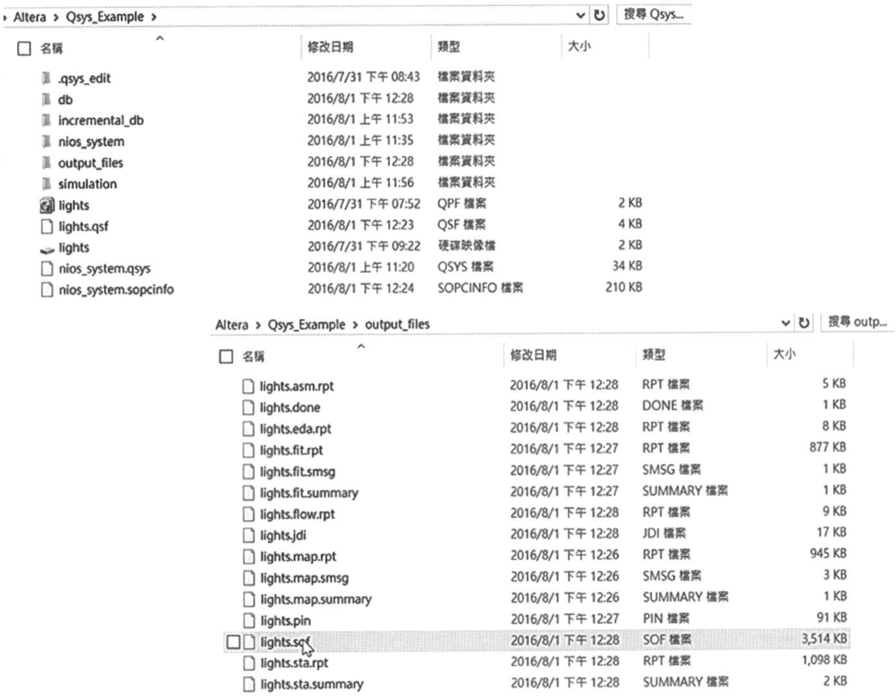

圖 11-22　lights.sof 檔在 output_files 中

Lights.sof 這個 Computer 雖然可以經由 Quartus II 的 Tools > Programmer 下載到 DE2-115 的 FPGA，但是一個 Memory 空白的 Computer，是毫無用處的。Computer 的 Memory 必須裝入適當的軟件，才能夠工作。如何將軟件裝入 Memory？它須要另外一個叫做 Monitor Program[1] 的軟件，將在第十二章中詳細介紹。

---

[1] Monitor Program 14.1 可從以下 download：
https://www.altera.com/support/traning/uiversity/materials-software.html

11-6 課外練習

(1) 試從圖 11-2：NIOS II 系統的方塊圖，和 nios_system.vhd，及圖 11-16：Nios II system 模塊的 lights.vhd。說明 lights project 的產生和完成的步驟。

(2) 在 DE2-115 板子上用 Qsys 建立的 nios_system.vhd，可否用於 DE0-Nano-SoC 上？試以本章的 Qsys_Example 為例，詳細說明其可否之實況。

第十二章 Intel/Altera "監控程序" 的介紹

12-1 簡介

　　Altera 的**監控程序** (Monitor Program) 可以讓使用者輕鬆地編譯和調試 ASM 語言和 C 語言程序。它顯示程序被執行時，如中央處理器、寄存器和系統存儲器的內容及 Nios II 處理器的狀態。它允許用戶通過程序設置斷點，和單一步驟等。

　　Altera 的 Quartus II Program 包含著 Quartus II、Nios II、Qsys 等 Programs。但不包含 Monitor Program，因此必須從網路上下載[註1]，而且下載的版本，必須與 Quartus II Program 的版本完全相同。

12-2 使用 Altera 的監控程序，來下載所設計完成的電路，再啟動應用程序

　　第十一章所設計完成的 Lights.sof，這個 Computer 電路，雖然可以經由軟件 Quartus II 的 Tools > Programmer 下載到 DE2-115 的 FPGA 硬體上。但是更簡單的方法，是使用 "監控程序" 直接下載電路到 FPGA 硬體上，同時啟動電路的應用程序。

　　由 QSYS 工具所產生的 I/O 並行接口，可以由儲存器來控制。依據

註1　Monitor Program 14.1 可從以下 download：
　　https://www.altera.com/support/traning/uiversity/materials-software.html

PIO 的設定，儲存器可能有 4 個之多，其中的一個儲存器稱為 "數據儲存器"。當 PIO 被設定為輸入接口，"數據儲存器" 的數據被讀到，便是讀到 PIO 輸入線上的數據。而當一個 PIO 的設定為輸出接口，數據被 Nios II 處理器寫到 "數據儲存器"，也就是來到 PIO 的輸出。又當一個 PIO 的設定為雙向接口時，那輸入和輸出合用同一組線，在這種情況下，就需要增加一個 "數據方向儲存器"，由它來決定輸入/輸出方向的轉換。在 Lights 這個例子，它是單一方向的 PIOs，祇須要一個 "數據儲存器"。Qsys 指定 switches 的 "數據儲存器" 地址為 0x00002010，LEDs 的 "數據儲存器" 地址為 0x00002000，如圖 12-1 所示。

我們的任務很簡單，祇要把 8 個滑動開關的設定圖案，顯示到 8 個 LEDs 上。我們首先將使用 Nios II 組合語言，然後再用 C 語言來展示它們的做法。

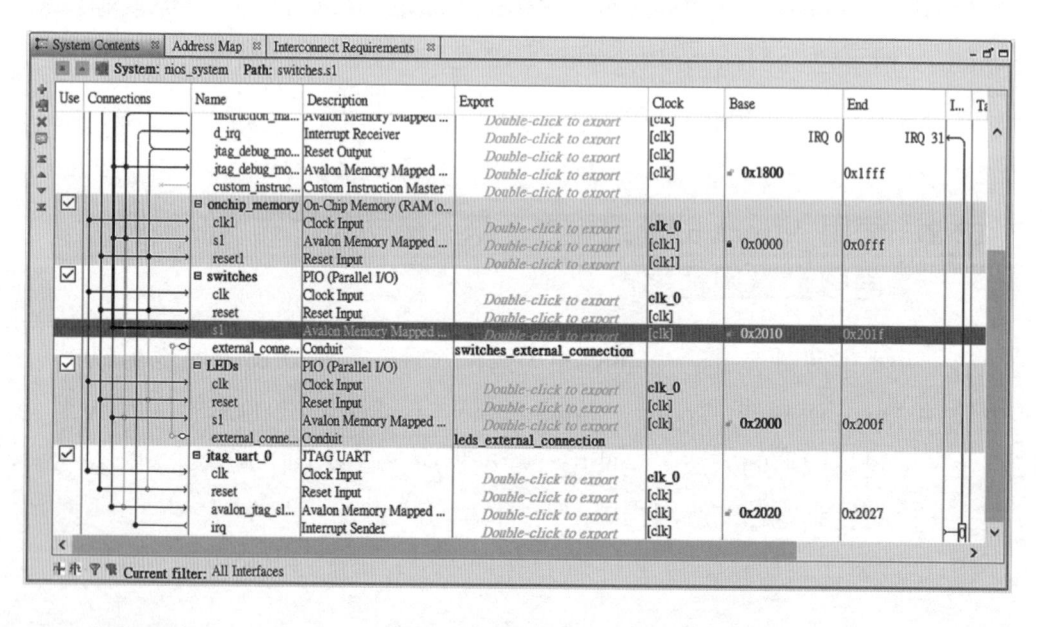

圖 12-1　Qsys 所指定 switches 和 LEDs 的 Data Register Adderss

12-3 Nios II 的組合語言程式

圖 8-2 為 Nios II 的組合語言程式，該程式加載 2 個 "數據儲存器" 的地址到 2 個 PIOs 到 Nios II 處理器的 registers r2 跟 r3。然後進入一個 "無限循環"，為的是可以隨時從 PIO/switches 轉移數據到 PIO/LEDs 上。

```
1  .equ switches, 0x00002010
2  .equ leds, 0x00002000
3  .global _start
4  _start: movia r2, switches
5          movia r3, leds
6          LOOP: ldbio r4, 0(r2)
7          stbio r4, 0(r3)
8          br LOOP
9  .end
```

◎ 圖 12-2 　用以控制 Lights 的組合語言程式

雙擊 Altera Monitor Program icon 如圖 12-3 所示，當有 Altera Monitor Program 視窗的出現。再選用 **File > New Project**。

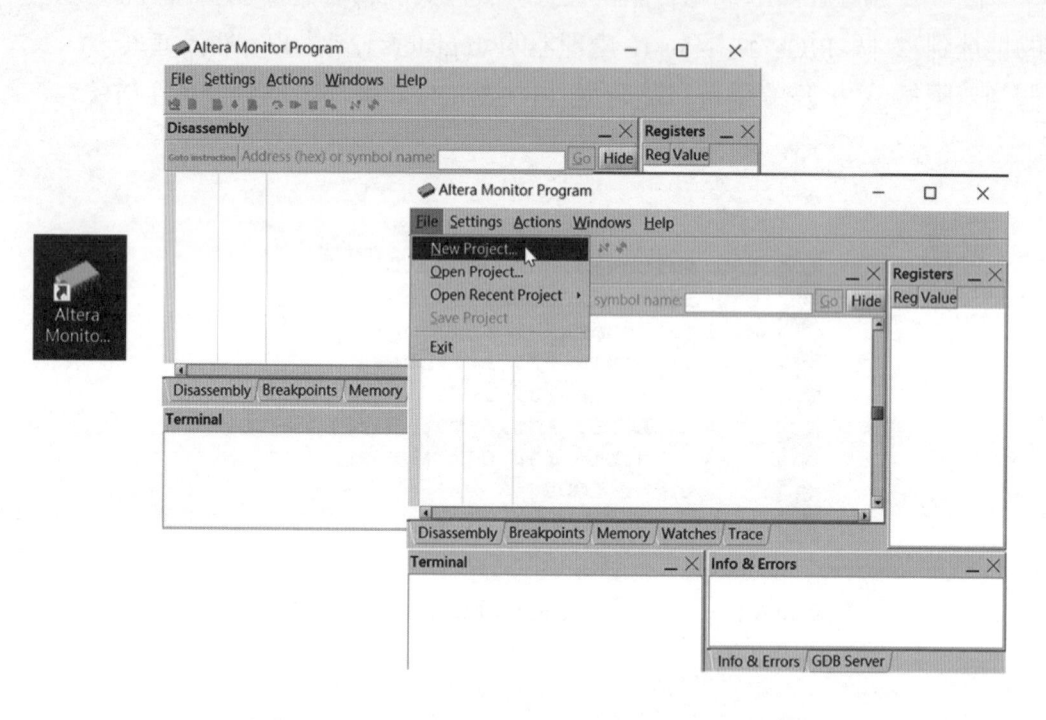

◉ 圖 12-3 ◉　　Altera Monitor Program 視窗

第十二章　Intel/Altera "監控程式" 的介紹　275

　　單擊 Altera Monitor Program 的 File > New Project 就有 New Project Wizard 視窗的出現。指定 C:\qsys_Example\App_software 為 Project directory，填入 Lights_Example 為 Project name 如圖 12-4 所示，再單擊 next。

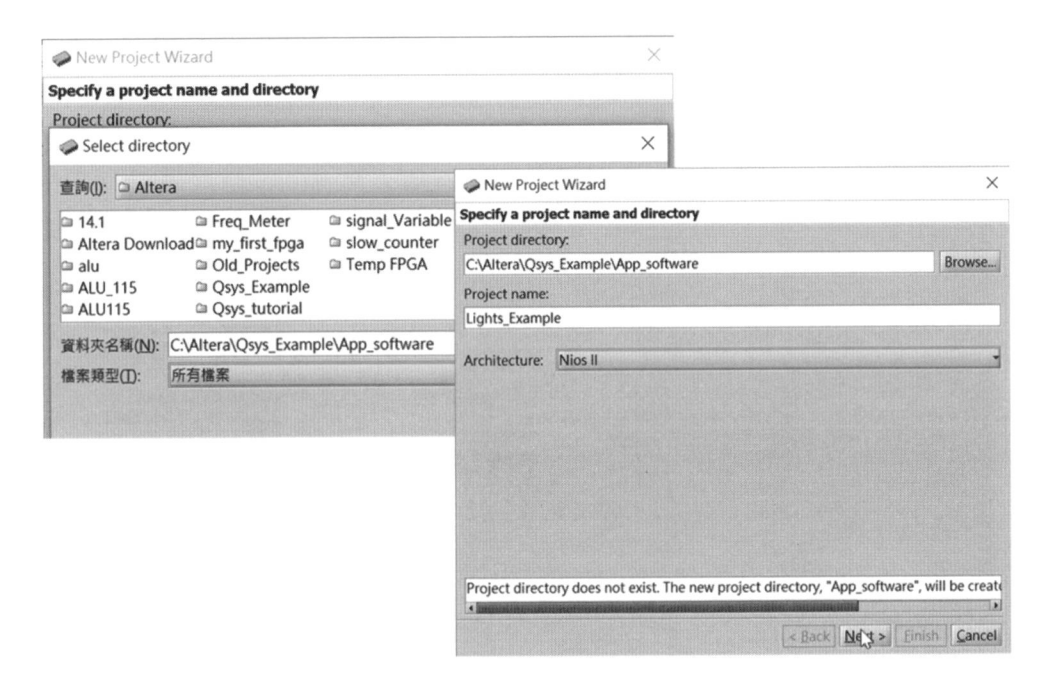

圖 12-4　指定 Project directory 和 Project name

接下來是指定 System 如圖 12-5 所示,選用 **<Custom system>**,
sopcinfo file,和 lights.sof file 的位置,再單擊 **next**。

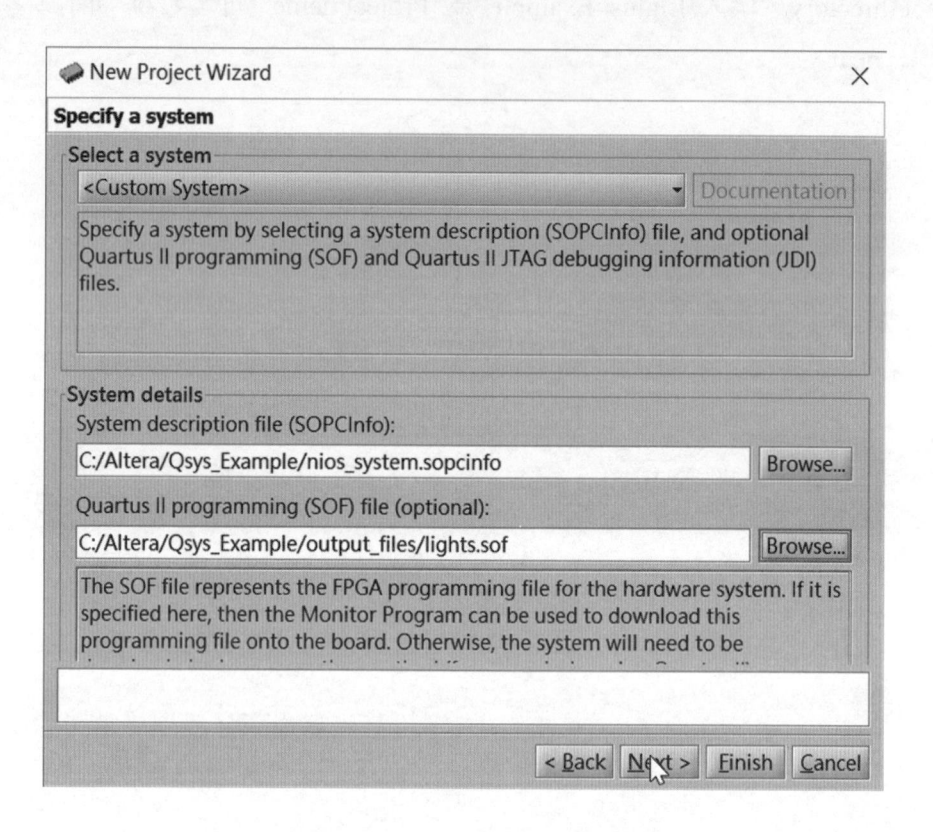

図 12-5 系統指定視窗

第十二章　Intel/Altera "監控程序" 的介紹　277

程序類別指定視窗如圖 12-6 所示。應選用 **Assembly Program**，再單擊 next。

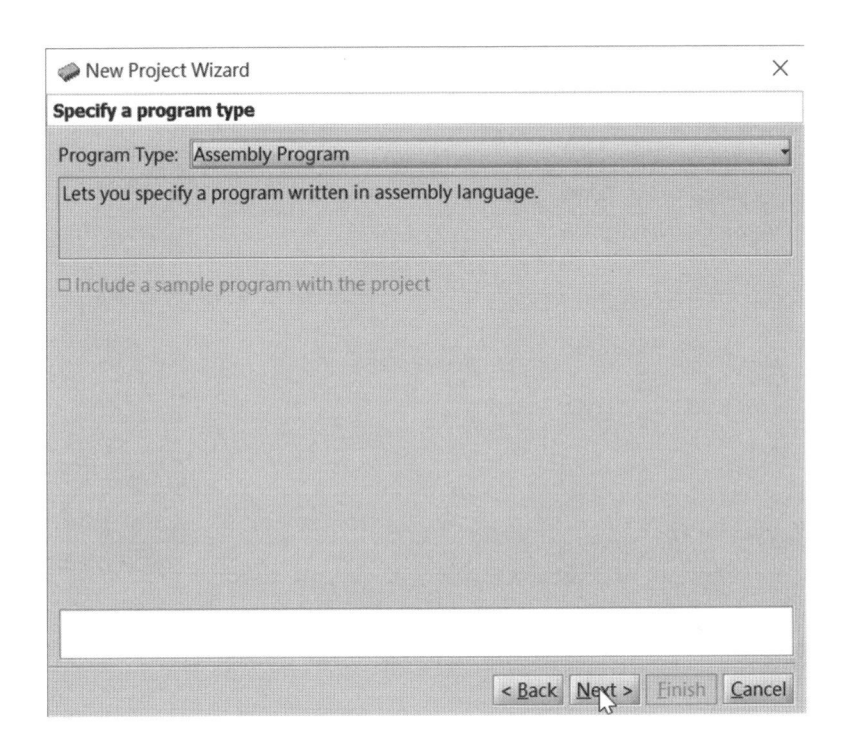

◦ 圖 12-6 ◦　　　程序類別指定視窗

圖 12-7 為指定使用那個應用程序的視窗，應填入 lights.s 及其來源，再單擊 next。

▓ 圖 12-7 ▓　　指定使用那個應用程序

第十二章　Intel/Altera "監控程序" 的介紹　279

　　系統參數的指定視窗，如圖 12-8 所示。這時候應當把 DE2-115 接上並開啓電源，同時連結上 PC 與 USB-Blaster 間的電纜，選用與 PC 連結的 USB-Blaster [USB-0]，nios2_processor，和 jtag。然後單擊 **next**。

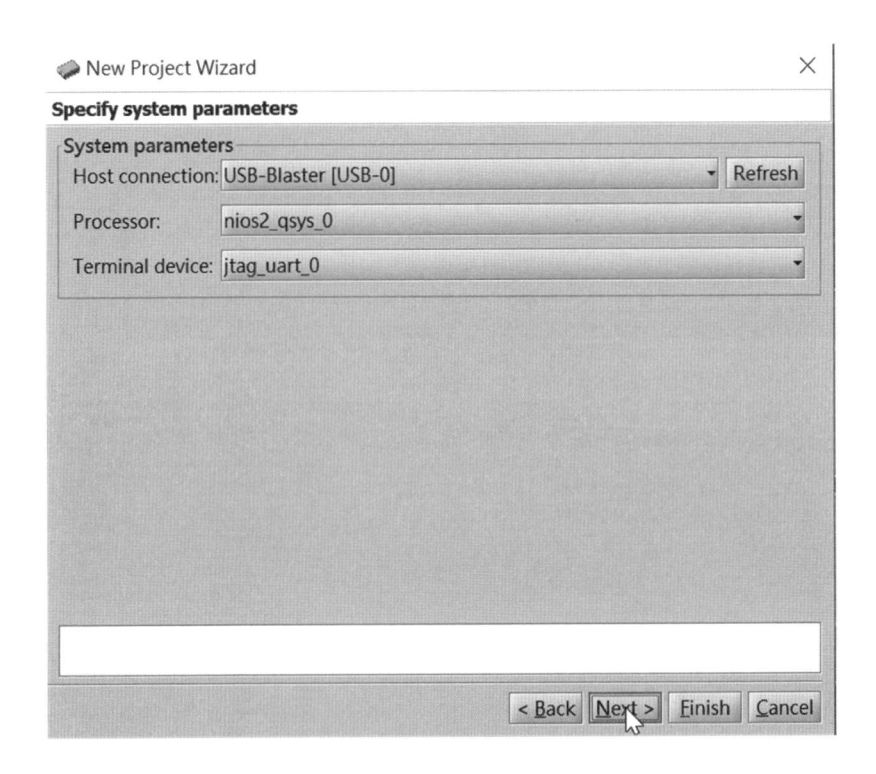

◦ 圖 12-8 ◦　　系統參數指定的視窗

程式儲存器指定的視窗，自動產生不必填寫，如圖 12-9 所示，全部設定完成，單擊 **Finish** 即可。

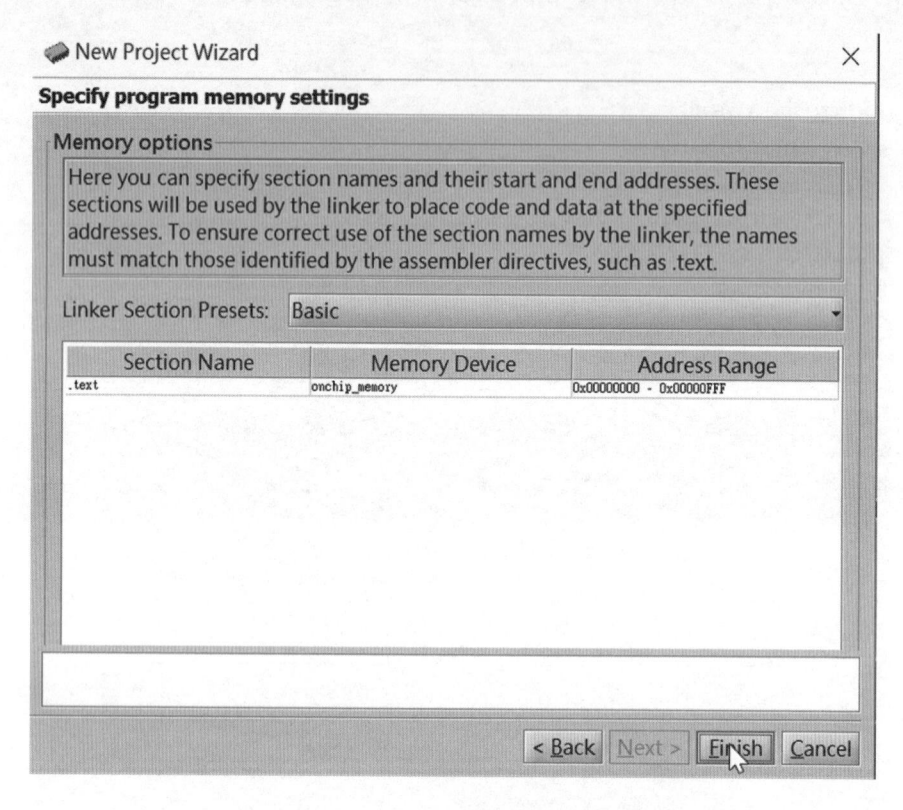

圖 12-9 程式儲存器指定的視窗

第十二章　Intel/Altera "監控程序" 的介紹　281

　　Monitor Program 提示，將所參與的 System file 成功地下載到 Monitor Program 內，如圖 12-10 所示。

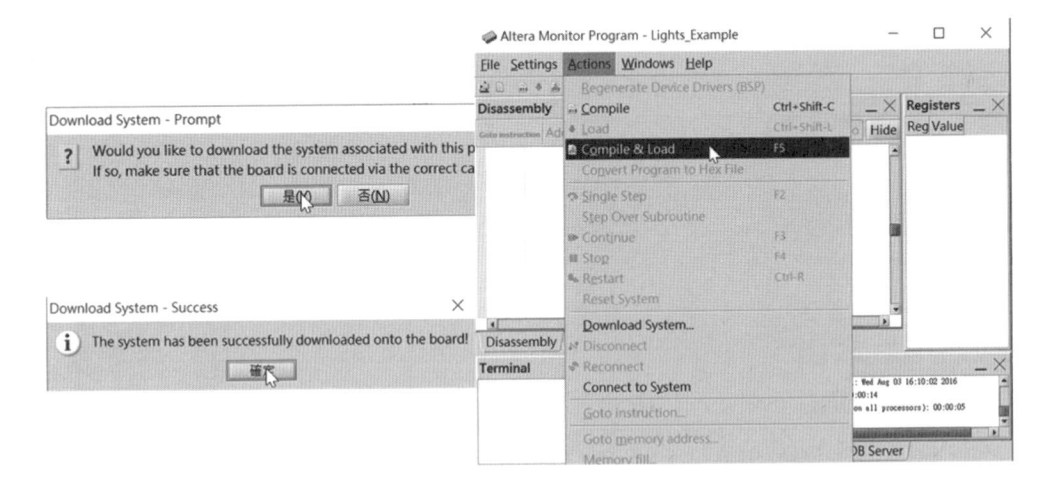

◦ 圖 12-10 ◦　　System file 成功地下載到 Monitor Program 中

下載完畢之後,可以做的項目,可以在 Altera Monitor Program 的 Actions 上列出,選用 **Compile & Load**,如果程式沒錯,當如圖 12-11A 所示。

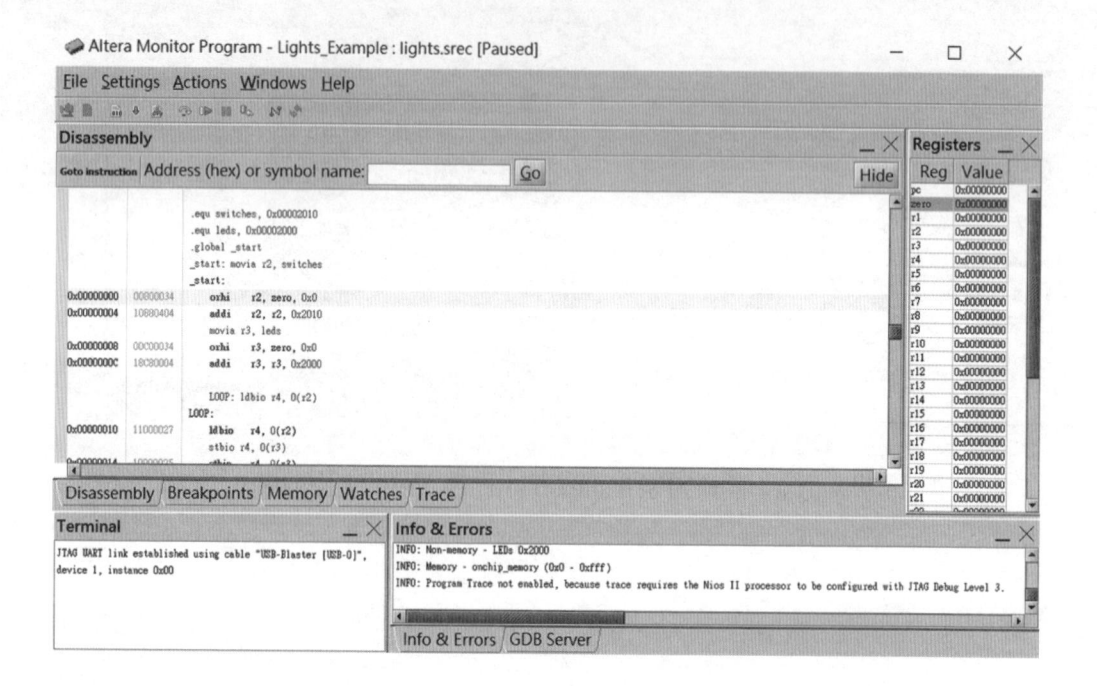

圖 12-11A Altera Monitor Program 的 Actions 種類

第十二章　Intel/Altera "監控程序" 的介紹　283

如果要 run 這個程式當如圖 12-11B 單擊 Actions > Continue，這時候 LEDGs 就會顯示 Slide Switches 的 ON/OFF 情況。

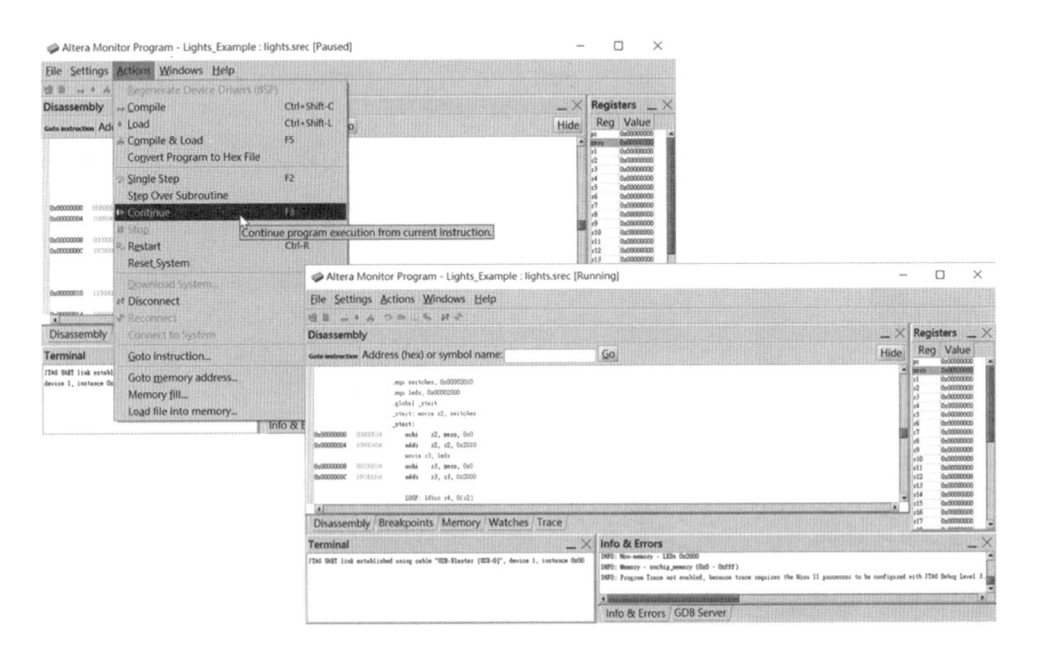

☙ 圖 12-11B ☙　　單擊 Actions > Continue LEDGs 顯示 Slide Switches 的 ON/OFF

12-4　Nios II 的 C 語言程式

圖 12-12 為 Nios II 的 C 語言程式，這個程式跟圖 8-2 用來控制 Lights 的組合語言程式，作用完全相同。

```
1 #define switches (volatile char *) 0x0002010
2 #define leds (char *) 0x0002000
3 void main()
4 { while (1)
5 *leds = *switches;
6 }
```

圖 12-12　用以控制 Lights 的 C 語言程式

接下來是開啟 Altera Monitor Program 視窗，如圖 8-13 所示。

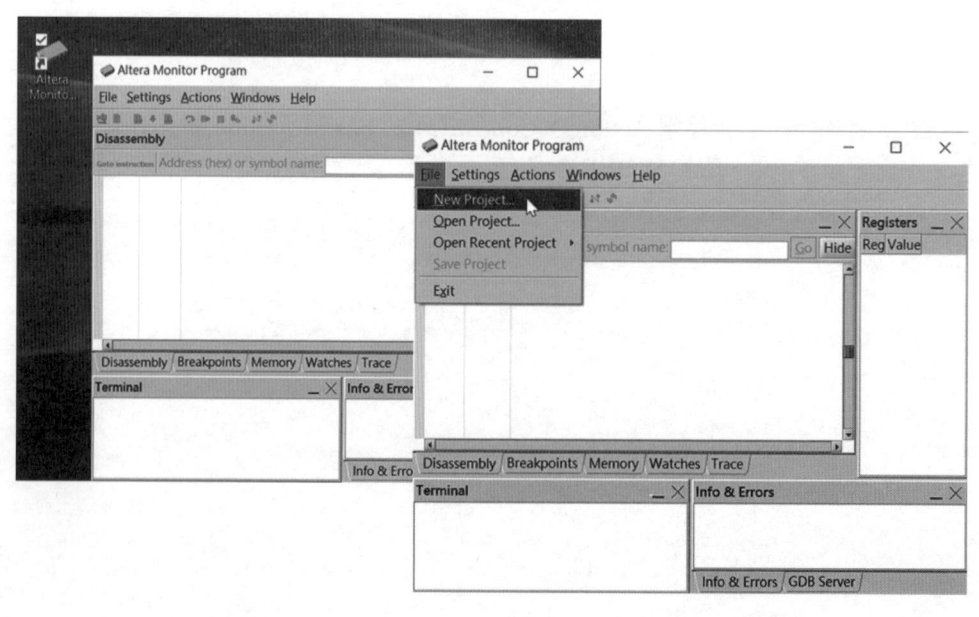

圖 12-13　Altera Monitor Program 視窗

單擊 **Altera Monitor Program** 的 File > New Project 就有 New Project Wizard 視窗的出現。指定 D:\qsys_tutorial\app_software2 為 Project directory，塡入 lights_example 為 Project name 如圖 12-14 所示，再單擊 next。

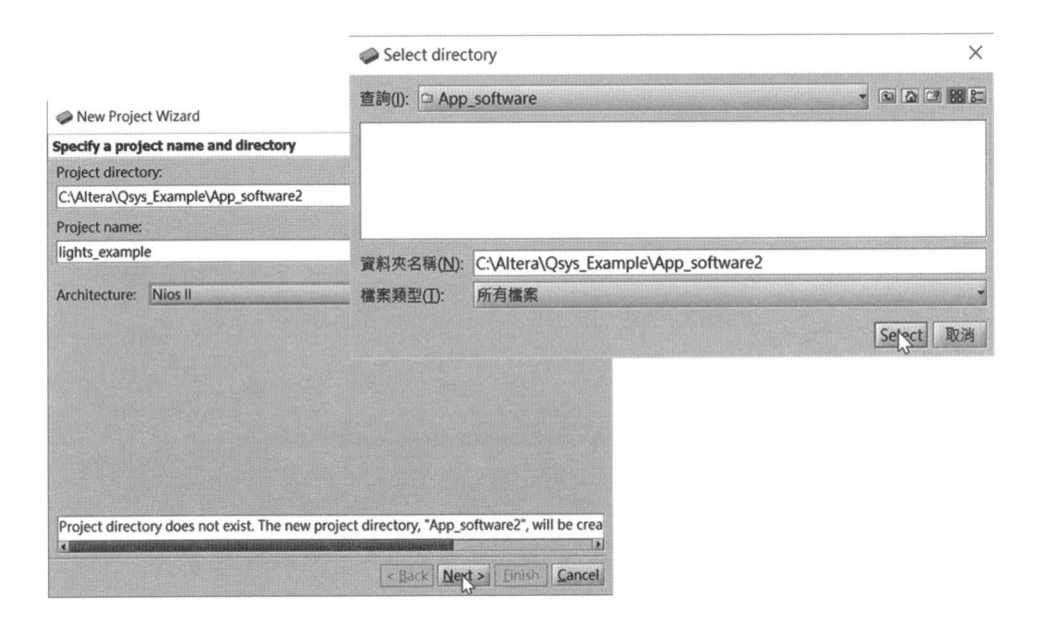

◦ 圖 12-14 ◦　　　　指定 Project directory 和 Project name

首先是選用 **Qsys_Example Project** 中的 sopcinfo 和 sof 檔，如圖 12-15A 所示。

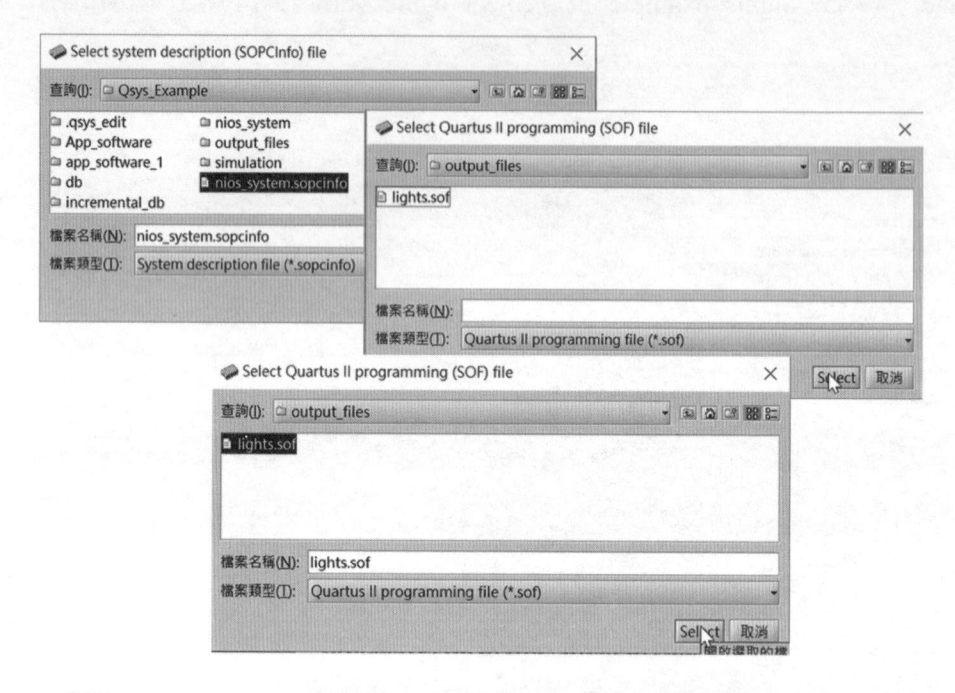

☞ **圖 12-15A** 選用 Qsys_Example Project 中的 sopcinfo 和 sof 檔

第十二章　Intel/Altera "監控程序" 的介紹　287

接下來是指定 System 如圖 12-15B 所示，選用 <Custom system>，SOPCinfo file，和 lights.sof file 的位置，再單擊 next。

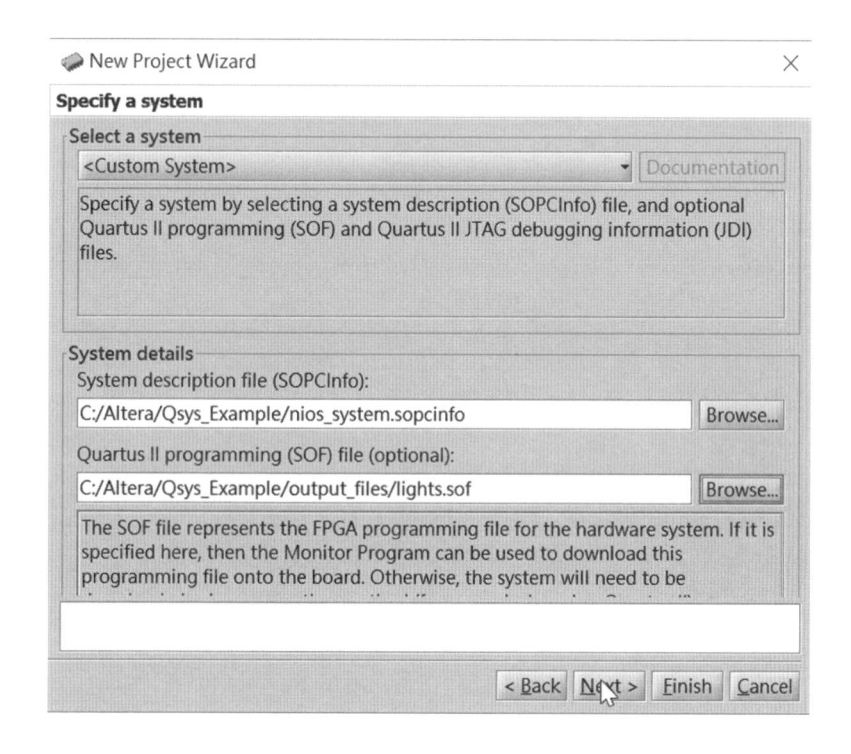

◦圖 12-15B◦　　指定 <Custom system> 的 nios_system.sopcinfo 和 lights.sof 檔

程序類別指定視窗如圖 12-16 所示。應選用 C Program，再單擊 **next**。

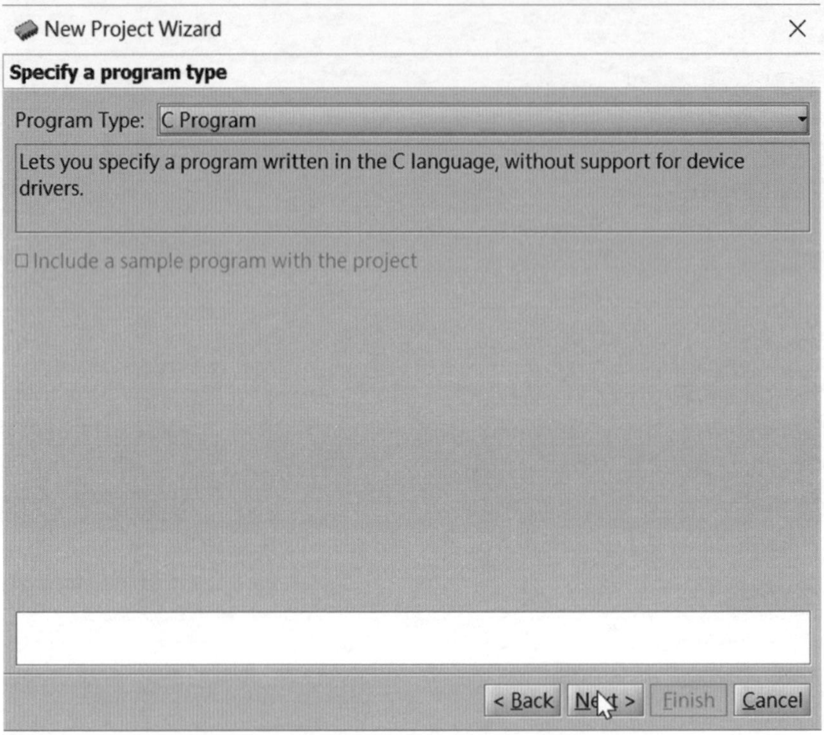

圖 12-16 程序類別指定視窗

第十二章 Intel/Altera "監控程序" 的介紹　289

　　圖 12-17 為指定使用那個應用程序的視窗，應選用的是 lights.c 及其來源，再單擊 next。

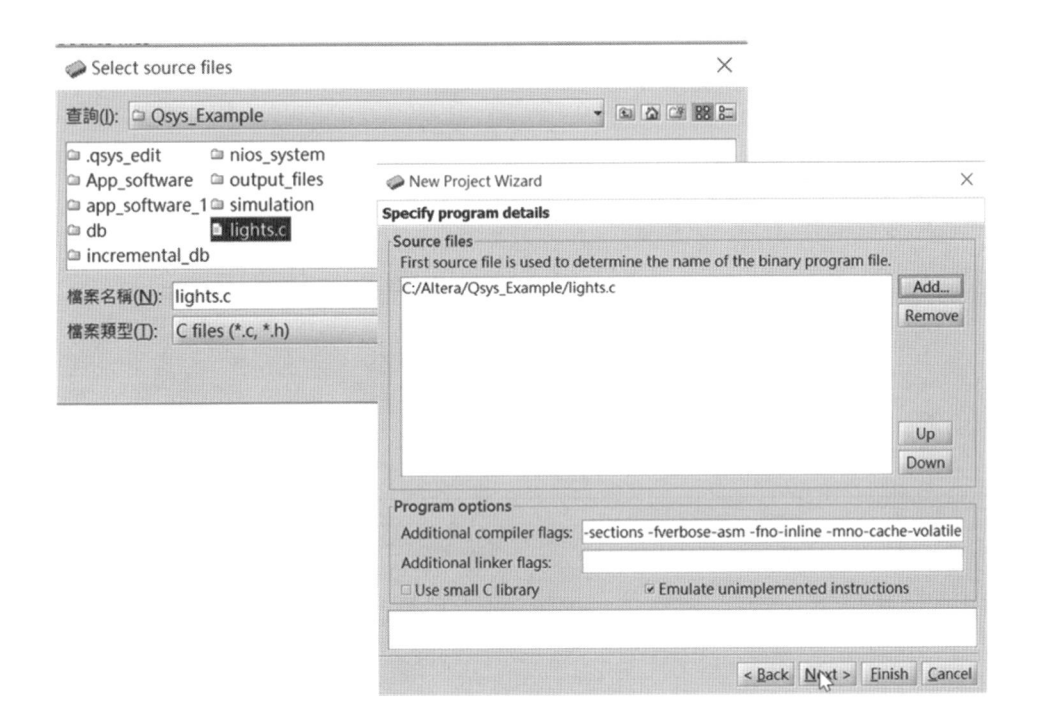

◌ 圖 12-17 ◌　　　　指定使用那個應用程序

系統參數的指定視窗，如圖 12-18 所示。這時候應當把 DE2-115 接上並開啟電源，同時連結上 PC 與 USB-Blaster 間的電纜，選用與 PC 連結的 USB-Blaster [USB-0]、nios2_processor，和 jtag。然後單擊 **Next**。

```
New Project Wizard                                           ×
Specify system parameters
┌─ System parameters ──────────────────────────────────────┐
│ Host connection: USB-Blaster [USB-0]          ▼ [Refresh]│
│ Processor:       nios2_qsys_0                          ▼ │
│ Terminal device: jtag_uart_0                           ▼ │
│                                                          │
│                                                          │
│                                                          │
│                                                          │
│                                                          │
│                                                          │
│                                                          │
│                                                          │
│                                                          │
│                                                          │
│                              < Back  Next >  Finish  Cancel│
└──────────────────────────────────────────────────────────┘
```

圖 12-18　系統參數指定的視窗

程式儲存器指定的視窗，是自動產生的，不必填寫。如圖 12-19 所示。全部設定已完成，單擊 Finish 即可。

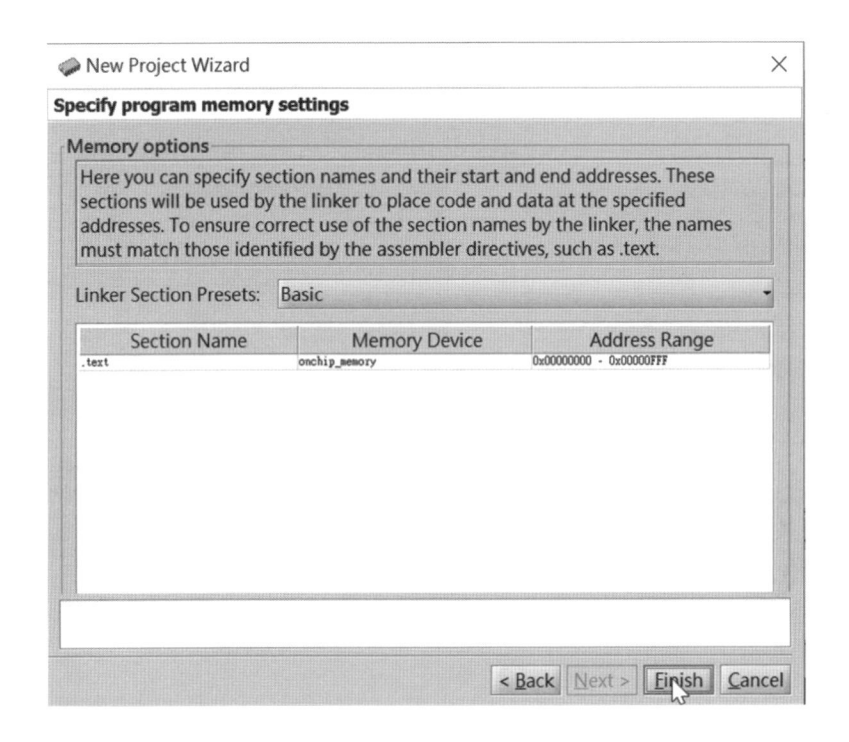

圖 12-19　程式儲存器指定的視窗

Monitor Program 提示，將所有參與的 System file 成功地下載到 Monitor Program 內，如圖 12-20 所示。

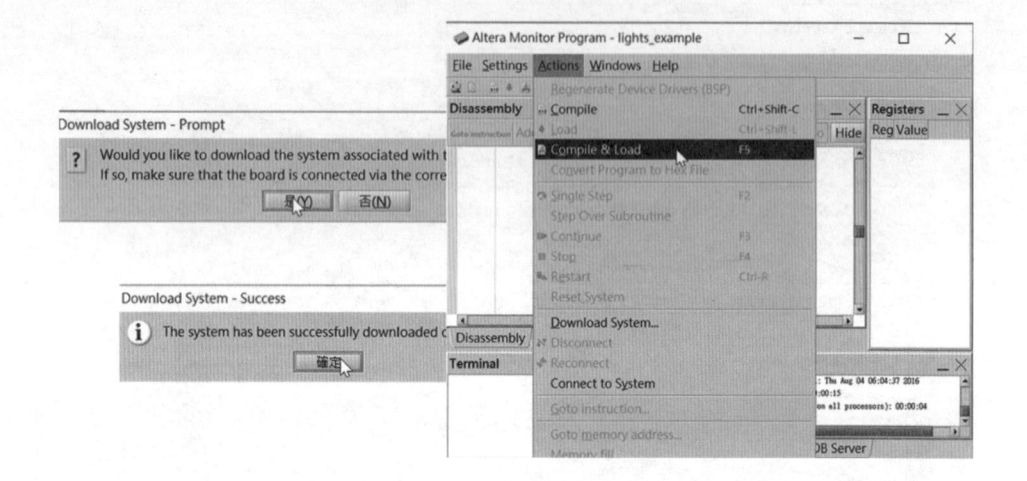

圖 12-20　System file 成功地下載到 Monitor Program 中

下載完畢之後，要做的項目，可以在 Altera Monitor Program 的 Actions 上列出，選用 **Compile & Load**，如果程式沒錯，當如圖 12-21 所示。視窗上顯示 **[Paused]**，為暫停。

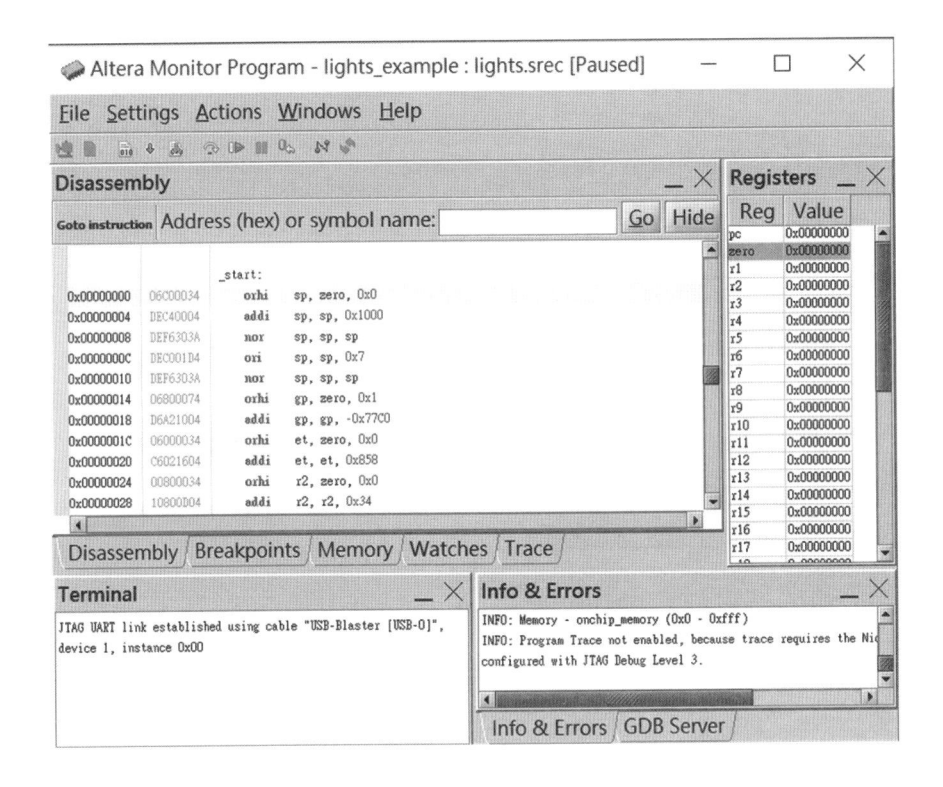

圖 12-21 Altera Monitor Program in Action

294 iLAB FPGA 數位系統設計、模擬測試與實體除錯

　　如果要 run 這個程式當單擊 **Actions** 中的 **Continue**,如圖 12-21 所示。這時候 LEDs 就會顯示 Slide Switches 的 ON/OFF 情況。

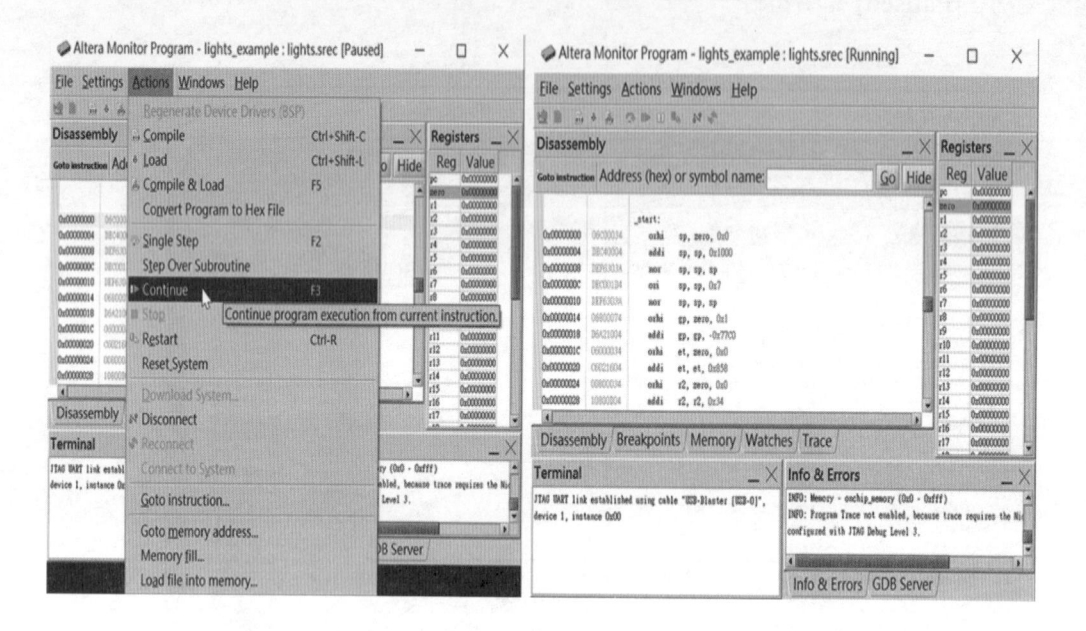

◎ 圖 12-22 ◎　　監控程序 continue run lights.c program

12-5 課外練習

(1) Altera "監控程序" 的主要用途是什麼？

(2) 試比較 ASM 和 C 程序在使用上的優劣點。

附錄 A　Altera 版模擬測試軟件 ModelSim VHDL Simulator 的介紹

VHDL [1] 是硬件設計的通用語言。它的目標是要縮短從概念到完成硬件設計的所需時間。Altera 版的 ModelSim VHDL Simulator，它不但被用來做電路設計、電路的功能和時序的測試，和 FPGA 的電路合成，都會用到 VHDL。

電路設計、電路的功能和時序的測試所使用的 VHDL 可以加入時間的語句。FPGA 的電路合成，則不可以有時間和波形產生的語句。

Altera 版的 ModelSim VHDL Simulator，雖然也可以編輯 VHDL，但是稍微長一點的 VHDL 檔（> 50 Lines）或多個 VHDL 檔要比較編輯，這時候最好使用離線的編輯器 [2]。

Altera 的網站上，有提供 Quartus II 和 ModelSim 的網路版軟件，免費下載。

---

[1] July 1983 DOD awarded Intermetrics, IBM, Texas Instruments to create VHDL
[2] 建議使用可以免費下載的 Crimson Editor。

以下為第一章簡單 ALU 電路的模擬測試的例子：

這個例子參與的檔為 alu_115.vhd 和它的 testbench.vhd。Crimson Editor 顯示如圖 A-1 所示。

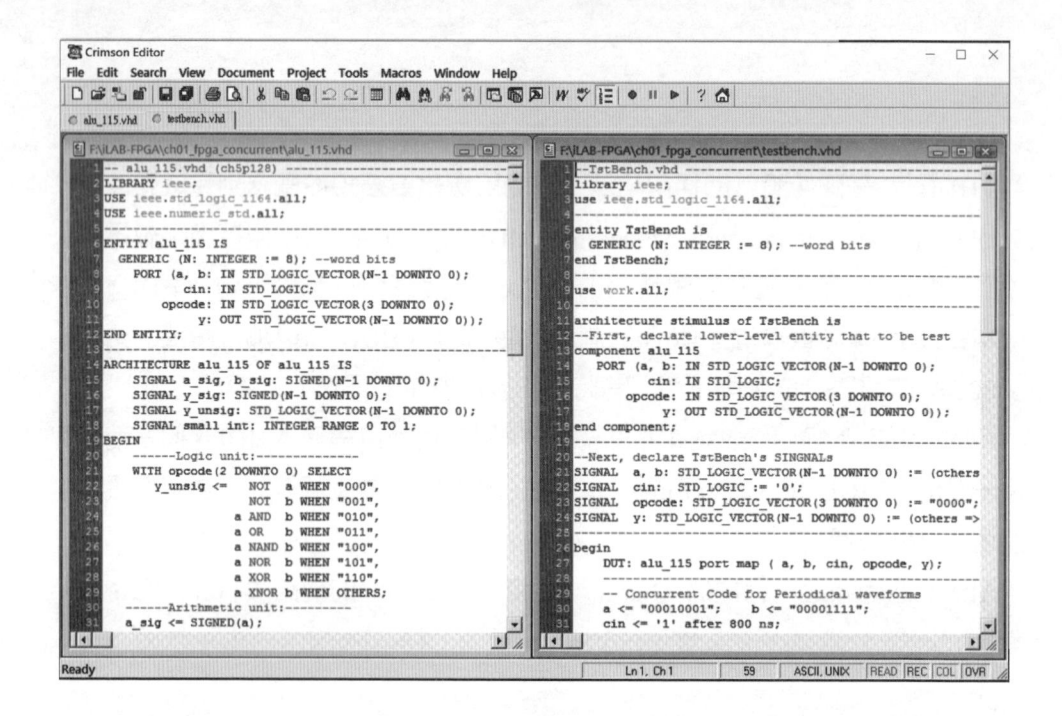

☜ 圖 A-1 ☞　　　Crimson Editor 顯示 alu_115.vhd 和它的 testbench.vhd 檔

附錄 A　Altera 版模擬測試軟件 ModelSim VHDL Simulator 的介紹　299

http://download.cnet.com/Crimson-Editor/3000-2352_4-10031858.html

啟動 ModelSim 由雙擊 ModelSim 捷徑開始，如圖 A-2 所示。

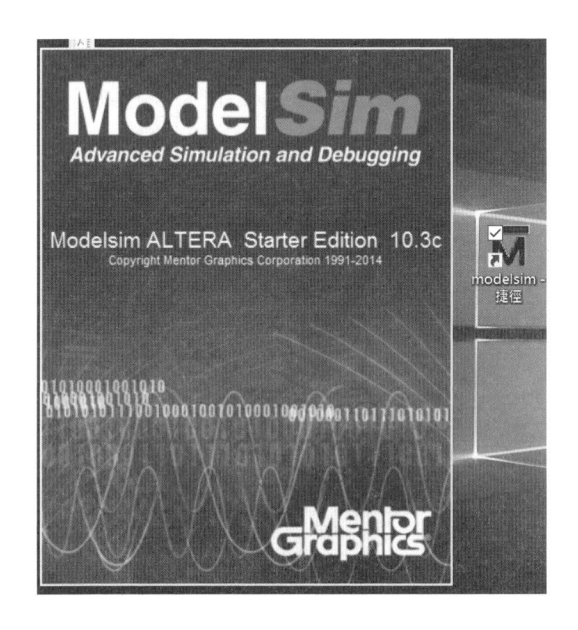

圖 A-2　啟動 ModelSim 由雙擊 ModelSim 捷徑開始

接下來再單擊 ModelSim Welcome 視窗的 Jumpstart 如圖 A-3 所示。

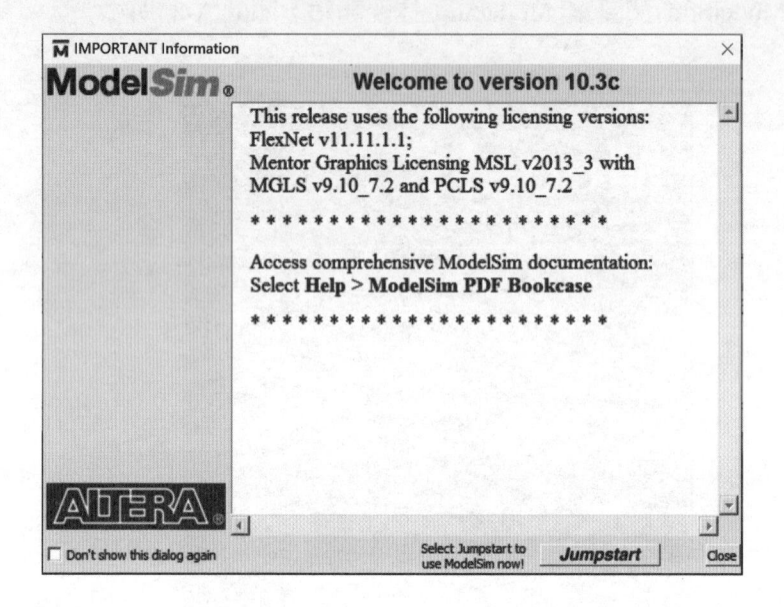

圖 A-3　單擊 ModelSim Welcome 視窗的 Jumpstar

再單擊 Welcome to ModelSim 視窗的 Create a Project 如圖 A-4 所示。

圖 A-4　單擊 Welcome to ModelSim 視窗的 Create a Project

在圖 A-5 Create Project Windows 的 Project Location 和 Project Name 填入圖 A-1 有關 Project 的名字及位置。

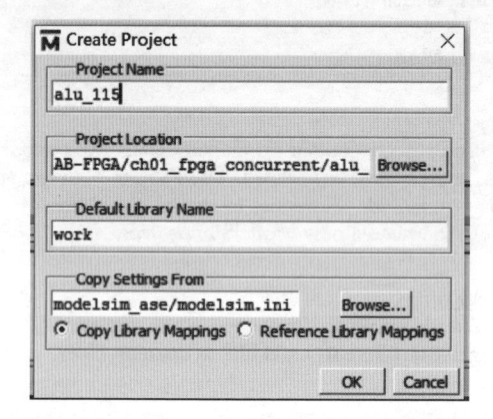

◦ 圖 A-5 ◦　　　填寫 Project 的名字及位置

Project 是否有已存在的 VHD 檔？有，則單擊 Add Existing File 如圖 A6 所示。

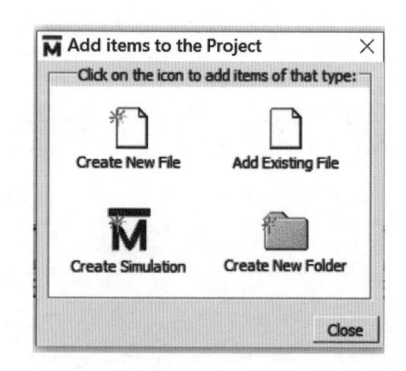

◦ 圖 A-6 ◦　　　單擊 Add Existing File 加入已存在的 VHD 檔

附錄 A　Altera 版模擬測試軟件 ModelSim VHDL Simulator 的介紹

在 Project alu_115 中，選用 alu_115.vhd 和 testbench.vhd，如圖 A-7 所示。

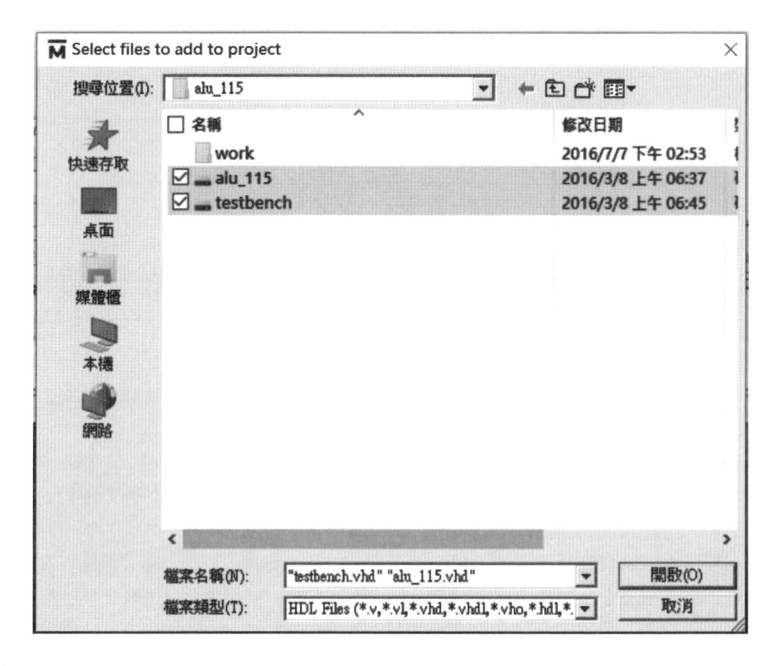

圖 A-7　在 Project alu_115 中，選用 alu_115.vhd 和 testbench.vhd

圖 A-8 顯示 alu_115.vhd 和 testbench.vhd 二檔，加入到 Project alu_115 中。請按 OK。

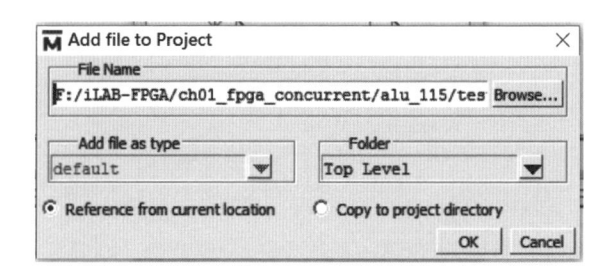

圖 A-8　alu_115.vhd 和 testbench.vhd 二檔，加入到 Project alu_115 中

已完成了 vhd 檔加入到 alu_115 Project，可以將 Add iteams to the Project 視窗 Close。alu_115.vhd 和 testbench.vhd 顯示在 Project 視窗內，二個檔的狀態都是 '?' 表示尚未 Compile，如圖 A-9 所示。

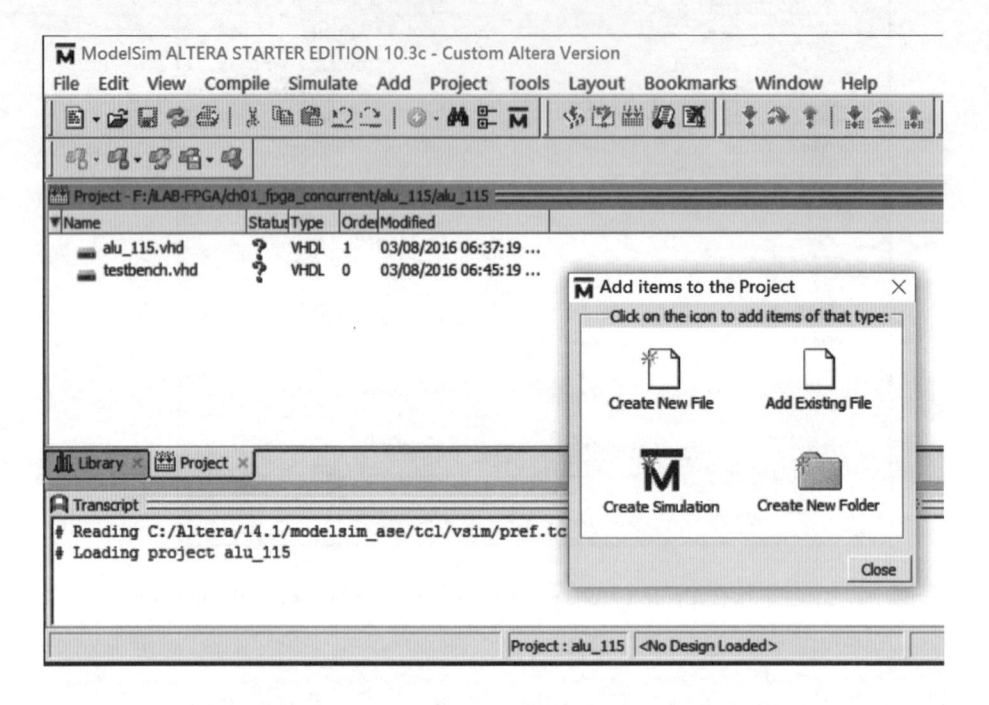

圖 A-9　關閉 Add iteams 視窗，Project 視窗中的二個 vhd 檔尚未 Compile

在圖 A-10 的工作欄選用 **Compile > Compile All**，ModelSim 開始編輯 Project 視窗中的二個 vhd 檔。

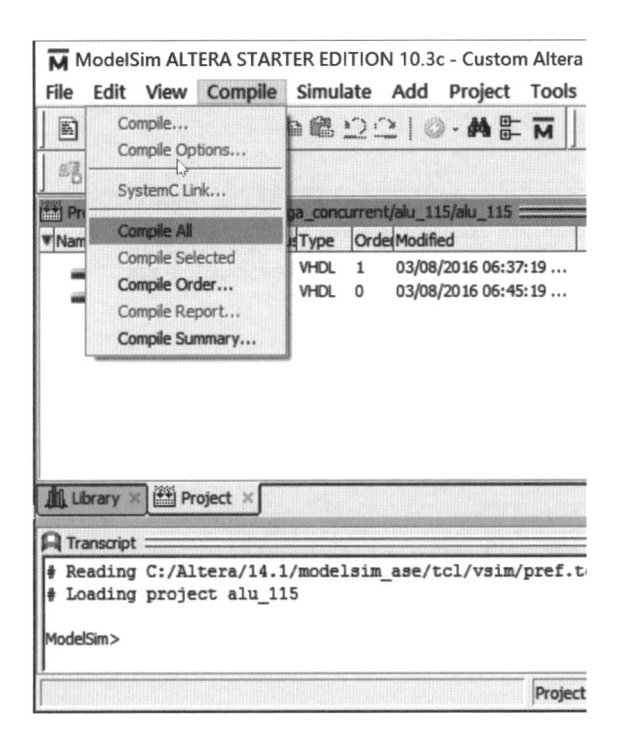

圖 A-10　ModelSim 開始編輯 Project 視窗中的二個 vhd 檔

圖 A-11 顯示圖 A-10 編輯的結果是完全成功。接下來才可以 Simulation。

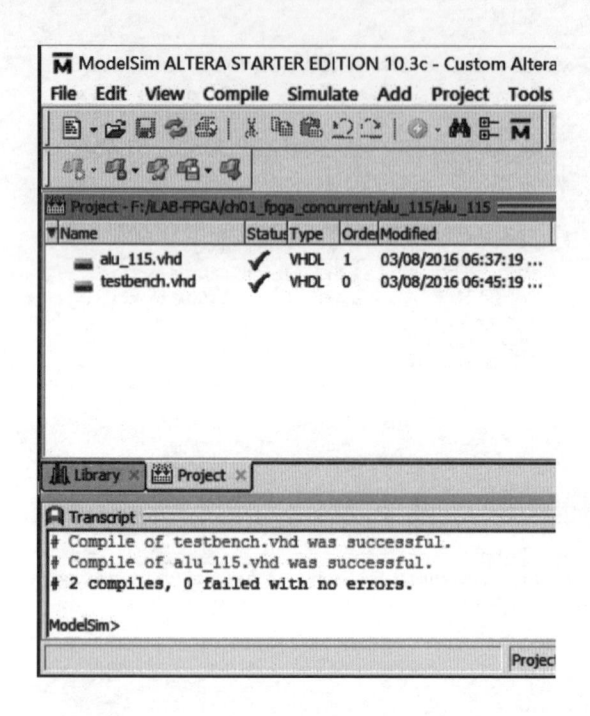

◎ 圖 A-11 ◎ 編輯的結果是完全成功

附錄 A Altera 版模擬測試軟件 ModelSim VHDL Simulator 的介紹

在圖 A-12 的工作欄選用 Simulation > Start Simulation，ModelSim 開啓 Start Smulation 視窗。

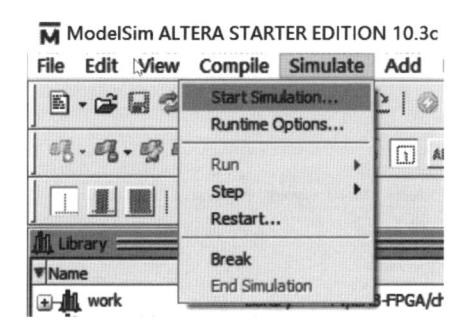

圖 A-12　模擬測試從工作欄選用 Simulation > Start Simulation 開始

圖 A-13 為 Start Simulation windows。在 Design 下選用 work > testbench，然後按 OK。

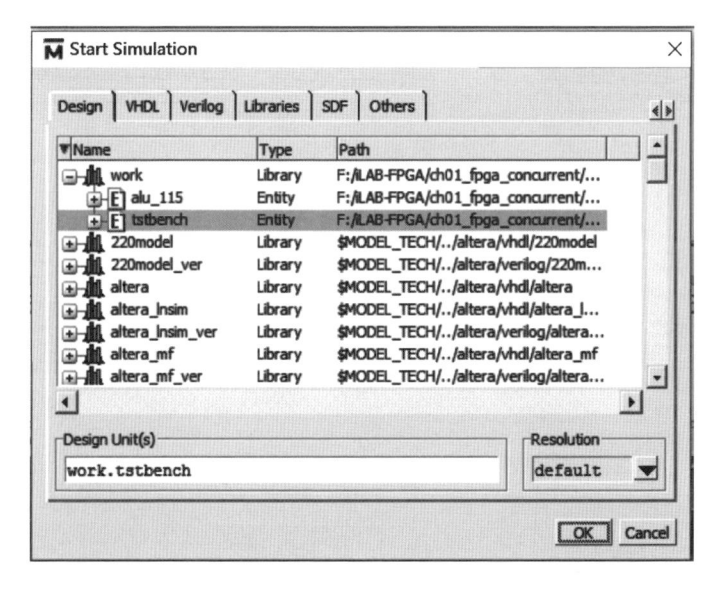

圖 A-13　Start Simulation windows 的選用

這個時候個會有 Objects windows 和 Wave windows 的出現，Objects windows 中有電路的輸入和輸出信號，這些信號必須複製到 Wave windows 上去，複製的方法是右擊 Objects windows 的空白處，然後選用 Add to > Wave > Signal in Region。如圖 A-14 所示。

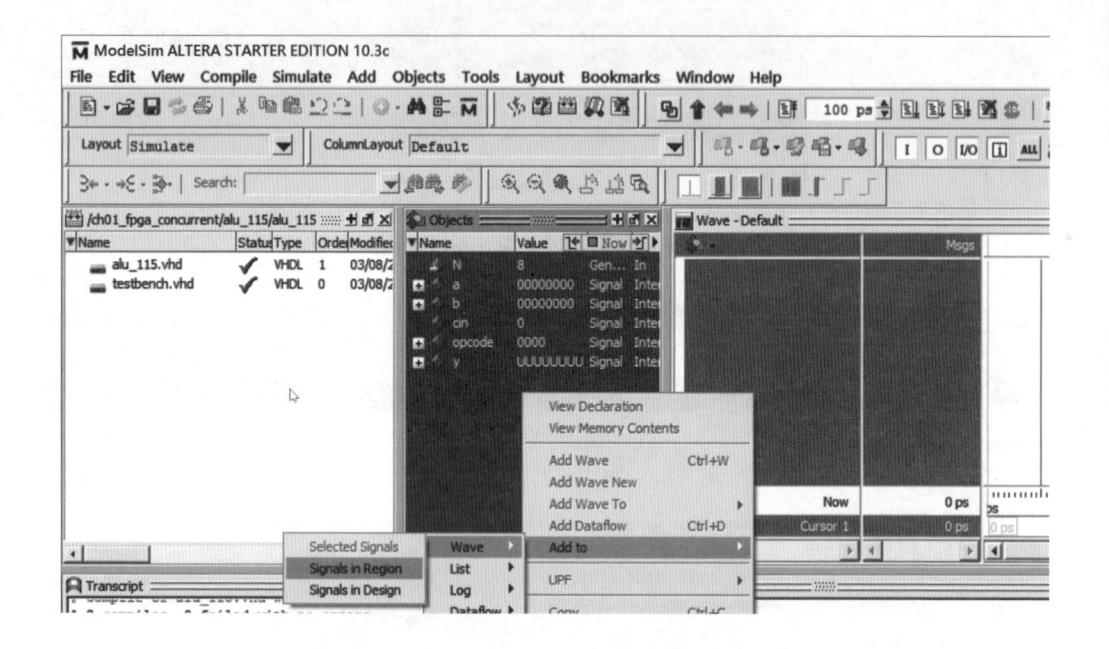

圖 A-14　複製 Objects windows 的信號到 Wave windows 上

附錄 A　Altera 版模擬測試軟件 ModelSim VHDL Simulator 的介紹　309

然後可以在 Transcript Windows 鍵入 run 1600 ns，這個 1600 ns 來自 Project 的 testbench.vhd。如圖 A-15 所示。

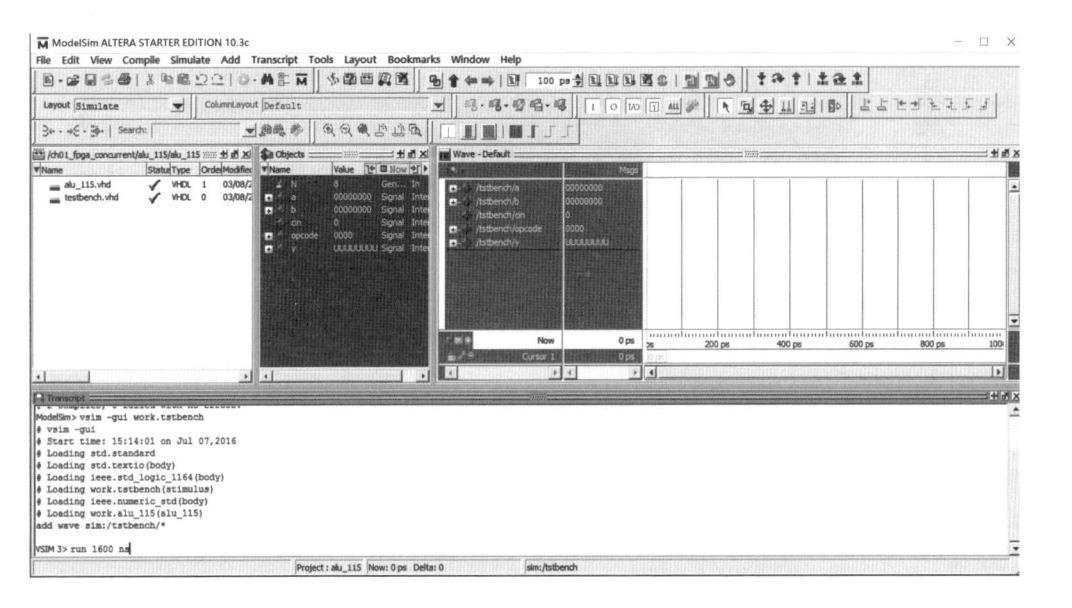

◎ 圖 A-15 ◎　　　Transcript Windows 鍵入測試的時間 run 1600 ns

Run 1600 ns 的結果，如圖 A-16 所示。

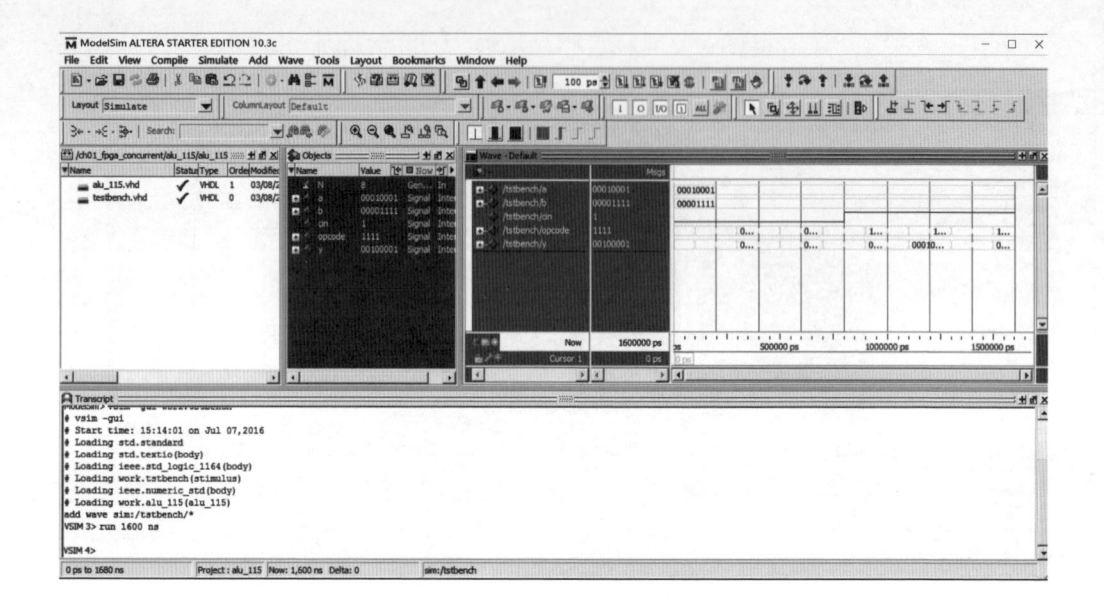

◦圖 A-16 ◦———— Test run 1600 ns 的結果

附錄 A　Altera 版模擬測試軟件 ModelSim VHDL Simulator 的介紹

為了能看到波形的全部，可以在工作欄選用 **Wave > Zoom > Zoom Full**。如圖 A-17 所示。

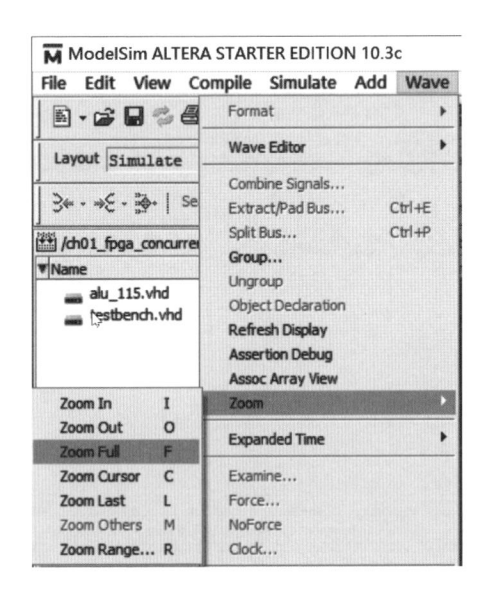

◦ 圖 A-17　　　在工作欄選用 Wave > Zoom > Zoom Full 能看到波形的全部

單擊 Wave windows 的右上角 '+' 號，並再次 Zoom Full，可以看到波形被放大後的全部。如圖 A-18 所示。

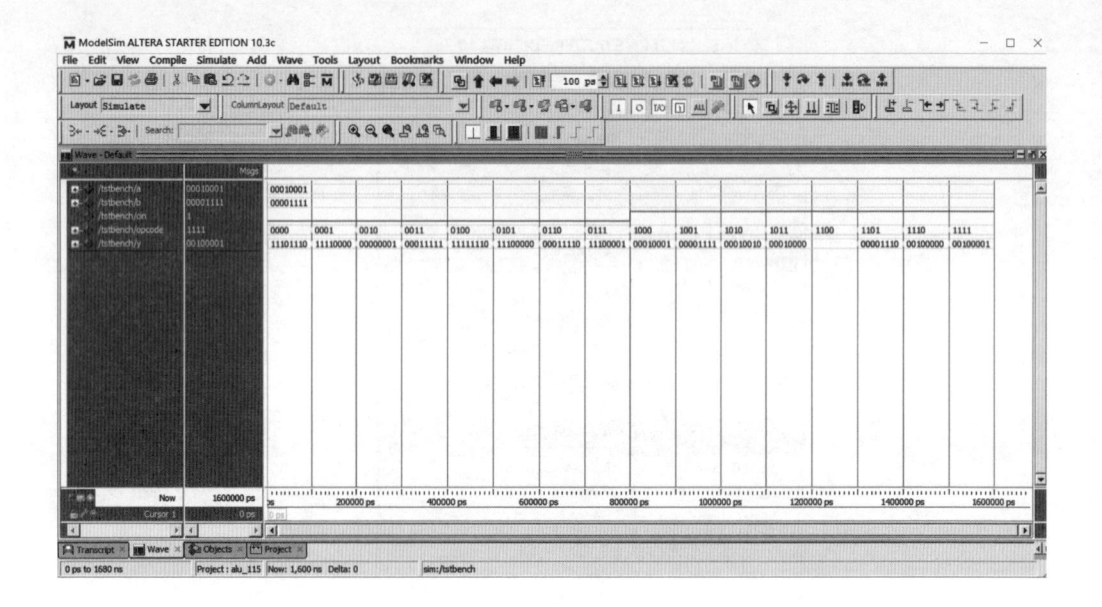

圖 A-18 波形被放大後的全部

　　模擬測試是電路在做 FPGA 合成前的必須，而且不可缺的步驟，模擬測試的輸入信號波形，可做 FPGA 合成後使用 Digital Pattern Generator 和 Logic Analyzer 設定的參考。

附錄 B　測試儀器 Analog Discovery 與其控制軟件 Waveforms 的介紹

　　數位系統的測試，除了測試其功能外，還需要測試其信號間的時序。教育用的 DE 型板子，提供了開關、LED 和 LCD 等顯示器，祇能觀測到部分的功能測試。對於測試信號間的時序，可以用 Analog Discovery 測試儀器內的 Digital Pattern Generator 和 Logic Analyzer 來獲得解決。

　　Analog Discovery 測試儀器是一個軟件定義的儀器，依靠控制軟件 Waveforms 在 PC 的環境下進行測試。Analog Discovery 測試儀器本身祇是接受來自 Waveforms 的命令，把信號輸出和接受信號的輸入。輸出和輸入信號的處理和顯示，由 Waveforms 在 PC 中進行。

　　Analog Discovery 和 Waveforms 軟件，都是美國 Digilent 公司的產品。Waveforms 軟件可以從該公司的網站上免費下載[註1]。

[註1] http://store.digilentinc.com/waveforms-2015-download-only/

Analog Discovery 測試儀器的硬體，如圖 B-1 所示。

圖 B-1　Analog Discovery 測試儀器的硬體

附錄 B　測試儀器 Analog Discovery 與其控制軟件 Waveforms 的介紹　315

　　測試儀器的接線和功能，如圖 B-2 所示。從接線和功能的名稱可以看出右面標示著 Digital I/O Signals 是供給數位電路 I/O 之用，共 16 條信號線。也就是說 Digital Pattern Generator 和 Logic Analyzer 的信號總和不得超過 16 條信號線。

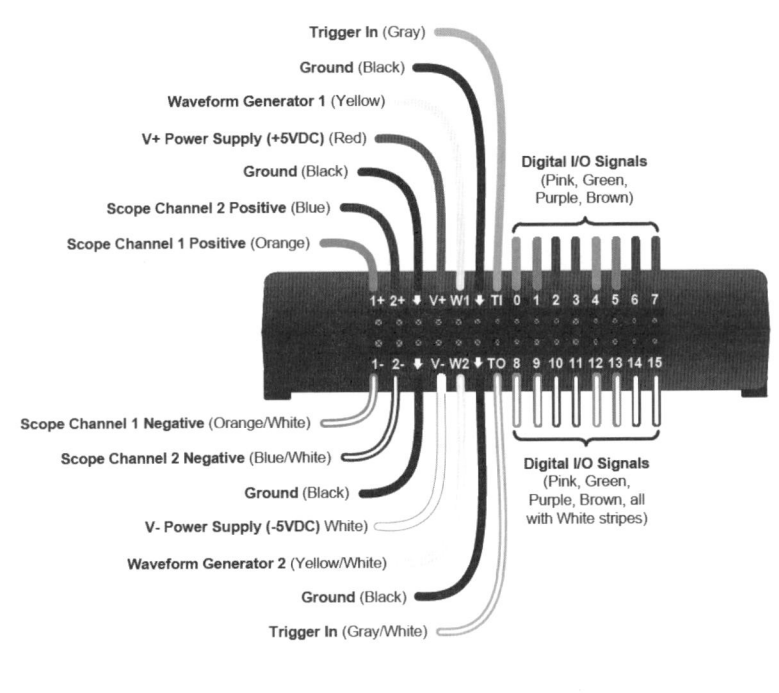

圖 B-2　測試儀器的接線和功能

　　左邊接線性所標示的，有正負 5 V 的二個電源、二個浮動的示波器、二個函數產生器、二條觸發信號輸入線，和四條接地用線。

啟動 Waveform 1 之後的畫面，如圖 B-3 所示。Digital 部分的 out，為 Pattern Generator 用來提供給被測試電路的輸入之用。Digital 部分的 in，為 Logic Analyzer 用來接收被測試電路的輸出之用。

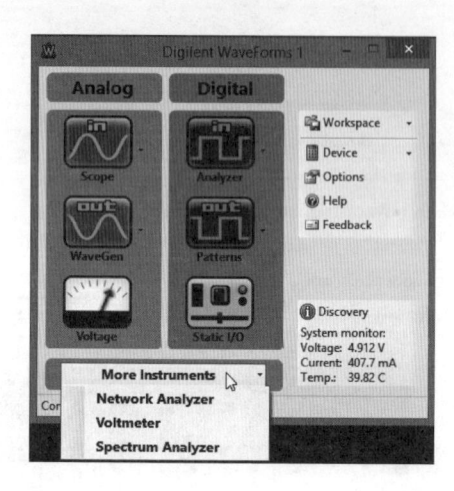

圖 B-3　啟動 Waveform 1 之後所顯示的畫面

　　Analog 部分的 out，為函數產生器，用來提供給被測試 Analog 電路的輸入之用。Analog 部分的 in，為示波器，用來觀察被測試 Analog 電路的輸出或輸入信號之用。

　　圖 B-3 的左下端的 More Instruments 中還有 Network Analyzer、Voltmeter 和 Spectrum Analyzer 三件儀器，是網絡分析和測試通訊電路，不可缺少的工具。

附錄 B 測試儀器 Analog Discovery 與其控制軟件 Waveforms 的介紹 317

範例：

使用 Analog Discovery 的 Digital Pattern Generator 和 Logic Analyzer 來測試第一章 ALU_115 的 Analog Discovery 與 DE2-115 的 J5 接線，如圖 B-4。

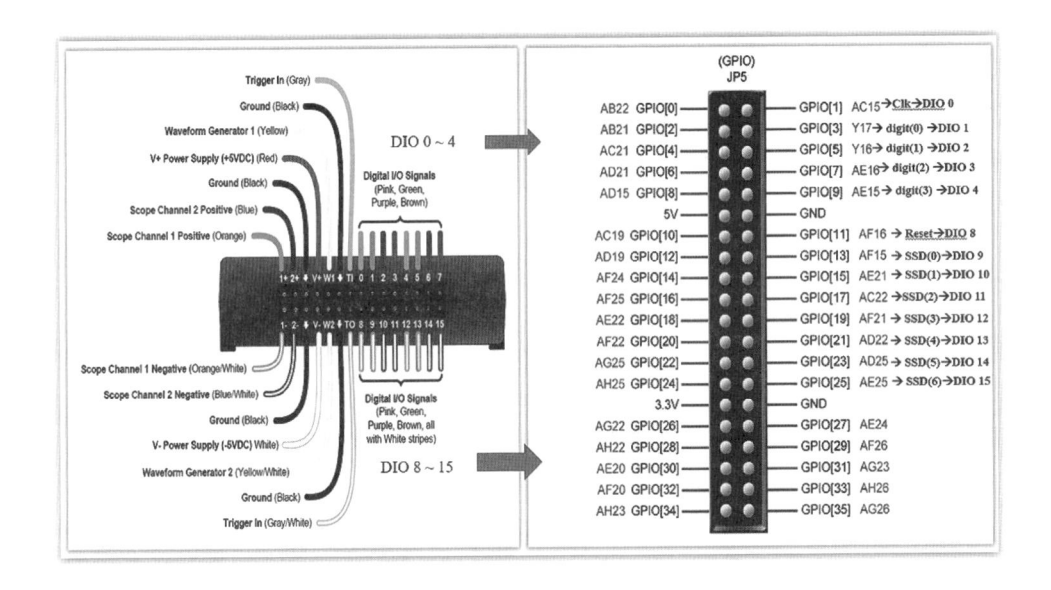

◎ 圖 B-4 ◎ 測試 ALU_115 的 Analog Discovery 與 DE2-115 的 J5 接線

圖 B-5 是它們的實體照相。

◯ 圖 B-5 ◯　　　測試 ALU_115 的實體照相

測試這個 ALU_115 電路，由於 Digital I/O Signals 總共祇有 16 條信號線，所以 8 個 bits 的 a() 和 b() 決定用 DE2-115 的 slide switches 來替代，如圖 B-6 所示。

	FPGA Pin No.	Description
a(0) →	PIN_AB28	Slide Switch[0]
a(1) →	PIN_AC28	Slide Switch[1]
a(2) →	PIN_AC27	Slide Switch[2]
a(3) →	PIN_AD27	Slide Switch[3]
a(4) →	PIN_AB27	Slide Switch[4]
a(5) →	PIN_AC26	Slide Switch[5]
a(6) →	PIN_AD26	Slide Switch[6]
a(7) →	PIN_AB26	Slide Switch[7]
b(0) →	PIN_AC25	Slide Switch[8]
b(1) →	PIN_AB25	Slide Switch[9]
b(2) →	PIN_AC24	Slide Switch[10]
b(3) →	PIN_AB24	Slide Switch[11]
b(4) →	PIN_AB23	Slide Switch[12]
b(5) →	PIN_AA24	Slide Switch[13]
b(6) →	PIN_AA23	Slide Switch[14]
b(7) →	PIN_AA22	Slide Switch[15]

◯ 圖 B-6 ◯　　　8 個 bits 的 a() 和 b() 輸入，使用 DE2-115 的 slide switches

附錄 B　測試儀器 Analog Discovery 與其控制軟件 Waveforms 的介紹　319

其它的輸入信號如 Cin，和 Opcode() 共 5 條信號線，由 Digital Pattern Generator 來提供。ALU_115 的輸出 Y() 共 8 條信號線，由 Logic Analyzer 來接收，如圖 B-4 的右邊所示。

Digital Pattern Generator的設定　在沒有連接 Analog Discovery 到PC之前，雙擊圖 B-3 的 **Digital out**，就會出現如圖 B-7 的 Digital Pattern Generator。視窗上會出現 Demo mode 的字樣，可以用來做編輯。按一下 **X**，選用 **Clear List**，就能清除整個視窗。如圖 B-7 所示。

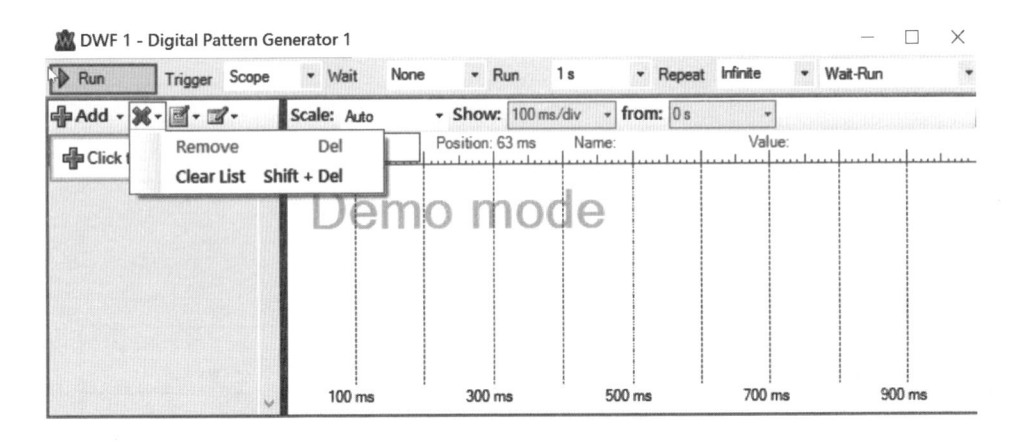

圖 B-7　Digital Pattern Generator 視窗清除的方法

接下來按一下 **+Add > Add Signal**，如圖 B-8 所示。在 Add Signal windows 中選用 **DIO 0**，再按 **Add**。

圖 B-8　Signal 的選用和設定之一

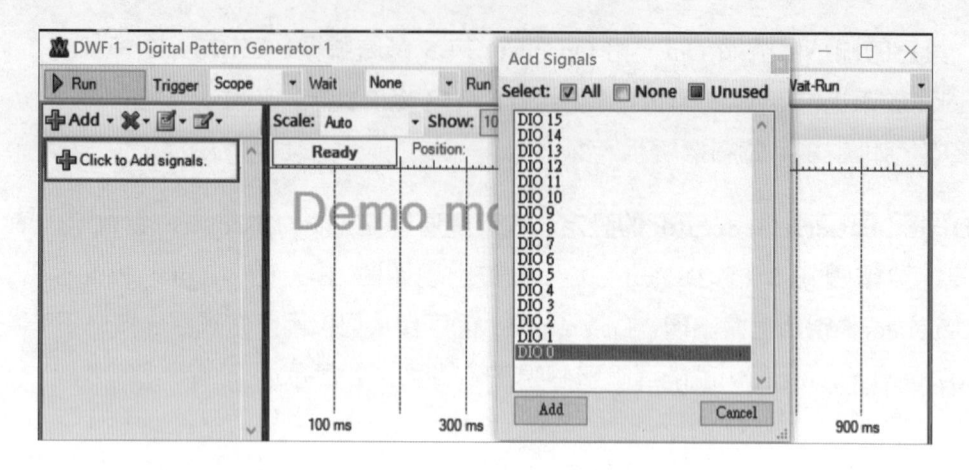

圖 B-9 為選用 DIO 0 為 Signal 所得的結果。接下來是 Name 的修正、信號 Type 的選用、Output 的型態，和參數等。

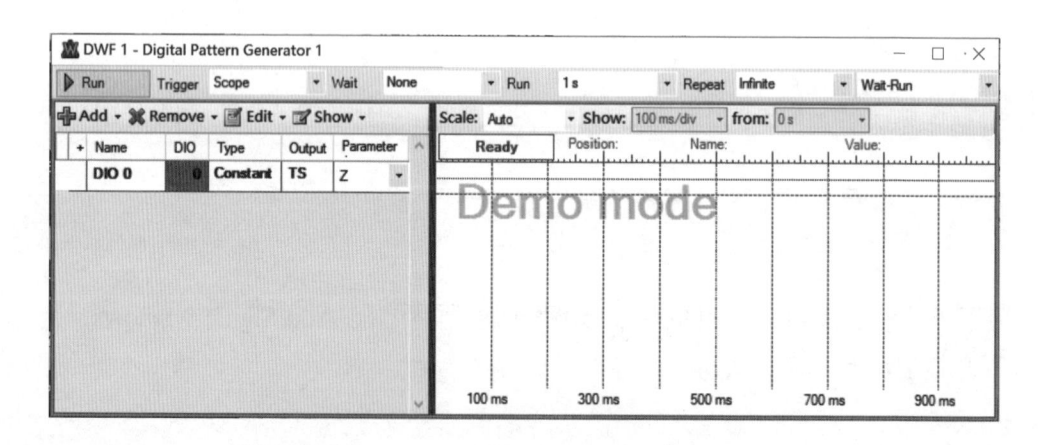

圖 B-9　選用 DIO 0 為 Signal 所得的結果

選用 DIO 0，如圖 B-10 所示。首先要做的是名稱的修改，選用 Edit Properties of "DIO 0"。

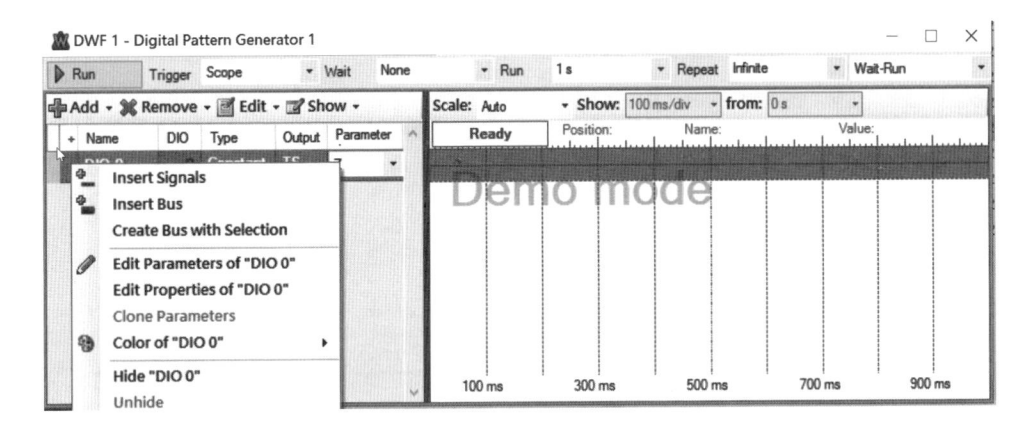

圖 B-10　選用 Edit Properties of "DIO0" 來做名稱的修改之二

選用 Edit Properties of "DIO 0" 的結果。如圖 B-11 所示。請在 Edit "DIO 0" 的 Name 項，填入 Cin。然候正再單擊 OK。

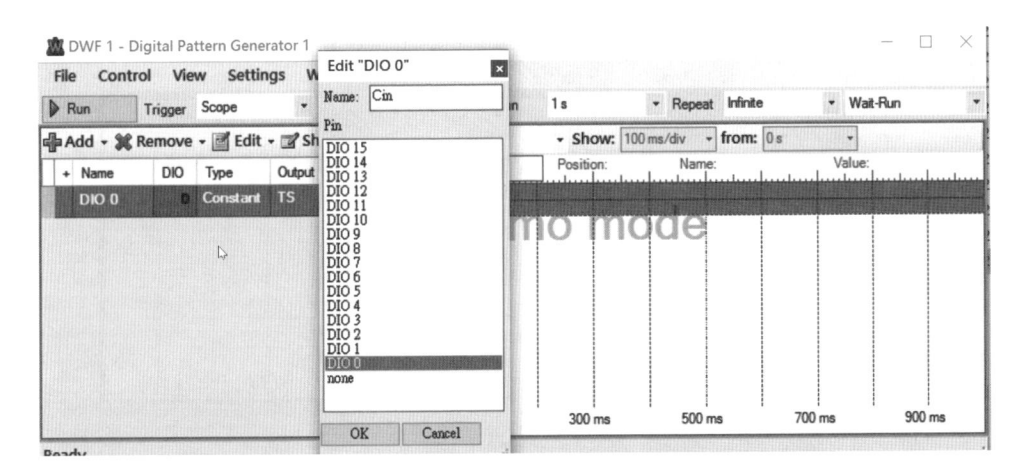

圖 B-11　Edit Properties of "DIO 0" 來做名稱的修改之三

信號 Type 的選用，如圖 B-12 所示，選用 Custom。

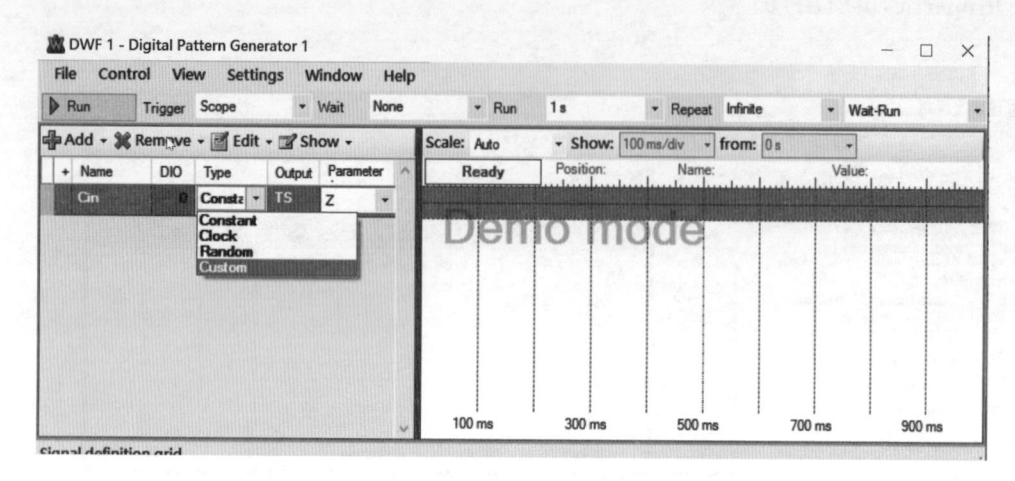

◖ 圖 B-12 ◗ 信號 Type 的選用 Custom 和設定之一

選用 Custom 信號的輸出型態，如圖 B-13 所示。選用 Peak to Peak，也就是 Logic '0' 跟 '1'。

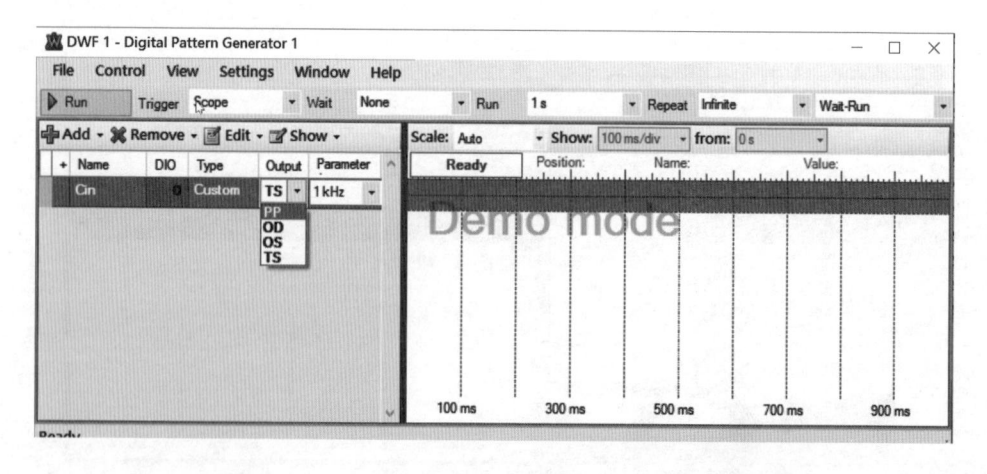

◖ 圖 B-13 ◗ 信號 Type 的選用 Custom 的輸出型態

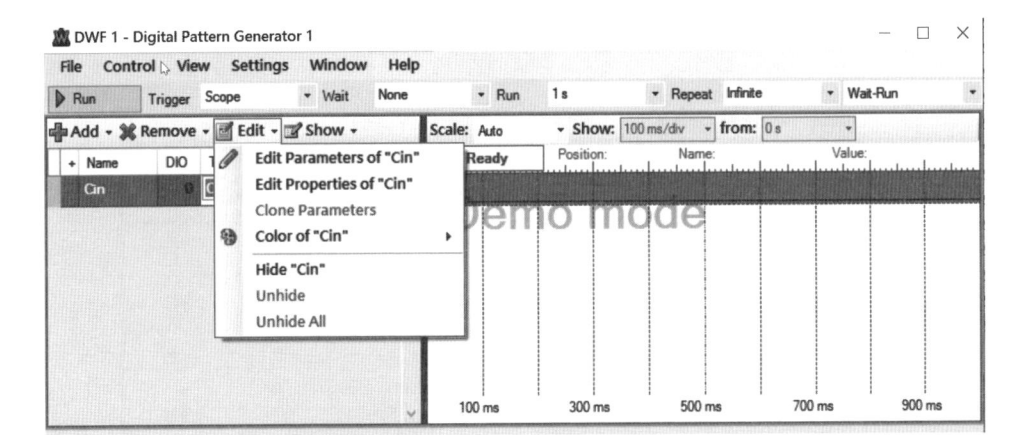

附錄 B　測試儀器 Analog Discovery 與其控制軟件 Waveforms 的介紹　323

最後是 Edit Parameters of "Cin" 的選用和設定，如圖 B-14 所示。

圖 B-14　Cin 信號的選用和設定之一

按下 Edit Parameters of "Cin" 來到了圖 B-15 的 Edit "Cin" windows。把屬於 Value 的 5 項填好，Buffer Size 跟 Opcode 有密切的關係，4 bits 最多為 16 項，因而選用 **20**，頻率 1 KHz 顯示全部當為 20 ms。Cin 的波形 0～8 ms 為 '0'，8～16 ms 為 '1'。最後一定要單擊右上方的 Lock 才算完成。

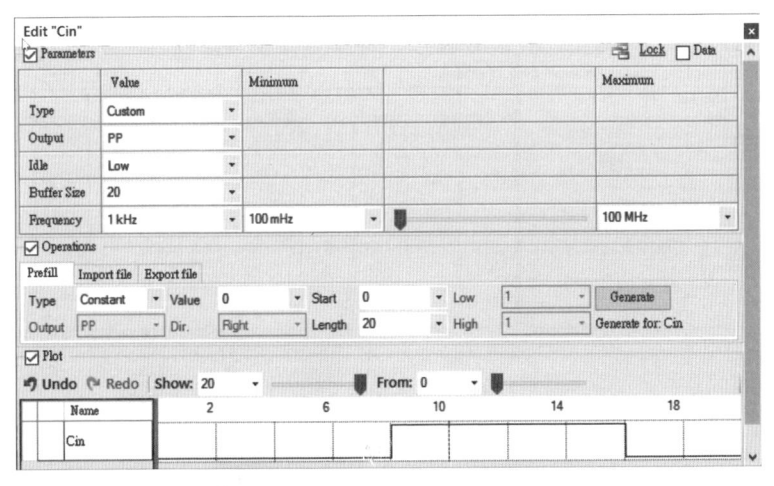

圖 B-15　Edit 信號 Cin

Edit 信號 Cin 完成後 Digital Pattern Generator 當如圖 B-16 所示。

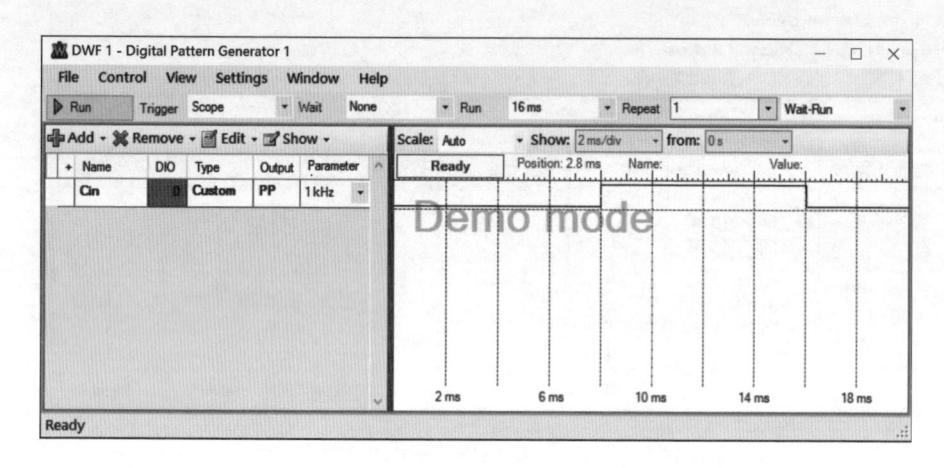

圖 B-16　Edit 信號 Cin 完成後的 Digital Pattern Generator

Opcode 可以用加入 BUS 的方法來處理，如圖 B-17 所示。

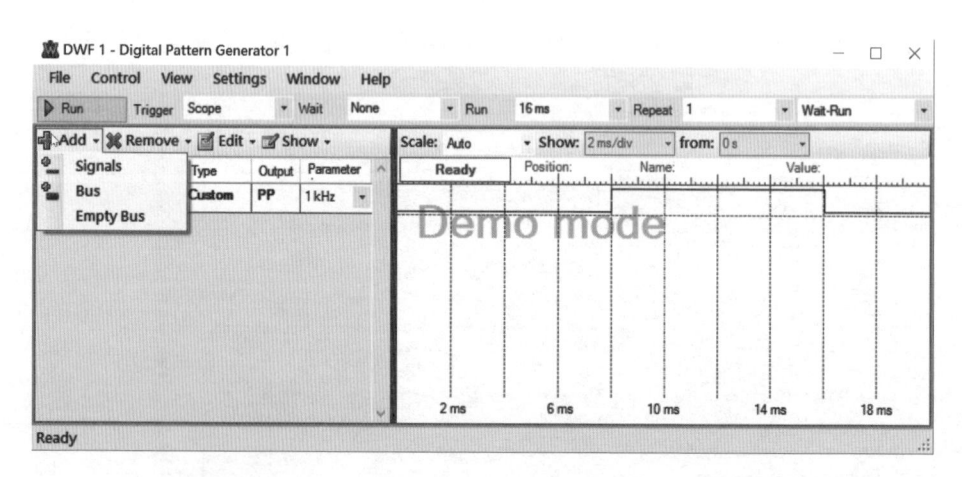

圖 B-17　Opcode 可以用加入 BUS 的方法來處理

該 BUS 名為 Bus Opcode，4 bits 取用 DIO1 ~ DIO4，Format 選用 Decimal。如圖 B-18 所示。最後要單擊 Add Bus 視窗上的 OK。

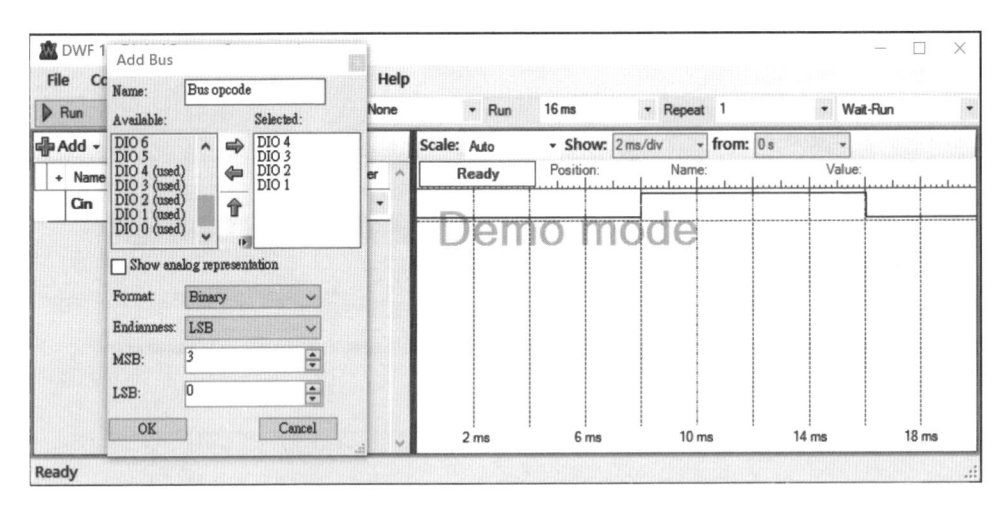

◎ 圖 B-18 ◎　　　Add Bus Opcode 設定之一

圖 B-18 的結果產生圖 B-19，接下來要設定是 Opcode Bus 的 Type 和 Output。

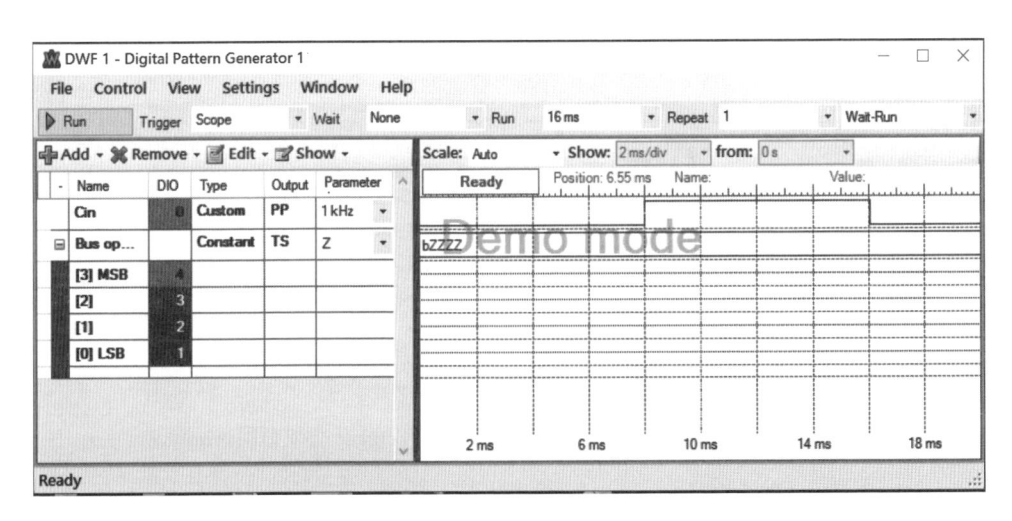

◎ 圖 B-19 ◎　　　Add Bus Opcode 設定之二

Opcode 的 Type，有多個現成的波形可以選用，這個例子是選用 **Binary Counter**。如圖 B-20 所示。

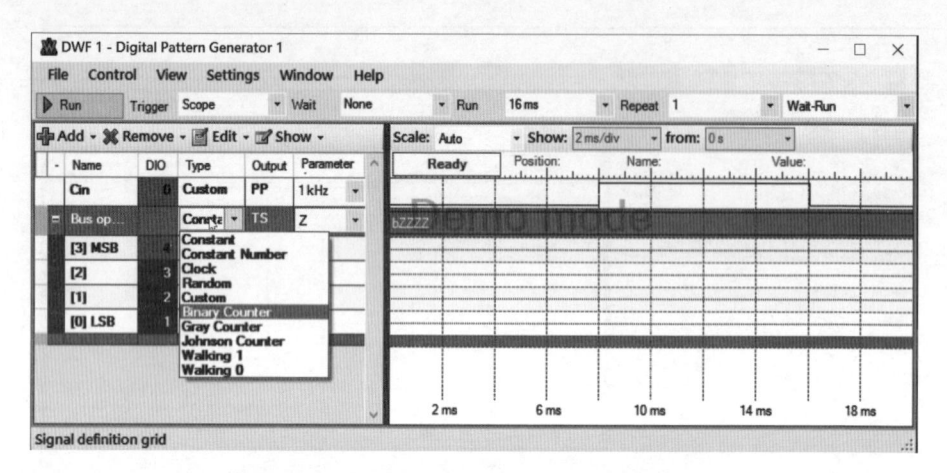

圖 B-20　Add Bus Opcode 設定之三

圖 B-21 顯示 Digital Pattern Generator 對 ALU_115 的輸入波形，其中 Trigger 這項，至為重要，必須選用 **Analyzer**。

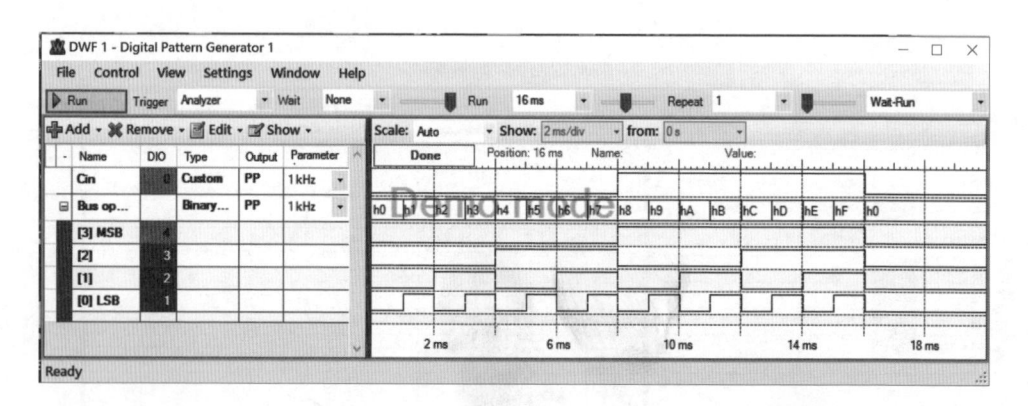

圖 B-21　Pattern Generator 的輸出波形，Trigger 必須選用 Analyzer

附錄 B　測試儀器 Analog Discovery 與其控制軟件 Waveforms 的介紹　327

Logic Analyzer 的設定　在沒有連接 Analog Discovery 到 PC 之前，雙擊圖 B-3 的 **Digital in**，就會出現 Logic Analyzer 視窗。同樣地會出現 Demo mode 的字樣，可以用來做編輯。按一下 **X**，選用 **Clear List**，就能清除整個視窗的舊有。如圖 B-22 所示。

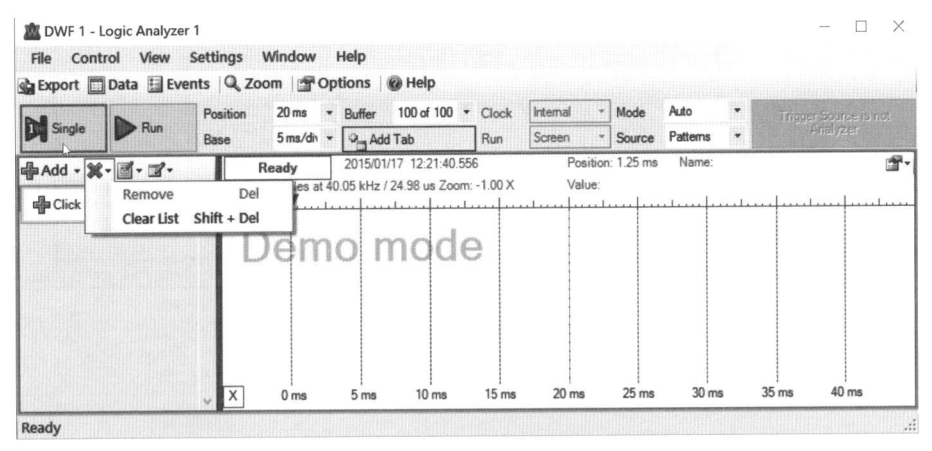

圖 B-22　未經設定的空白 Logic Analyzer 視窗

為了能觀測輸出間的時序關係，Logic Analyzer。圖 B-22 就把輸出的信號都放到 Logic Analyzer 內。

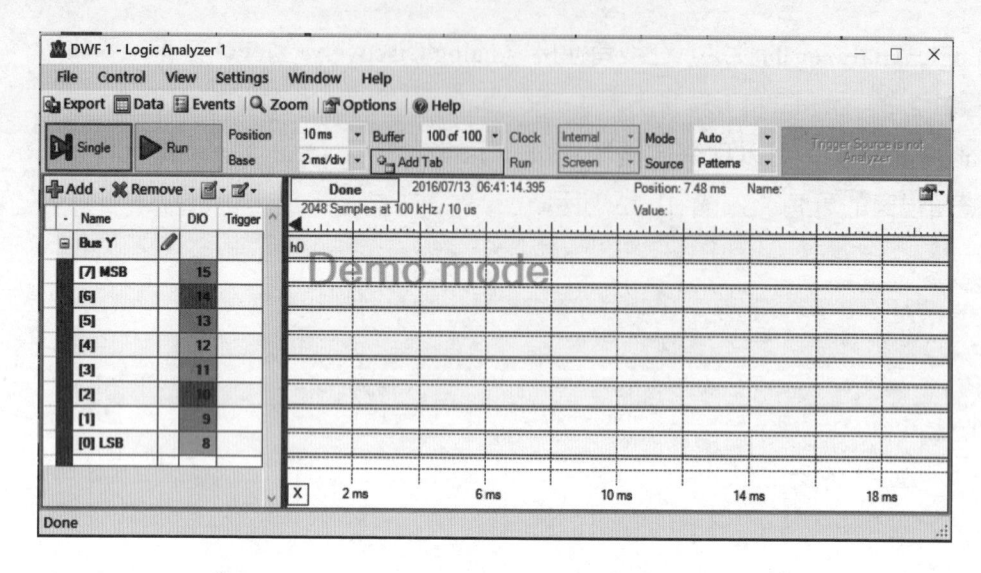

圖 B-23　輸出的信號放到 Logic Analyzer 內

Waveform 準備工作大致完畢。在測試 ALU_115 電路之前，必先完成以下幾點：

1. 連接 Analog Discovery 與 PC 間的接線。此時 Digital Pattern Generator 和 Logic Analyzer 視窗上的 Demo mode 字樣，便會自動消失。
2. 連接 Analog Discovery 與 DE2-115 J5 間的接線，同時也將 DE2-115 的電源打開。
3. Quartus/Programmer 選用 alu_115.sof，將 FPGA Program 好。
4. 將 DE2-115 的 Slide Switch 7:0 設定為 "00010001"。Slide Switch 15:8 設定為 "00001111"。

附錄 B　測試儀器 Analog Discovery 與其控制軟件 Waveforms 的介紹　329

這時候的 Digital Pattern Generator 當如圖 B-24 所示。

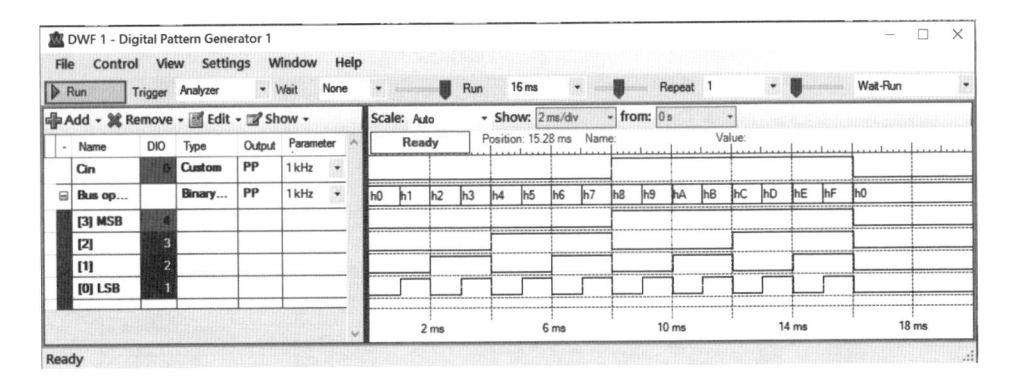

⦿ 圖 B-24 ⦿　　Analog Discovery 連接到 PC 後的 Digital Pattern Generator

按下 Pattern Generator 左上角的 **Run**，waveform 視窗上的 Ready 將因為 Trigger 是 Analyzer 而改變為 Armed。等到 Logic Analyzer 按下 **Single** 之後才改變為 **Done**，待 Logic Analyzer 讀得信號後，再轉為 Ready。

圖 B-25 為 Logic Analyzer 讀得的信號，比較 ModelSim 模擬測試的結果，如圖 B-26 所示之 Timing Diagram，完全相同。

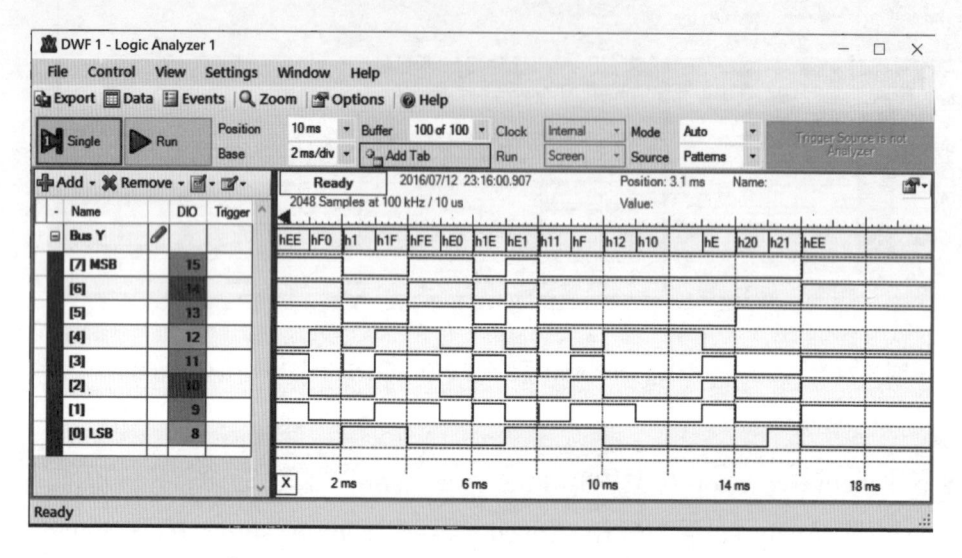

圖 B-25　Logic Analyzer 所讀得的信號

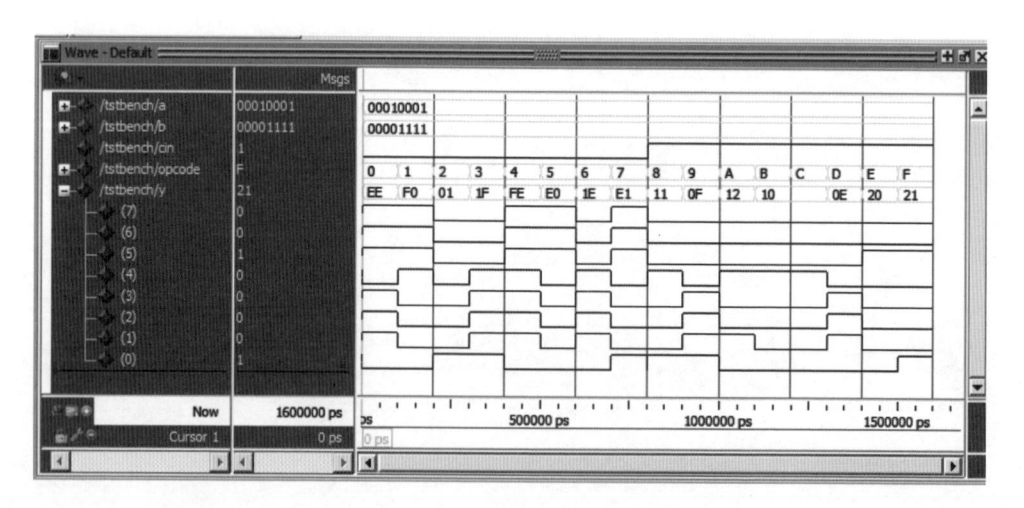

圖 B-26　ModelSim 模擬測試的結果之 Timing Diagram